TECHNIQUES OF CHEMISTRY

ARNOLD WEISSBERGER, *Editor*

VOLUME VIII

SOLUTIONS AND SOLUBILITIES

TECHNIQUES OF CHEMISTRY

ARNOLD WEISSBERGER, *Editor*

TECHNIQUES OF CHEMISTRY

VOLUME VIII

SOLUTIONS AND SOLUBILITIES

Edited by

MICHAEL R. J. DACK

Research School of Chemistry
The Australian National University
Canberra, Australia

PART I

A WILEY-INTERSCIENCE PUBLICATION

JOHN WILEY & SONS

New York • London • Sydney • Toronto

Library of Congress Cataloging in Publication Data:

Dack, Michael R   J   1943-
  Solutions and solubilities.

  (Techniques of chemistry ; v. 8)
  "A Wiley-Interscience publication."
  Includes bibliographical references.
  1. Solution (Chemistry) 2. Solubility. I. Title.

QD61.T4 v. 8 [QD541]   540'.28s [541'.34]
ISBN 0-471-93266-3   75-2331

Printed in the United States of America

10 9 8 7 6 5 4 3 2 1

AUTHORS OF PART I

EDWARD S. AMIS

Department of Chemistry, University of Arkansas, Fayetteville, Arkansas

RUBIN BATTINO

Chemistry Department, Emory University, Atlanta, Georgia

A. BEN-NAIM

Department of Physical Chemistry, Hebrew University, Jerusalem, Israel

SHERRIL D. CHRISTIAN

Department of Chemistry, The University of Oklahoma, Norman, Oklahoma

H. LAWRENCE CLEVER

Chemistry Department, Wright State University, Dayton, Ohio

CHARLES A. ECKERT

Department of Chemical Engineering, School of Chemical Sciences, University of Illinois, Urbana, Illinois

ROMESH KUMAR

Chemical Engineering Division, Argonne National Laboratory, Argonne, Illinois

EDWIN H. LANE

Department of Chemistry, The University of Oklahoma, Norman, Oklahoma

P. LASZLO

Institut de Chemie, Université de Liège, Belgium

THOMAS L. McMEEKIN

Formerly Research Professor in Biology, University of South Carolina, Columbia, South Carolina

J. M. PRAUSNITZ

Department of Chemical Engineering, University of California, Berkeley, California

INTRODUCTION TO THE SERIES

Techniques of Chemistry is the successor to the Technique of Organic Chemistry Series and its companion—Technique of Inorganic Chemistry. Because many of the methods are employed in all branches of chemical science, the division into techniques for organic and inorganic chemistry has become increasingly artificial. Accordingly, the new series reflects the wider application of techniques, and the component volumes for the most part provide complete treatments of the methods covered. Volumes in which limited areas of application are discussed can easily be recognized by their titles.

Like its predecessors, the series is devoted to a comprehensive presentation of the respective techniques. The authors give the theoretical background for an understanding of the various methods and operations and describe the techniques and tools, their modifications, their merits and limitations, and their handling. It is hoped that the series will contribute to a better understanding and a more rational and effective application of the respective techniques.

Authors and editors hope that readers will find the volumes in this series useful and will communicate to them any criticisms and suggestions for improvements.

ARNOLD WEISSBERGER

Research Laboratories
Eastman Kodak Company
Rochester, New York

PREFACE

Most chemical processes of industrial and biological importance occur in solution. The role of solvent is so great that million-fold rate changes take place in some reactions simply by changing the reaction medium. Our bodies contain some 65 to 70% water which acts as a lubricant, as an aid to digestion, and more specifically, as a stabilizing factor to the double helix conformation of DNA. But what do we know about the detailed structure of water and the other solvents? What happens at a microscopic and macroscopic level when species are dissolved in these solvents? How can we investigate such phenomena?

It would be presumptuous to suppose that *Solutions and Solubilities* answers all these questions. The title itself is so all-embracing that a book many times its size could only discuss the many facets of the subject in a superficial manner. Specialized topics such as electrochemistry, photochemistry in solution, the electron in solution, and molten salts were all considered for inclusion in the volume. The fact that they, and other topics, have been omitted does not downgrade their importance to solution chemistry. In the selection of topics, we have attempted to cut across traditional divisions in chemistry, and provide the basis for integrated lecture courses at graduate level. Each chapter contains a wealth of background material on its subject and, justifying the volume's place in the *Techniques of Chemistry* series, describes the most recent experimental procedures. We have been especially fortunate with our contributing authors, so many of whom are recognized leaders in their field of study. To them, a debt of gratitude is owed for the care and enthusiasm with which they entered this project.

The advice and encouragement of Professors R. P. Bell, K. J. Laidler, and R. W. Taft during the planning stage of the volume are gratefully acknowledged. It is also a pleasure to acknowledge the innumerable suggestions made by colleagues over the years which have undoubtedly helped to remedy some of my own inadequacies and gaps of knowledge. Production of *Solutions and Solubilities* is testimony to the help freely given by Dr. Arnold Weissberger, and to the patience and skill of staff at John Wiley & Sons.

MICHAEL R. J. DACK

Canberra, Australia
August 1974

CONTENTS

TECHNIQUES OF CHEMISTRY

ARNOLD WEISSBERGER, *Editor*

VOLUME VIII

SOLUTIONS AND SOLUBILITIES

Chapter I

MOLECULAR THERMODYNAMICS OF REACTIONS IN SOLUTIONS

Charles A. Eckert

There exists a close relationship between solution thermodynamics and the kinetics of chemical reactions in solution. It is the purpose of this chapter to discuss some of the methods that have been used in applying the techniques of solution theory and molecular thermodynamics to problems in chemical kinetics. Further mention is made of how some of the problems arising in such applications may be dealt with, and discussion is presented of those areas in which further research is clearly needed.

A series of three interrelated steps is involved in applying a molecular solution theory to solve a chemical kinetic problem, whether it occurs in a gaseous, liquid, or solid solution. In the first step, the rates of reactions are related to the thermodynamic properties in solution of the reacting species and the reaction transition state through the classic transition state theory, which yields the enormous advantage of reducing a rate problem to an equilibrium problem. Both the theory and experimental techniques of solution thermodynamics are better understood and more accurate than those for kinetics. Moreover, advances in molecular thermodynamics and applications of statistical mechanics will surely provide new approaches to phase equilibria, which can be applied in conjunction with existing knowledge about intermolecular forces to characterize chemical reactions.

1

Next, one employs some theory of solutions, some correlation, frequently borrowed from molecular thermodynamics, or preferably when possible, as many actual experimental thermodynamic data as are available, to relate reaction rates (or equilibria) to known quantities. It must be emphasized that neither the methods discussed here, nor any other practical approach now available, can ever predict a priori rate constants. They can and do relate the rates to the rates of the same reaction or a similar reaction under other thermodynamic conditions.

Often, however, in order to predict the solution properties of the transition state, a molecule not isolable for study by its very definition, one must have some knowledge of its structure and properties. Frequently, it is necessary to characterize the transition state, generally by the thermodynamic determination of its activation parameters and inferences drawn from these. The energy and entropy of activation may be determined by well-known conventional techniques, and high-pressure kinetics has been used to determine activation volumes. These values may often provide information about transition states not available otherwise—such as their structure (the mechanism of a reaction), their dipole moment, their specific solvent interactions, their degree of solvation, or even the details of their interaction with catalyst molecules. While such techniques are highly useful, a detailed discussion of them is beyond the scope of this chapter, and ample examples are available (see, for instance, Refs. 19, 46, 64). It is sufficient to state here that these studies often can provide the detailed information about a transition state needed to use molecular thermodynamics to solve problems in chemical kinetics.

Probably the most useful area of application of these methods has been in rational solvent selection and design for chemical reactions. Commonly, solvent effects on chemical reactions are a factor of several orders of magnitude in the rate, and cases have been reported with factors as high as 10^9 [8, 11]. Thus a proper choice of a reaction solvent can have a very large effect on reaction rates, and molecular solution theories have been shown to be useful in designing solvents or solvent mixtures to optimize yields or selectivities of reactions or sets of reactions [19, 70]. For instance, often a solvent mixture is better than either pure component for separation processes [28] and, correspondingly, examples exist in which solvent mixtures result in markedly different reaction rates than is possible with pure solvents [42, 43], sometimes resulting in a rate enhancement of orders of magnitude [63].

1 TRANSITION STATE THEORY

The transition state theory [23, 27, 73] provides a thermodynamic framework within which the interpretation, correlation, and prediction of the rates of chemical reactions are feasible. It postulates for any elementary reaction the existence of an intermediate species M, called the transition state of the reaction:

$$A + B \rightarrow M \rightarrow \text{product}$$

The complex, an association of A and B resulting from the collision, is a normal molecule, except for its instability with respect to motion along the reaction coordinate. This activated complex is considered to be in strict thermodynamic equilibrium with the original reactants:

$$A + B \rightleftharpoons M$$

and the reaction rate constant is assumed to be proportional to the concentration of complex M, found from chemical equilibrium considerations:

$$[M] = K[A][B]\frac{\gamma_A \gamma_B}{\gamma_M} \qquad (1.1)$$

where K is the equilibrium constant, and the γ's are activity coefficients in the mixture. Equation (1.1) does not provide an absolute value of the chemical reaction rate constant k, but yields relative values of k for differing conditions.

The application of this approach does not really require the existence of the transition state. It has proved repeatedly to be a useful method for calculating kinetic behavior, no matter whether the transition state is regarded as a configuration through which the system passes or as a real molecule. In any event, Marcus [48] recently showed that transition state theory is not dependent on an equilibrium assumption; rather the same end result may be reached by assuming only the adiabatic postulate, that is, that as the molecules move along the reaction coordinate and as the internuclear distances change, the electronic energy levels may shift but the quantum states remain constant.

Transition state theory was first applied to solution kinetics by Brønsted [7] and Bjerrum [6], who showed that the rate of reaction in a thermodynamically nonideal system is given by

$$\text{Rate} = k_0 \frac{\gamma_A \gamma_B}{\gamma_M}[A][B] \qquad (1.2)$$

where k_0 is the rate constant in the Raoult's law reference system (where $\gamma_A = \gamma_B = \gamma_M = 1$). Then, if the activity coefficients are evaluated in terms of some solution theory and the basic rate constant k_0 at atmospheric pressure is a function of temperature only, the apparent rate constant k in any real solution is given by the Brønsted-Bjerrum equation:

$$k = k_0 \left(\frac{\gamma_A \gamma_B}{\gamma_M}\right) \qquad (1.3)$$

The original example used to demonstrate this approach was the study of reactions between ions in dilute aqueous solutions, using the Debye-Hückel [14] treatment.

According to this approach, discussed in many common texts [38, 47], the activity coefficient of a spherical ion of diameter a in solution is expressed in terms of the charge number Z and the dielectric constant of the solvent ϵ as

$$\ln \gamma_i = -\frac{Z_i^2 e^2 \kappa}{2\epsilon k_B T(1 + \kappa a)} \tag{1.4}$$

where k_B is Boltzmann's constant,

$$\kappa = \sqrt{\frac{8\pi e^2}{\epsilon k_B T} I} \tag{1.5}$$

and I is the familiar ionic strength:

$$I = \frac{1}{2} \sum_j n_j Z_j^2 \tag{1.6}$$

If the solution is dilute, as it must be for the theory to be valid,

$$1 >> \kappa a$$

Using this result in conjunction with (1.3), and noting that

$$Z_M = Z_A + Z_B$$

we find the variation in the rate constant of an ionic reaction with ionic strength to be

$$\log \frac{k}{k_0} = Z_A Z_B A' \sqrt{I} \tag{1.7}$$

where A' is a constant for a given solvent, equal to 1.018 for water at 25°C. This result has been compared with experiment on many occasions, and is in both qualitative and quantitative agreement with experiment (see Fig. 1.1). Unfortunately, for practical purposes the range of applicability is small; in general it is not valid at concentrations in excess of 0.01 M.

Nonetheless, this demonstrates that the transition state theory is the bridge between thermodynamics and kinetics; it permits one to apply the methods of solution theory to rate problems. In order to do this successfully, several steps

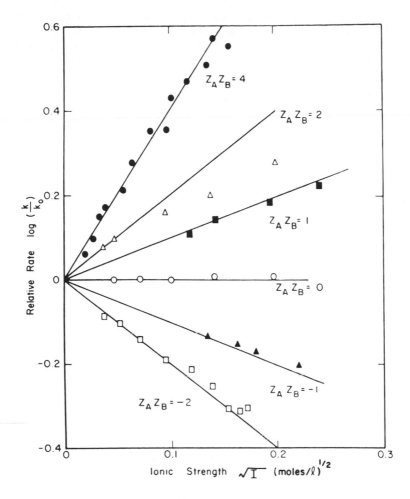

Fig. 1.1. Application of the Br∮nsted-Bjerrüm equation to ionic reactions by the Debye-Hückel theory [51].

are necessary. One must have a knowledge of the behavior of all species in the solution, including those in the transition state, plus a mathematical framework within which to correlate them. In some cases, such as the Debye-Hückel solutions mentioned above, or in gas-phase solutions (Section 2), this is relatively simple. However, most liquid mixtures are more complicated, and no exact prediction is possible. Ideally, actual thermodynamic experimental data may be available and applicable to the kinetics problem. More often, some approximate theory of solutions may have to serve. Moreover, occasionally the solution behavior of the activated complex cannot be predicted until its structure and properties have been determined, requiring an inferential investigation of this transient species to permit prediction of its solvent interactions.

2 GAS-PHASE REACTIONS

Although this book deals primarily with liquids, gas-phase solutions do of course exist, and often solvent properties can be quite important in the dense gas (or fluid) region. Moreover, reactions in the gas phase provide an outstanding example of the use of a theory of solutions in kinetic problems.

For a homogeneous gas-phase reaction, the rate constant is a function of temperature only for the ideal gas; for a real gas it varies with both density and composition. For a bimolecular gas-phase reaction [20], the Brønsted-Bjerrum equation gives

$$\frac{k}{k_0} = \frac{\phi_A \phi_B z}{\phi_M} \tag{1.8}$$

where the ϕ's are fugacity coefficients, and z is the usual compressibility factor of the gaseous mixture. If one applies the virial equation of state,

$$\frac{Pv}{RT} = z = 1 + \frac{B}{v} + \frac{C}{v^2} + \cdots \tag{1.9}$$

where B and C are the second and third virial coefficients, respectively; the rate constant, relative to that in the ideal gas, is given by

$$\ln \frac{k}{k_0} = \frac{2}{v} \sum_j y_j \triangle B_j + \frac{3}{2v^2} \sum_j \sum_k y_j y_k \triangle C_{jk} + \cdots \tag{1.10}$$

where the summations are over all components in the mixture, and

$$\triangle B_j = B_{Aj} + B_{Bj} - B_{Mj}$$

$$\triangle C_{jk} = C_{Ajk} + C_{Bjk} - C_{Mjk}$$

Equation (1.10) is general, but limited to the region of convergence of the virial equation, as well as to the number of coefficients that can be evaluated.

Equation (1.10) was applied to the pyrolysis of hydrogen iodide:

$$2HI \rightarrow H_2 + I_2$$

using a model of the complex based on that of Wheeler, Topley, and Eyring [67], and virial coefficients estimated by corresponding states theory. The

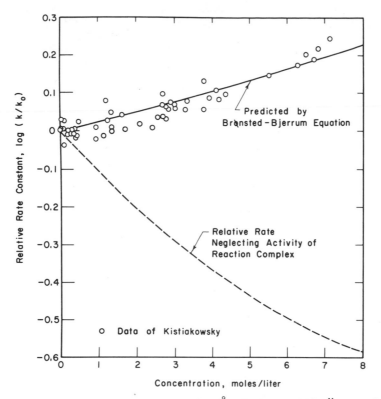

Fig. 1.2. Rate constant for HI pyrolysis at 321.4°C. The Brønsted-Bjerrüm equation gives an excellent prediction of the variation with gas density.

results for the variation in the rate constant with density $(1/v)$ are in excellent agreement with the experimental results of Kistiakowsky [41], as shown in Fig. 1.2. By contrast, an alternate prediction is also shown in this figure for the variation in rate based on the assumption that the rate is proportional to the activity of the reactants only, neglecting the activity of the complex, as suggested previously by other investigators. Such an assumption clearly predicts effects in the wrong direction.

Gas-phase solvent effects on the same reaction have been calculated [50] using a modified Redlich-Kwong equation [54]. This has two distinct advantages over the truncated virial equation. First, it is applicable at far higher densities, up to and well beyond the critical density. Second, only the critical temperature and pressure and the acentric factor are required; the necessity of using these or other data to estimate virial coefficients is thus avoided. When it is assumed that there are no competing reactions with the other species present, the types of results obtained are shown in Fig. 1.3. The figure presents calculated rate

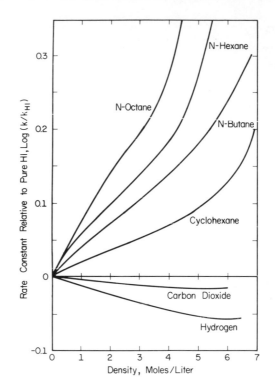

Fig. 1.3. Solvent effects on a gas-phase reaction; the pyrolysis of HI dilute in various gases at 300°C.

constants relative to the rate constant in pure HI at the same overall mole density, as a function of density. From this plot it is evident that, although the rate constant is augmented by increasing density, a light diluent such as hydrogen, nitrogen, or even CO_2 slows the reaction relative to the rate constant for pure HI. However, higher-boiling diluents such as medium-weight hydrocarbons cause sizable increases in the rate constant. Such results are readily explained in terms of stabilization of the large, moderately acentric complex by diluents exhibiting stronger intermolecular forces, such as butane, hexane, and octane. Lighter gases, with lesser forces, stabilize the complex less than does pure HI, and thus attenuate the rate (Fig. 1.3). These conclusions agree with the results of Sortland and Prausnitz [60].

In general, the greatest effects were observed in the region closest to the critical point of the mixture. Such a result is hardly surprising, since dense gases can often have excellent solvent properties, interacting strongly with dilute species. This result was shown most dramatically by Simmons and Mason [58], who investigated the gas-phase 1,2-cycloaddition dimerization of chlorotrifluoroethylene at pressures up to 100 atm, and also made careful measurements of the PVT properties of the monomer. They used their results to develop a dense-gas equation of state which permitted them not only to predict the rate of the reaction at high pressures, but also to interpret some of the properties of the transition state.

These results demonstrate that gas-phase nonidealities may be treated by the Brønsted-Bjerrum expression also, although in general the effects are small unless the gas is quite dense. Success in making quantitative predictions depends strongly on the validity and applicability of the equation of state used.

3 SOLVENT EFFECTS ON LIQUID-PHASE REACTIONS

Kinetic solvent effects on the rates of reactions can be enormous. A change in solvent often causes variations in the rate constant of 100 to 1000 for reactions between nonionic species, and often of factors as great as 10^5 in reactions between ions. As mentioned earlier, in extreme cases kinetic solvent effects as great as a factor of 10^9 have been observed. Many discussions are available covering varied aspects of this topic, for example, those by Baekelmans, Gielen, and Nasielski [3] and by Amis [1]. One attempt has been reported [26] in which corresponding states theory was used to correlate solvent effects, with moderate success. In addition, many investigators have proposed linear free-energy-type relationships for kinetic solvent effects (see, e.g., Refs. 1, 9, 12, 16, 24, 30, 65, 66, 72). However, such methods are discussed in other chapters of this volume. Rather, here we wish to demonstrate the application of the Brønsted-Bjerrum equation [see (1.3)].

This expression relates kinetic solvent effects to the activities in solution, so in principle a knowledge of solution activities permits rigorous prediction of the effect of solvents on reaction rates. However, this requires experimental data on the solution behavior not only of the reacting species, but also of the transition state, and the latter data are rarely available. It is always preferable to use actual data to estimate from various correlations or approximations, and these are frequently available for reactants. However, activity data for transition states are generally those back-calculated from kinetic measurements such as, for example, the studies of Parker [53] on ion-molecule transition states, or those of Wong and Eckert [71] on a molecular reaction.

Thus, to apply the Brønsted-Bjerrum expression, we must use some approximate theory of solutions and, as discussed in Section 1, generally when such methods are to be useful for predictive purposes, they are strictly applicable only to one particular type of solution, such as dispersion-force solutions, dipolar interactions, and ionic solutions. However, the effects often overlap, such as a nonpolar reaction in a highly dipolar solvent. Nonetheless, in considering solvent effects on reaction rates, reactions are considered in three broad categories:

nonpolar reactions, polar reactions, and ionic reactions.

For nonpolar reactions, which show the smallest solvent effects, the most reliable approach is avialable in the regular solution theory of Hildebrand and Scott [36, 37]. Although this approach is not the only possible one, it has certain advantages. It is simple, requiring only pure-component data, and it is readily extended to multicomponent mixtures. It has the limitation of applying only to solutions of nonionic, nonpolar, or slightly polar molecules. Applying this technique to kinetics was first proposed by Glasstone, Laidler, and Eyring [27].

According to the regular solution theory, the activity coefficient of component i in a multicomponent mixture is given in terms of its liquid molar volume v_i, solubility parameter δ_i (the square root of the cohesive energy density), and the volume-fraction average solubility parameter of the solvent $\overline{\delta}$.

$$RT \ln \gamma_i = v_i(\delta_i - \overline{\delta})^2 \qquad (1.11)$$

By combining (1.10) and (1.11), the relative rate expression for an elementary reaction is derived [70]:

$$\ln \frac{k_x}{k_{x_0}} = \frac{v_A(\delta_A - \overline{\delta})^2 + v_B(\delta_B - \overline{\delta})^2 - v_M(\delta_M - \overline{\delta})^2}{RT} \qquad (1.12)$$

Equation (1.12) is valid under conditions in which the assumption that there is no excess entropy of mixing is justified. For dilute solution the average solubility parameter can be approximated by the solubility parameter of the solvent.

The values of the molar volume and solubility parameter for the reactants in (1.12) are readily available pure-component properties. However, the properties of the transition state may also be obtained from independent measurements. Its volume is available from the effect of pressure on the reaction rate in the liquid phase, that is, the volume of activation data. For the solubility parameter we combine this volume with the energy of vaporization, found from a thermodynamic cycle similar to that suggested by Harris and Prausnitz [34],

$$
\begin{array}{ccc}
A_g & + \; B_g & \rightarrow M_g \\
\uparrow & \uparrow & \uparrow \\
A_l & + \; B_l & \rightarrow M_l
\end{array}
$$

The energy of vaporization of the complex is equivalent to the liquid-phase dissociation of M to species A and B, vaporization of A and B, and the gas-phase recombination to give M. To a good approximation the activation energies of the dissociation and the recombination are equal in the gas and liquid phases for most nonpolar reactions. Then δ_M is given by

$$\delta_M = \frac{\nu_A \delta_A{}^2 + \nu_B \delta_B{}^2}{\nu_M}$$ (1.13)

Thus the combination of (1.12) and (1.13) provides a predictive expression for kinetic solvent effects for nonpolar reactions with no adjustable parameters.

The Diels-Alder reaction provides a good class of reactions to use as examples. It consists of the addition of a compound containing a double or triple bond to the 1,4-positions of a conjugated diene system, with the formation of a six-membered hydroaromatic ring. A number of such reactions have been studied, and the results are available. In general, there are few side reactions, and the reactants and products are not too highly polar.

It should be pointed out that for application of (1.13) the units used for the rate constant and for the thermodynamic properties must be self-consistent. For example, in the Debye-Hückel treatment, activity coefficients are expressed in volumetric concentration units (moles per liter), and so are the rate constants. For most other expressions for activity coefficients, the concentration unit is the dimensionless mole fraction, so we use consistent rate constants in mole fraction units k_x. Unfortunately, this point has been overlooked too often. For reactions in dilute solution, (1.3) is valid, with k and k_0 replaced by k_x and k_{x_0}, respectively. For a bimolecular reaction with volumetric concentrations in moles per milliliter, these quantities differ from each other by a constant factor, the molar volume of the solvent. Normally, in applying (1.12) it is convenient to eliminate k_{x_0} by choosing one solvent as a reference, and this has been done in the examples shown here.

Figures 1.4 to 1.6 demonstrate the use of regular solution theory to predict kinetic solvent effects on the Diels-Alder reactions of maleic anhydride with chloroprene [48] and isoprene [15, 29, 59], and acrylonitrile with isoprene [61]. This technique was applied to many other cases (see, e.g., Ref. 21), and in general the data were found to be in good agreement with the theory even for some relatively polar solvents such as acetone. As might be expected, the most polar solvents, acetonitrile and nitromethane, tend to show the largest deviations.

In these results the regular solution theory works surprisingly well, even in relatively polar mixtures. Such results can be explained by the independent determination of the activity coefficient in solution of the transition state for the Diels-Alder reaction of maleic anhydride with 1.3-butadiene. Wong and Eckert [72] made direct measurements of the reactant activities, plus the kinetic solvent effects. They characterized the multicomponent mixture with the excellent Wilson [69] equation and back-calculated γ_M. They showed that the ratio of activity coefficients for maleic anhydride and the transition state γ_A/γ_M is nearly constant, even though the individual γ's vary by orders of magnitude. Since the Brønsted-Bjerrum relation involves only this ratio of activity coeffi-

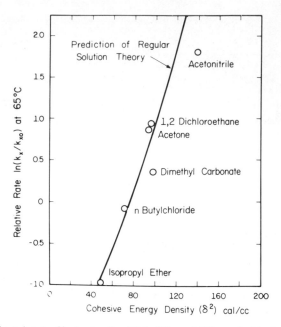

Fig. 1.4. Kinetic solvent effects on the Diels-Alder addition of chloroprene to maleic anhydride [48].

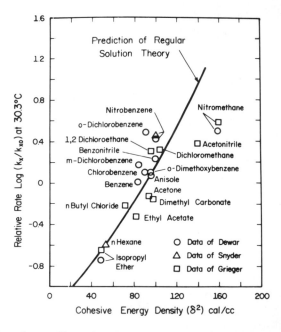

Fig. 1.5. Kinetic solvent effects for the Diels-Alder cycloaddition of isoprene-maleic anhydride [15, 29, 59].

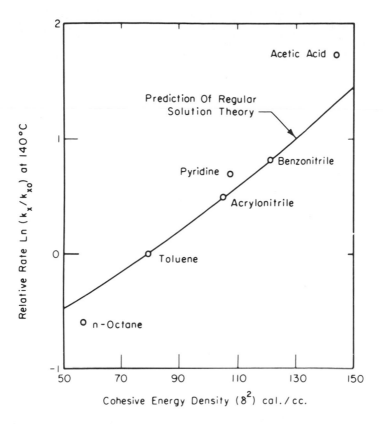

Fig. 1.6. Kinetic solvent effects on the Diels-Alder condensation of isoprene with acrylonitrile [61].

cients, it appears that the polar effects cancel out. In other words, this is truly a nonpolar reaction, in spite of the presence of relatively polar species, and this accounts for the success of regular solution theory. The discrepancies in extreme cases, such as for nitromethane as in Fig. 1.5 and for acetic acid as in Fig. 1.6, are not caused by failure of this cancellation effect, but rather because the regular solution theory is inadequate to predict the variation of activity of the nonpolar diene in these solvents.

Regular solution theory has also been applied to polar reactions, but more for correlative purposes than for prediction. Moreover, what success it has enjoyed in such applications should be viewed as surprising. For example, Stefani [62] showed such a result for the solvent effect on the rates of two Menschutkin reactions—the reaction of an alkyl halide with a tertiary amine (both highly polar) to form an ionic quaternary ammonium salt:

$$R_3N + R'X \rightarrow [R_3N\cdots R'\cdots X]^{\ddagger} \rightarrow R_3R'N^+ + X^-$$

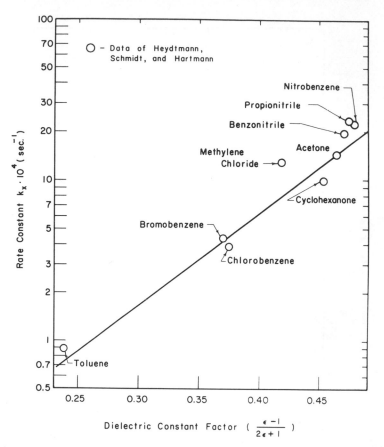

Fig. 1.7. A Kirkwood-type correlation of the solvent effects on the rate of reaction of a-picoline with ω-bromoacetophenone at 30°C [35].

For this reaction the transition state is considered to be extremely polar, yet still nonionic, and numerous examples of solvent effects on such reactions are available [22, 32, 33, 35, 39, 42, 43].

For the reaction of pyridine with methyl iodide in 13 solvents ranging in solubility parameter from 7 to 11 (cal ml^{-1})$^{1/2}$ (including hydrocarbons, ethers, organic halides, and nitro compounds), Stefani found a good linearity of the logarithm of the rate constant with respect to the square root of internal pressure (equivalent to the solubility parameter). Stefani's correlation corresponds to a solubility parameter of the Menschutkin complex of about 14 (cal ml^{-1})$^{1/2}$.

Also, (1.12) was applied [70] to compare with Stefani's correlation for the triethylamine-ethyl iodide reaction in a similar range of solvents. The result

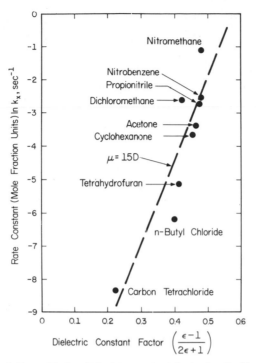

Fig. 1.8. Rate correlation with the dielectric constant factor for the Menschutkin reaction of tri-*n*-propylamine with methyl iodide at 30°C [22].

corresponds to a solubility parameter for this Menschutkin complex of about 13.5 (cal ml^{-1})$^{1/2}$. Since the transition state for the Menschutkin reaction is known to be nonionic but highly polar, the two values found for the solubility parameter of the complexes are qualitatively reasonable, but this approach should not be considered quantitatively reliable.

An approach specifically designed for polar solutes is that derived from the Kirkwood [40] expression for the work (Gibbs energy at constant T and P) required to place an ideal dipole μ at the center of a spherical cavity of radius r within an isotropic medium of dielectric constant ϵ:

$$\Delta G = -\frac{\mu^2}{r^3}\left(\frac{\epsilon - 1}{2\epsilon + 1}\right) \tag{1.14}$$

This expression neglects, of course, any nonelectrostatic interactions, such as van der Waals forces, hydrogen bonds, or charge-transfer complexes. When (1.14) is used to calculate activity coefficients and combined with (1.3), the kinetic solvent effect for a bimolecular reaction is given as

$$\ln \frac{k_x}{k_{x_0}} = \frac{1}{k_B T} \frac{\epsilon - 1}{2\epsilon + 1} \left(\frac{\mu_M^2}{r_M^3} - \frac{\mu_A^2}{r_A^3} - \frac{\mu_B^2}{r_B^3} \right) \qquad (1.15)$$

where the rate constant is linear in the dielectric constant factor, with the slope given by the term in parentheses, which might be termed a change in dipole moment density.

In general, the success of (1.15) in predicting kinetic solvent effects for polar reaction has been rather sporadic. Many investigators have pointed out [3, 4] that the nonelectrostatic contributions are rarely negligible. Nonetheless, the results show that (1.15) may be qualitatively useful, although quantitatively its success may be rather variable. For example, Fig. 1.7 and 1.8 show examples of the application of the Kirkwood method to rate data on Menschutkin reactions, and in aprotic solvents agreement is good.

Thus the Kirkwood approach is known to be an oversimplification, and in some cases (especially for associating solvents) it gives poor results, but it generally gives a reasonable if not highly quantitative picture of the polar interactions. Its predictions should be viewed cautiously, but it is significantly better than any other solution theory that now exists for polar reactions.

In addition to nonpolar and polar reactions, there are many important reactions involving ionic species. The original application of solution thermodynamics to chemical kinetics was the application of Debye-Hückel theory to reactions between ions, as discussed above. Although the electrostatic theory provides an exact solution and the assumptions are probably more reasonable than those in the Kirkwood treatment, it is applicable only in the limit of infinite dilution, in practice at very low concentrations. Even at relatively modest concentrations, in general above 0.01 M, other effects, such as ionic solvation, may become dominant. In spite of the great importance of this class of reactions, there are relatively few exact methods available for predicting variations in rate constants, primarily because of the limited quantitative understanding of existing ionic solution theories. Both the effects of nonidealities and the problems involved in characterizing them are much more severe for solutions containing ionized species than for solutions of nonelectrolytes.

In addition to the salt effect in dilute solution, Debye-Hückel theory has also been used to predict solvent effects, also in the limit of infinite dilution [55]. By differentiation of (1.7) the rate constant at infinite dilution in various solvents may be shown to be a linear function of the reciprocal of the dielectric constant [13, 17, 44]. An example of such a plot of solvent effects is presented in Fig. 1.9.

Unfortunately, however, there is little practical interest in kinetics in the very dilute Debye-Hückel region. Moreover, this theory applies only to ion-ion reactions, and predicts nothing about ion-molecule reactions. In most reactions

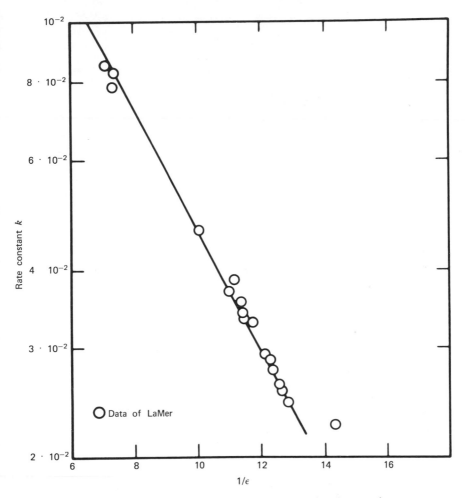

Fig. 1.9. The effect of the solvent dielectric constant on reactions between ions.

involving ions in solution at finite concentrations, the dominant factor is the relative solvation of ionic reactants and transition state, and virtually all other effects are second-order. For example, by including solvation by the method of Scatchard [56, 57], Clements and Eckert [10] extended predictions of the salt effect on the rate of an ion-ion reaction in several solvents up to ionic strengths of about unity, as shown for the bromoacetate-thiosulfate reaction in water in Fig. 1.10.

Ion-molecule reactions provide an example of the importance of solvation effects and the difficulty of predicting them. The kinetics of such reactions

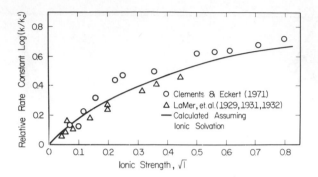

Fig. 1.10. The rate of reaction as a function of ionic strength for the bromoacetate-thiosulfate ion-ion reaction at $25\overset{\circ}{}C$.

depends on the Gibbs-energy difference between the reactants and the transition state; the problem of the solvent effect on reaction rates is therefore again the determination of the Gibbs energy, entropy, and enthalpy of solvation of the reactants and of the transition state [2, 45, 68].

For the bimolecular nucleophilic substitution reaction of an anion with an alkyl halide,

$$Y^- + RX \rightarrow [Y\cdot\cdot\cdot R\cdot\cdot\cdot X]^- \rightarrow RY + X^-$$

the Brønsted-Bjerrum expression becomes

$$\frac{k_s}{k_0} = \frac{^0\gamma_{Y^-}^s \cdot {}^0\gamma_{RX}^s}{^0\gamma_{\ddagger}^s} \tag{1.16}$$

where the solvent activity coefficient $^0\gamma_i^s$ reflects the change in the standard chemical potential of species i on transfer from an arbitrary reference solvent o to another solvent s.

Most reactions of this type are considerably faster in dipolar aprotic solvents than in protic solvents. Protic solvents such as water, ammonia, alcohols, and amides are hydrogen-bond donors, while aprotic solvents such as acetone, nitromethane, acetonitrile, and dimethylformamide are highly polar but no more than weak hydrogen-bond donors. The much faster rates of bimolecular nucleophilic substitution reactions in dipolar aprotic solvents relative to protic solvents can be attributed to the enhanced solvation of the nucleophiles in the protic solvent, or to greater solvation of the transition states in the dipolar aprotic solvents, since solvation of the neutral molecule is considered to be relatively small. Parker [53] determined in various solvents the activity coefficients of anions, neutral molecules, and transition states relative to a reference solvent.

In the determination of the solvent activity coefficient of anions, he used extra-thermodynamic assumptions to divide the solvent activity coefficient of electrolytes into those of the individual ions. From the resulting values he concluded that the reactions are faster in dipolar aprotic solvents, because the reactant anion Y^- is more solvated by protic than by dipolar aprotic solvents, and this outweighs any effects due to transition-state anion (a larger, charge-dispersed species) or neutral molecule solvation. The main source of the solvation of anions by protic solvents is hydrogen bonding, since normal anions are strong hydrogen-bond acceptors.

An alternate approach for interpreting the solvent effects on the rate of such reactions was attempted by Haberfield, Clayman, and Cooper [31], who determined the enthalpy of transfer of a transition state from one solvent to another. The enthalpy of transfer of transition states from a protic to an aprotic solvent was found to be exothermic, and in most cases the enthalpies of transfer of the reactants were also exothermic. This means that both the transition state and the reactants are generally more solvated in an aprotic solvent than in a protic solvent. However, the effect is much greater for the transition state than for the nucleophiles, which results in a lower enthalpy of activation in the aprotic solvent.

Thus one can see that, for even this relatively simple reaction, two quite different approaches to the interpretation have been proposed, and both are basically qualitative. One method emphasizes solvation of nucleophiles, and the other approach deals primarily with solvation of the transition state, as the primary factor in determining the reaction rate.

Unfortunately, the large variations in solvation, reflected both in the rate data as well as in calorimetric data, are not in general predictable. Some factors are understood qualitatively [12, 52], and clearly some recent advances in modern solvation theory appear promising [25], but a significantly better quantitative understanding of solvation phenomena will be required before it will be possible to make accurate and reliable predictions of solvent effects on reactions involving ionic species.

4 APPLICATIONS TO REACTION DESIGN

Once the solution thermodynamics for a reaction can be written in any sort of analytical form, one may apply it to optimum design of a solvent or solvent mixtures [19]. In many cases, as is evident from Fig. 1.4 to 1.8, the optimum solvent may merely be one with extreme properties, such as the greatest (or lowest) value of the solubility parameter or of the dielectric constant. However, such is not always the case. From the Gibbs-Duhem equation, it has been shown that in any ternary system one component must exhibit an extremum in its activity coefficient [28]. This means that since the relative rate k_x/k_{x_0} is

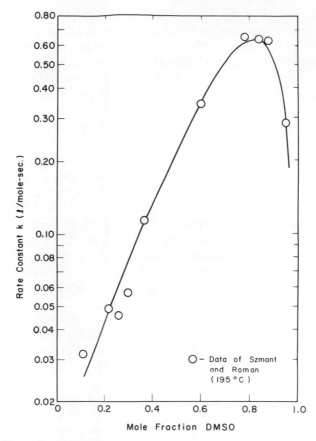

Fig. 1.11. Rate of a Wolff-Kishner reaction in a mixed solvent of dimethyl sulfoxide-butyl carbitol [63].

proportional to the ratio of the individual activity coefficients by (1.3), the rate of reaction may also show an extremum. In many cases then, a properly chosen mixture of solvents may maximize or minimize the rate of reaction better than any individual solvent alone. An extreme example of such a reaction is shown in Fig. 1.11, in which the Wolff-Kishner reaction of benzophenone hydrazone in dimethyl sulfoxide-butyl carbitol mixtures exhibits a rate maximum in a mixed solvent that is orders of magnitude faster than in either pure solvent.

Although no predictive scheme is available to handle such a complex reaction, an analytical form is available for nonpolar reactions. For such a reaction in a binary solvent mixture in which the regular solution behavior is followed and the reactants are dilute, (1.12) can be used with the average solubility parameter given by

$$\overline{\delta} = \Phi_1\delta_1 + \Phi_2\delta_2 \tag{1.17}$$

where Φ is the volume fraction, and the subscripts 1 and 2 refer to the two solvents. Then the extremum of the ratio of the rate constants k_x/k_{x_0} with respect to $\overline{\delta}$ is given by

$$\overline{\delta}_{extremum} = \frac{\nu_A\delta_A + \nu_B\delta_B - \nu_M\delta_M}{\nu_A + \nu_B - \nu_M} \tag{1.18}$$

The denominator of (1.18) is no more than the negative of the volume of activation for the reaction. Further differentiation indicates that whether the $\overline{\delta}_{extremum}$ given is for a maximum or minimum depends on the sign of the volume of activation. For a Diels-Alder reaction in which the volume of activation is generally negative, the $\overline{\delta}_{extremum}$ given is for a minimum in the rate.

The Diels-Alder dimerization of isoprene provides an example of this phenomenon. Calculations [70] based on regular solution theory, assuming a value of $\delta_M = 7.1$ (cal ml^{-1})$^{1/2}$, show that a mixture of about 30% (by volume) of neopentane with CS_2 gives a minimum in the rate constant. The rate in pure CS_2 is about 6% faster than in the mixture, and in pure neopentane more than 35% faster. The results are shown in Fig. 1.12. Although the effect in this case

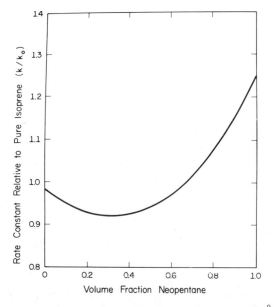

Fig. 1.12. The rate of the Diels-Alder dimerization of isoprene at 65°C in mixtures of neopentane-carbon disulfide, as calculated from regular solution theory.

is not terribly large, it illustrates the extremum possible in a mixture. For more polar mixtures (for which regular solution theory is less applicable) the effect would be much greater.

A similar analysis has been applied to the calculation of the relative rates of two competing reactions. Consider two bimolecular reactions:

$$A_1 + B \xrightarrow{k_1} M_1$$

$$A_2 + B \xrightarrow{k_2} M_2$$

where k_1 and k_2 are the rate constants for the two reactions, A_1 and A_2 are the reactants competing for B, and M_1 and M_2 are the corresponding activated complexes. The variation in the relative rate from solvent to solvent is given by

$$\frac{k_1}{k_2} = \left(\frac{k_1}{k_2}\right)_0 \frac{\gamma_{A_1}}{\gamma_{M_1}} \frac{\gamma_{M_2}}{\gamma_{A_2}} \tag{1.19}$$

where $(k_1/k_2)_0$ again refers to the ideal reference system defined previously. If sufficient thermodynamic data are known, the relative rates of the two competing reactions can in principle be calculated. Otherwise, the regular solution theory can be applied:

$$\ln \frac{k_1}{k_2} = \ln \left(\frac{k_1}{k_2}\right)_0 +$$

$$\frac{\nu_{A_1}(\delta_{A_1} - \overline{\delta})^2 + \nu_{M_2}(\delta_{M_2} - \overline{\delta})^2 - \nu_{A_2}(\delta_{A_2} - \overline{\delta})^2 - \nu_{M_1}(\delta_{M_1} - \overline{\delta})^2}{RT} \tag{1.20}$$

An example of a pair of competing reactions is the Diels-Alder condensation of methyl acrylate with cyclopentadiene, which yields both an *endo* product and an *exo* product. The separate reactions, through separate transition states to form the two possible stereoisomeric products, constitute a pair of competing reactions.

Since the reactants are the same for both reactions, the two terms in A drop out of (1.20), and further it is reasonable to assume that the molar volumes of the two isomeric transition states are equal, $\nu_{M_N} = \nu_{M_X}$. Then (1.20) becomes

$$\ln \frac{k_N}{k_X} = \ln \left(\frac{k_N}{k_X}\right)_0 + \frac{\nu_{M_X}(\delta_{M_N} + \delta_{M_X} - 2\overline{\delta})(\delta_{M_X} - \delta_{M_N})}{RT} \tag{1.21}$$

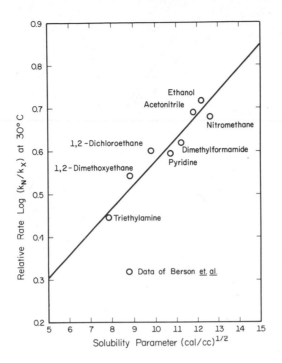

Fig. 1.13. The prediction of regular solution theory for the relative rates of the formation endo (N) and exo (X) Diels-Alder adducts of cyclopentadiene with methyl acrylate [5].

The results, using the data of Berson, Hamlet, and Mueller [5], are shown in Fig. 1.13, and the agreement is remarkably good.

One further example of the use of an analytical equation for reaction is the expression for Diels-Alder additions to maleic anhydride developed by McCabe and Eckert [48]. This involves application of the Hammett equation along with regular solution theory to predict simultaneously both solvent and substituent effects on the rate for this class of reactions. Although intended to correlate reactions and equilibria involving aromatic side chains, the Hammett linear free-energy equation has been found to be useful for Diels-Alder reactions, probably because of the highly conjugated quasi-aromatic nature of the transition state.

Some typical results for Diels-Alder reactions are shown in Fig. 1.14, and the rates are linear for all solvents investigated. Well within the accuracy of most literature data, such lines are parallel for many solvents and, if the intercept is considered to be a solvent parameter d, there results a general equation for both solvent and substituent effects:

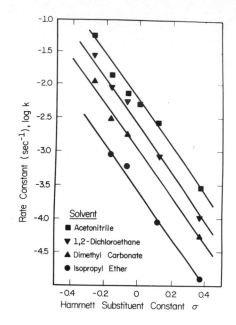

Fig. 1.14. Comparison of the predictions of the modified Hammett expression (1.22) with experimental data for the rate of addition of substituted butadienes with maleic anhydride in four solvents [48].

$$\log \frac{k}{k_0} = \rho(\sigma - \sigma_0) + (d - d_0) \qquad (1.22)$$

where the subscript 0 refers to the reference reaction in the reference solvent. As might be expected, the intercepts d correlate quite closely with the solubility parameter and, for hundreds of data over a very wide range of rates, solvents, and substituents, (1.22) has been shown to predict rates to an accuracy of about ±25%. Considering that the precision of the data is often no better than this, (1.22) represents a remarkably good expression, with no adjustable parameters, for making rough predictions of a wide variety of rates.

Similarly, any valid analytical expression for the activity coefficients in (1.3) may be used for solvent design, either in predicting kinetic solvent effects or in optimizing solvents or solvent mixtures to maximize or minimize rates. Further, one may design optimum solvent systems for reaction sets to maximize either yields or selectivities. The real need is for accurate experimental data and theories of mixtures to characterize the solution behavior of reactants and transition states.

References

1. E. S. Amis, *Solvent Effects on Reaction Rates and Mechanisms,* Academic, New York, 1966.
2. E. M. Arnett, W. G. Bentrude, J. J. Burke, and P. McC. Duggleby, *J. Am. Chem. Soc.,* **87,** 1541 (1965).
3. P. Baekelmans, M. Gielen, and J. Nasielski, *Ind. Chim. Belg.,* **29,** 1265 (1964).
4. S. W. Benson, *The Foundations of Chemical Kinetics,* McGraw-Hill, New York, 1960.
5. J. A. Berson, Z. Hamlet, and W. A. Mueller, *J. Am. Chem. Soc.,* **84,** 297 (1962).
6. N. Bjerrum, *Z. Phys. Chem.* (Leipzig), **108,** 82 (1924).
7. J. M. Brønsted, *Z. Phys. Chem.* (Leipzig), **102,** 169 (1922).
8. H. C. Brown and Y. Okamoto, *J. Am. Chem. Soc.,* **79,** 1913 (1957).
9. S. Brownstein, *Can. J. Chem.,* **38,** 1590 (1960).
10. L. D. Clements and C. A. Eckert, *Ind. Eng. Chem., Process Des. Develop.,* **10,** 401 (1971).
11. D. J. Cram, B. Rukborn, C. A. Kingsbury, and P. Haberfield, *J. Am. Chem. Soc.,* **83,** 3678 (1961).
12. M. R. J. Dack, *Chem. Brit.,* **6,** 347 (1970); *Chem. Tech.,* **1,** 108 (1971).
13. H. G. Davis and V. K. LaMer, *J. Chem. Phys.,* **10,** 585 (1942).
14. P. J. W. Debye and E. Hückel, *Phys. Z.,* **24,** 185 (1923).
15. M. J. S. Dewar and R. S. Pyron, *J. Am. Chem. Soc.,* **92,** 3098 (1970).
16. Y. Drougard and D. Decroocq, *Bull. Soc. Chim. Fr.,* **1969,** 2972.
17. S. Eagle and J. C. Warner, *J. Am. Chem. Soc.,* **58,** 2335 (1936).
18. C. A. Eckert, *Ind. Eng. Chem.,* **59**(9), 20 (1967).
19. C. A. Eckert, *The Effect of High Pressure on Reactions in Solution, Ann. Rev. Phys. Chem.,* **23,** 239 (1972).
20. C. A. Eckert and M. Boudart, *Chem. Eng. Sci.,* **18,** 144 (1963).
21. C. A. Eckert, C. K. Hsieh, and J. R. McCabe, *AIChE J.,* **20,** 20 (1974).
22. C. A. Eckert, S. P. Sawin, and C. K. Hsieh, in press.
23. M. G. Evans and I. M. Polanyi, *Trans. Faraday Soc.,* **31,** 875 (1935).
24. A. H. Fainberg and S. Winstein, *J. Am. Chem. Soc.,* **78,** 2770 (1956).
25. H. L. Friedman, *Chem. Brit.,* **9,** 300 (1973).
26. H. L. Frisch, T. A. Bak, and E. R. Webster, *J. Phys. Chem.,* **66,** 2101 (1962).
27. S. Glasstone, K. J. Laidler, and H. Eyring, *The Theory of Rate Processes,* McGraw-Hill, New York, 1941.
28. R. A. Grieger and C. A. Eckert, *Ind. Eng. Chem., Process Des. Develop.,* **6,** 250 (1967).
29. R. A. Grieger and C. A. Eckert, *Trans. Faraday Soc.,* **66,** 2579 (1970).
30. E. Grunwald and S. Winstein, *J. Am. Chem. Soc.,* **70,** 846 (1948).
31. P. Haberfield, L. Clayman, and J. S. Cooper, *J. Am. Chem. Soc.,* **91,** 787 (1969).

32. H. Hartmann, H. D. Brauer, H. Kelm, and G. Rinck, *Z. Phys. Chem.* (Frankfurt am Main), **61**, 53 (1968).
33. H. Hartmann, H. Kelm, and G. Rinck, *Z. Phys. Chem.* (Frankfurt am Main), **44**, 335 (1965).
34. H. G. Harris and J. M. Prausnitz, *Ind. Eng. Chem., Fundam.,* **8**, 180 (1969).
35. H. Heydtmann, A. P. Schmidt, and H. Hartmann, *Ber. Bunsenges. Phys. Chem.,* **70**, 444 (1966).
36. J. H. Hildebrand and R. L. Scott, *Regular Solutions,* Prentice-Hall, Englewood Cliffs, N.J., 1962.
37. J. H. Hildebrand and R. L. Scott, *The Solubility of Nonelectrolytes,* 3rd ed., Dover, New York, 1964.
38. T. L. Hill, *Introduction to Statistical Thermodynamics,* Addison-Wesley, Reading, Mass., 1960.
39. J. C. Jungers, et al., *L'Analyse Clinetique de la Transformation Chimique,* Vol. 2, Technip, Paris, 1969.
40. J. G. Kirkwood, *J. Chem. Phys.,* **2**, 351 (1934).
41. G. B. Kistiakowsky, *J. Am. Chem. Soc.,* **50**, 2315 (1928).
42. Y. Kondo and N. Tokura, *Bull. Chem. Soc. Jap.,* **37**, 1148 (1964).
43. Y. Kondo and N. Tokura, *Bull. Chem. Soc. Jap.,* **40**, 1433, 1438 (1967).
44. V. K. LaMer and M. E. Kamner, *J. Am. Chem. Soc.,* **57**, 2662 (1935).
45. J. E. Leffler and E. Grunwald, *Rates and Equilibria of Organic Reactions,* Wiley, New York, 1963.
46. W. J. leNoble, *Kinetics of Reactions in Solutions under Pressure, Progr. Phys. Org. Chem.,* **5**, 207 (1967).
47. G. N. Lewis, M. Randall, K. S. Pitzer, and L. Brewer, *Thermodynamics,* 2nd ed., McGraw-Hill, New York, 1961.
48. J. R. McCabe and C. A. Eckert, *Ind. Eng. Chem., Fundam.,* **13**, 168 (1974).
49. R. A. Marcus, *J. Chem. Phys.,* **46**, 959 (1967).
50. T. R. Mills and C. A. Eckert, *Ind. Eng. Chem., Fundam.,* **7**, 327 (1968).
51. A. R. Olson and T. R. Simonson, *J. Chem. Phys.,* **17**, 1167 (1949).
52. A. J. Parker, *Quart. Rev. Chem. Soc.,* **16**, 163 (1962).
53. A. J. Parker, *Chem. Rev.,* **69**, 1 (1969).
54. O. Redlich, F. J. Ackerman, R. D. Gunn, M. Jacobson, and S. Lau, *Ind. Eng. Chem. Fundam.,* **4**, 369 (1965).
55. G. Scatchard, *Chem. Rev.,* **10**, 229 (1932).
56. G. Scatchard, *Phys. Z.* **33**, 22 (1932).
57. G. Scatchard, *Chem. Rev.,* **19**, 309 (1936).
58. G. M. Simmons and D. M. Mason, *Chem. Eng. Sci.,* **27**, 89, 2307 (1972).
59. R. B. Snyder and C. A. Eckert, *AIChE J.,* **19**, 1126 (1973).
60. L. D. Sortland and J. M. Prausnitz, *Chem. Eng. Sci.,* **20**, 847 (1965).
61. J. C. Soula, D. Lumbroso, M. Hellin, and F. Coussement, *Bull. Soc. Chim. Fr.,* 2059, 2065 (1966).
62. A. P. Stefani, *J. Am. Chem. Soc.,* **90**, 1694 (1968).
63. H. H. Szmant and M. N. Roman, *J. Am. Chem. Soc.,* **88**, 4034 (1966).
64. K. E. Weale, *Chemical Reactions at High Pressures,* Spon, London, 1967.

65. P. R. Wells, *Chem. Rev.*, **63**, 171 (1963).
66. P. R. Wells, *Linear Free Energy Relationships*, Academic, New York, 1968.
67. A. Wheeler, B. Topley, and H. Eyring, *J. Chem. Phys.*, **4**, 178 (1936).
68. K. B. Wiberg, *Physical Organic Chemistry*, Wiley, New York, 1964.
69. G. M. Wilson, *J. Am. Chem. Soc.*, **86**, 127 (1964).
70. K. F. Wong and C. A. Eckert, *Ind. Eng. Chem., Proc. Des. Develop.*, **8**, 568 (1969).
71. K. F. Wong and C. A. Eckert, *Trans. Faraday, Soc.*, **66**, 2313 (1970).
72. K. F. Wong and C. A. Eckert, *Ind. Eng. Chem.*, **62**(9), 16 (1970).
73. W. F. K. Wynne-Jones and H. Eyring, *J. Chem. Phys.*, **3**, 493 (1935).

Chapter II

MOLECULAR ORIGIN OF IDEAL SOLUTIONS AND SMALL DEVIATIONS FROM IDEALITY

A. Ben-Naim

1 INTRODUCTION

The concept of ideal solutions accompanies the chemist from the very early stage of studies when he is first exposed to equilibrium constants, solubilities, distribution coefficients, and so on, up to the highest level of research work, when the processing and interpreting of experimental data are required.

In spite of this apparently ubiquitous application, this concept is subject to a considerable amount of confusion, misinterpretation, and often misleading usage. The source of the trouble may be the existence of various kinds of ideality, a distinction among which is difficult to make on the basis of classical thermodynamics alone.

The general idea of an *ideal state* appears in many branches of physics and chemistry. Such are ideal gases, characterized by the absence of interactions among particles, or ideal crystals having perfectly ordered structures.

In solution chemistry the concept of *ideal solutions* refers simultaneously to several completely different ideal states, which of course is an obvious source of confusion. Moreover, not only the term ideal solution, but also the thermodynamic formulation of the various ideal states have a unified appearance:

$$\mu_i = \mu_i^0 + kT \ln x_i \tag{2.1}$$

where μ_i is the chemical potential of the ith component, x_i is its mole fraction, k is the Boltzmann constant, T is the absolute temperature, and μ_i^0 is a constant independent of x_i. Relation (2.1) is a very useful one, as it provides explicit dependence of the chemical potential on composition.

The central purpose of this chapter is to analyze the various cases of ideal solutions and the molecular origin of each one. It seems that there exist three basically different kinds of ideal solution: (1) ideal gas mixtures, (2) symmetric ideal solutions, and (3) dilute ideal solutions. The first and simplest of the three is applicable to any mixture, provided the total density (or pressure) is very low. In other words, it is characterized by either the absence, or negligence, of intermolecular forces. The second and third are of more interest to the chemist, as there is no restriction on the total density of the system. The second ideality arises when the components are similar (in a sense which is more precisely stated in Section 4), for example, two isotopically different components. The third ideality is not restricted to any requirement of similarity, and applies whenever one component is very diluted in the rest of the system. (In all the discussions in this chapter, we omit the cases of ionic solutions and solutes that dissociate in the solution.)

Thus we see that the mere claim that a given solution obeys relation (2.1) offers no information as to the type of ideality under which the system exists. The distinction among the various kinds of ideality becomes even more important when deviations from ideality are considered. In a very superficial way, one may modify relation (2.1) by appending an activity coefficient γ_i so that

$$\mu_i = \mu_i^0 + kT \ln x_i \gamma_i \tag{2.2}$$

where γ_i is presumed to correct for nonideality effects. But again, the mere statement of (2.2) is meaningless if one does not specify the type of ideality being corrected. A given mixture, when looked on as deviating from a symmetric ideal solution, may have a small γ_i, and yet the same mixture, viewed as deviating from a dilute ideal solution, may require a very large γ_i. This point may be magnified by the following illustration. Consider a two-component mixture at a given temperature T, pressure P, and composition x_A (the mole fraction of component A). The chemical potential of A may be expanded in either one of the forms:

$$\mu_A = c_1(T) + kT \ln x_A + \sum_i A_{1i} \rho^i \tag{2.3a}$$

$$= c_2(P,T) + kT \ln x_A + \sum_i A_{2i} \Delta^i \tag{2.3b}$$

$$= c_3(P,T) + kT \ln x_A + \sum_i A_{3i} \rho_A{}^i \tag{2.3c}$$

where c_i are constants, independent of x_A. In (2.3a) we consider deviation from ideal gas mixtures, that is, the expansion in the total density ρ of the system accounts for deviations due to interactions among the various components. In (2.3b) we consider deviations from symmetric ideal solutions, and the expansion is carried out in terms of a dissimilarity parameter Δ (for more details see Section 4). Finally, in (2.3c) we refer to deviations from dilute ideal solutions, hence expansion is carried out in terms of the concentration of component A.

Although any of these expansions may be used legitimately for any system, there are cases in which one choice is preferable to another. For instance, a mixture of $CHCl_3$ and $CHBr_3$ with $x_A \sim \frac{1}{2}$ is very close to a symmetric ideal solution, hence the expansion (2.3b) should give a very small correction to the ideal case. However, the same system when viewed as deviating from a dilute ideal solution results in a relatively large and complicated expansion of the (2.3c) type. A second example may be seen in a solution of benzene in water. Since benzene is diluted in water, it may behave very much like a dilute ideal solution, thus expansion of the form (2.3c) is most suitable.

In order to make the best choice of the expansion in (2.3), we should have some knowledge of the system. Even so, thermodynamics alone does not offer any information on the contents of the expansion coefficients. Furthermore, thermodynamics in general does not specify the meaning of the constants c_i, often referred to as a standard chemical potential. The important observation that can be made from this rudimentary discussion is that once we have made the choice of our reference ideal state, that is, once we have chosen the appropriate expansion for our system, the meaning of the activity function thus defined is committed to a particular kind of deviation. This cannot be reinterpreted in terms of deviations from other kinds of ideality. Failure to make this distinction is a very common source of error.

Relations (2.3) were based on a specific choice of a set of thermodynamic variables, namely, P, T, and x_A. Other choices are also in use, for example, T, V, x_A or T, μ_B, x_A, and so forth, for which the corresponding expansions in (2.3) give different expansion coefficients.

Thus the deviation from certain ideal states depends not only on the kind of ideality chosen, but also on the choice of thermodynamic variables describing the system, a matter often overlooked in current textbooks on thermodynamics. In addition, there is a widespread practice of using different concentration scales such as mole fraction, molarity, or molality. Again, different expansions of the kind (2.3) result from different choices of concentration scale.

In realizing the plethora of cases, there is no surprise that confusion is sometimes the rule rather than the exception. In this chapter we have undertaken to spell out in detail the various kinds of ideality and their molecular origins, in the hope of disentangling the apparent complexity and creating a systematic classification of cases.

Most of the subject matter of this chapter is not new. In fact many authors [2, 4, 12] have already dealt with this problem and suggested various ways of classifying the different concepts of ideality. Here, however, the subject is presented from a somewhat more unified point of view, which uses as the basic ingredients the notion of molecular distribution functions. For this purpose we have summarized in Section 2 some fundamental properties of these functions on an elementary level. Several relations between thermodynamic quantities and molecular distribution functions are discussed in Section 3.

In Section 4 we begin our systematic exploration of the symmetric ideal solutions. The molecular origin of the necessary and the sufficient conditions for such an ideality is discussed in detail. This is followed, in Section 5, by a discussion of small deviations from symmetric ideal solution. Sections 6 and 7 are devoted to the case of dilute ideal solutions and small deviations from it. Here we focus attention on the various correction terms, or activity coefficients, resulting from the different choices of the independent thermodynamic variables. In Section 8 a simple example is presented in which explicit expressions for all kinds of activity coefficients may be given. We conclude the chapter with a discussion of the thermodynamics of transfer, a topic of general importance in solution thermodynamics.

2   MOLECULAR DISTRIBUTION FUNCTIONS

Nowadays the notion of molecular distribution functions commands a central

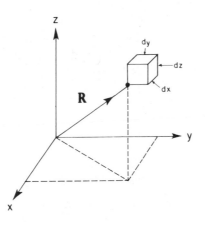

Fig. 2.1. An infinitesimal element of volume, $d\mathbf{R} = dx\,dy\,dz$ located at the point **R**.

role in the theory of fluids. These functions contain important information on the local mode of packing of the molecules on the one hand, and form a bridge between molecular properties and thermodynamic quantities on the other.

In this section a brief and a rather rudimentary presentation of these functions is given. We are concerned mainly with their definitions and some of their basic properties. We hope that ample illustrations will suffice to provide a good understanding of the properties of these functions. A more detailed treatment of them may be found in standard texts on fluids [6, 9, 11, 13]. For the sake of simplicity we restrict ourselves to systems composed of spherical molecules only, although some of the results obtained are valid for more general systems.

Consider a system of N identical spherical molecules contained in a volume V at a temperature T. Let $d\mathbf{R}$ denote an infinitesimal element of volume located at the point **R** (Fig. 2.1), that is,

$$d\mathbf{R} = dx\,dy\,dz \tag{2.4}$$

The singlet molecular distribution function is defined as the average local (number) density of particles at point **R** and is denoted by $\rho(\mathbf{R})$, that is, the average number of particles found in an element of volume $d\mathbf{R}$ at **R** is $\rho(\mathbf{R})\,d\mathbf{R}$. By average we mean average over all configurations of the system subjected to macroscopic characterization by the variables T, V, and N. A particle is counted in $d\mathbf{R}$ whenever its *center* falls within the volume $d\mathbf{R}$. For spherical molecules one may identify the center of the molecule with the center of mass. For more complex molecules one must choose a point to serve as the center of the molecule.

Clearly, since the total number of molecules in the system is fixed, we must arrive at the normalization condition

$$\int_V \rho(\mathbf{R}) \, d\mathbf{R} = N \tag{2.5}$$

where the integration extends over the entire volume of the system under consideration.

In a homogeneous fluid we expect that $\rho(\mathbf{R})$ has the same value at any point $\mathbf{R}$ within the system. (This is true apart from a very small region near the surface of the system, which we always neglect in macroscopic systems.) Thus from (2.5) we obtain

$$\rho(\mathbf{R}) \int_V d\mathbf{R} = \rho(\mathbf{R})V = N \tag{2.6}$$

or

$$\rho(\mathbf{R}) = N/V = \rho \tag{2.7}$$

The last equation states that the local density at any point $\mathbf{R}$ within the fluid is simply the bulk density ρ, a result that could have been anticipated on intuitive grounds as well. Such a relation obviously would not hold for an inhomogeneous system. For instance, if the system was subjected to the influence of an external field, the local density at various points would in general be dependent on the specific position.

One particular kind of inhomogeneity may be imposed on the system by fixing the position of one particle at, say, $\mathbf{R}$. The average local density at a point $\mathbf{R}'$ will be affected by the "external" field of force produced by the particle fixed at $\mathbf{R}$. We denote the local density at $\mathbf{R}'$, given a particle at $\mathbf{R}$, by $\rho(\mathbf{R}'/\mathbf{R})$, and the relevant normalization condition is now

$$\int_V \rho(\mathbf{R}'/\mathbf{R}) \, d\mathbf{R}' = N - 1 \tag{2.8}$$

(which is true if we have selected one of the original N particles of the system to be placed at $\mathbf{R}$. In computing the density at $\mathbf{R}'$, we do not count the particle at $\mathbf{R}$.) The pair distribution function is defined as the quantity $\rho(\mathbf{R}'/\mathbf{R})\rho(\mathbf{R})$. However, for our purposes we are concerned only with the pair correlation function defined below, rather than with the pair distribution function.

We now make use of the following physical reasoning. The function $\rho(\mathbf{R}'/\mathbf{R})$ should not depend on the specific point at which we have chosen to place the particle (which again is true except for a negligibly small region near the boundaries of the system). In addition, because of the spherical symmetry of the field of force produced by the particle at $\mathbf{R}$, we expect that $\rho(\mathbf{R}'/\mathbf{R})$ depends only on the scalar distance between $\mathbf{R}$ and $\mathbf{R}'$ (Fig. 2.2), which we denote by

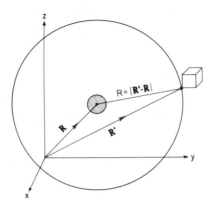

Fig. 2.2. Distribution of particles around a fixed particle at **R**. Because of the spherical symmetry of the field of force produced by this particle, any point on the spherical shell of radius R will have the same local density $\rho g(R)$.

$$R = |\mathbf{R'} - \mathbf{R}| \tag{2.9}$$

Then the local density at any point on the spherical shell of radius R around the point **R** will have the same value. Furthermore, we expect that for large distances $\mathbf{R} \to \infty$, the perturbation due to the particle fixed at **R** will be negligible, hence

$$\rho(\mathbf{R'}/\mathbf{R}) = \rho, \quad R \to \infty \tag{2.10}$$

that is, the local density at $\mathbf{R'}$ is the bulk density that would have been measured in the absence of an external field of force. We now introduce the pair correlation function $g(R)$ defined by

$$\rho(\mathbf{R'}/\mathbf{R}) = \rho g(R), \quad R = |\mathbf{R'} - \mathbf{R}| \tag{2.11}$$

where the dependence on the scalar distance R has been explicitly introduced in the notation $[g(R)$ and not $g(\mathbf{R},\mathbf{R'})]$.

The quantities $\rho(\mathbf{R})$ and $\rho(\mathbf{R'}/\mathbf{R})$ may also be assigned probabilistic meanings. However, for our purposes in this chapter it is sufficient to adopt the interpretation of these quantities in terms of local densities. We now turn to survey some of the most important features of the function $g(R)$.

1. At $R \to \infty$ we expect that $g(R)$ tends asymptotically to unity. This follows formally from the argument leading to (2.10). We say that at very

large separations between $\mathbf{R}$ and $\mathbf{R}'$ there is no correlation between the two points, thus

$$g(R) \xrightarrow{\ R \ \to \ \infty \ } 1 \tag{2.12}$$

(The limiting behavior in (2.12) is valid in the limit of infinitely large systems. For finite systems this is not exactly true, but we are not concerned with this problem here. For more details see Refs. 6 and 9).

2. At very short distances, say, $\mathbf{R} \lesssim \sigma$, where σ is the effective molecular diameter of the particles, we expect that

$$g(R) \xrightarrow{\ R \lesssim \sigma \ } 0 \tag{2.13}$$

This relation simply expresses the fact that the local density at very short distances becomes zero because the presence of the fixed particle at $\mathbf{R}$ excludes the approach of other particles at a distance smaller than σ.

3. For an ideal gas, that is, for a system of noninteracting particles, we have

$$g(R) \equiv 1, \text{ ideal gas} \tag{2.14}$$

which follows from the fact that correlation between particles is due to the existence of interaction. If the latter are required to vanish for an ideal gas, then there will be no correlation between the particles. In other words, fixing the position of one particle does not produce any inhomogeneity in the system, and the local density at each point is simply ρ.

4. At very low density the function $g(R)$ approaches what is essentially the Boltzmann distribution law, that is,

$$g(R) = \exp \ [-\beta U(R)], \quad \rho \to 0 \tag{2.15}$$

where $\beta = (kT)^{-1}$, and $U(R)$ is the pair potential operating between two particles.

The general form of $U(R)$ for simple and spherical molecules is depicted in Fig. 2.3. There are a few analytical functions that have essentially the same form as the one shown in Fig. 2.3. The most commonly used is the so-called Lennard-Jones pair potential [7], which reads

$$U(R) = 4\epsilon [(\frac{\sigma}{R})^{12} - (\frac{\sigma}{R})^6] \tag{2.16}$$

The parameter σ is conveniently regarded as the effective diameter of the molecules. At $R \leqslant \sigma$ the two particles exert strong repulsive forces on each other. The

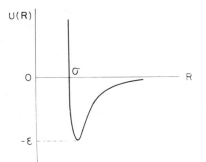

Fig. 2.3. General form of the pair potential $U(R)$ operating between two simple and spherical particles.

minimum of $U(R)$ occurs at $R_{min} = 2^{1/6}\sigma$ and has the value $-\epsilon$. Therefore ϵ may serve as a measure of the strength of the interaction between the two particles.

Note that (2.15) reduces to (2.14) when $U(R) \equiv 0$, that is, when no interaction is operative between the particles. Furthermore, from the form of $U(R)$ and relation (2.15) we expect that the function $g(R)$ at very low density will have a single peak at $R = 2^{1/6}\sigma$, as indeed is the case (see the left-hand side of Fig. 2.4).

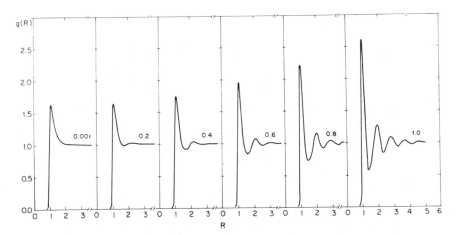

Fig. 2.4. Dependence of the pair correlation function $g(R)$ on the density of the system. The corresponding (number) densities ρ are indicated next to each curve. The functions were computed by numerical solution of the Percus-Yevick equation for a system of Lennard-Jones particles with parameters $\sigma = 1.0$ and $\epsilon/kT = 0.5$. The densities are in dimensionless units $\rho\sigma^3$ (for more details on the computational procedure, see Appendix A).

5. At successively higher densities new peaks evolve in $g(R)$, which become more pronounced as the density increases. The general density dependence of $g(R)$ is demonstrated in Fig. 2.4, and further details are summarized in Table 2.1 (All the illustrations of $g(R)$ in this chapter have been computed by a numerical solution of the Percus-Yevick equation, for a system of Lennard-Jones particles. For more details on the numerical calculations, the reader is referred to Appendix A.)

Table 2.1 Some Features of the Pair Correlation Function and Its Density Dependence[a]

ρ	R_1	$g(R_1)$	R_2	$g(R_2)$	R_3	$g(R_3)$	R_4	$g(R_4)$	G
0.01	1.1	1.63	–	–	–	–	–	–	2.57
0.1	1.1	1.64	2.2	1.02	–	–	–	–	2.34
0.2	1.1	1.65	2.2	1.03	–	–	–	–	1.43
0.3	1.1	1.69	2.2	1.03	3.3	1.003	–	–	0.22
0.4	1.1	1.75	2.2	1.04	3.2	1.004	4.3	1.000	-0.69
0.5	1.1	1.84	2.1	1.05	3.2	1.006	4.2	1.001	-1.13
0.6	1.1	1.95	2.1	1.08	3.1	1.012	4.1	1.002	-1.25
0.7	1.1	2.08	2.1	1.12	3.0	1.022	4.0	1.005	-1.21
0.8	1.1	2.22	2.0	1.16	3.0	1.038	3.9	1.011	-1.10
0.9	1.1	2.33	2.0	1.23	2.9	1.063	3.7	1.016	-1.01
1.0	1.0	2.69	2.0	1.29	2.8	1.093	3.7	1.038	-1.00

[a] Data are for Lennard-Jones particles with parameter $\sigma = 1.0$ and $\epsilon/kT = 0.5$. The densities given in the first column are in dimensionless units $\rho\sigma^3$. R_i is the location of the ith maximum of $g(R)$. The quantity G is defined in (2.20). The details in this table correspond to the curves of Fig. 2.4.

As is clearly seen from Fig. 2.4 and Table 2.1, the various peaks of $g(R)$ occur roughly at integral multiples of σ, that is, at $R \cong \sigma, 2\sigma, 3\sigma....$ This feature reflects the propensity of the spherical molecules to be packed, at least locally, in concentric and nearly equidistant spheres about a given molecule. The first maximum is almost invariably at $R \cong 2^{1/2}\sigma$, which is the location of the minimum of the Lennard-Jones potential [see (2.16)] and reflects the relatively large probability of finding a pair of particles at this distance. The second peak corresponds to the distance between the two particles, in such a way that a third particle can fit perfectly into the space between the two; similarly, higher peaks correspond to filling the space between the two particles by two, three, and so on, particles.

The typical mode of packing around a given molecule is illustrated in Fig. 2.5. On the left-hand side of the figure are two spherical shells of width $d\sigma$ at distances σ and 2σ from the center of a given molecule. The same two spherical

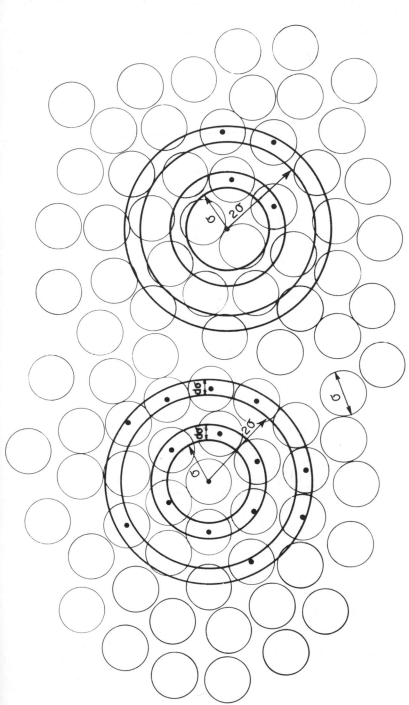

Fig. 2.5. An illustration showing a random distribution of spherical particles of diameter σ. Two spherical shells of width $d\sigma$ are drawn with radius σ and 2σ. On the left-hand side, the center of the sphere coincides with the center of one particle, whereas on the right-hand side the center has been chosen at a random point. It is clearly observed that the two shells on the left-hand side are filled, by center of particles, to a larger extent than the corresponding shells on the right-hand side. This excess of occupation is manifested in the various peaks of $g(R)$.

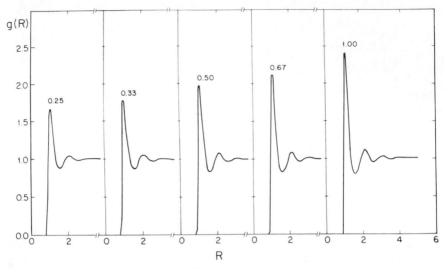

Fig. 2.6. Dependence of $g(R)$ on ϵ/kT. The computations have been carried out for Lennard-Jones particles at one density $\rho = 0.6$ ($\sigma = 1.0$) and various values of ϵ. The values of ϵ/kT are reported next to each curve.

shells are shown on the right-hand side from an arbitrary center. One clearly sees that the shells on the left-hand side are filled with many more particles than the shells on the right-hand side. This excess of particles at distances σ, 2σ, and so on, from a given particle is manifested in the various peaks of the function $g(R)$. As is clearly seen from Fig. 2.4, the correlation function decays quite rapidly to unity at distances of a few molecular diameters. One can see from Table 2.1 that for the reported cases $g(R)$ is practically unity at $R \gtrsim 5\sigma$.

Figure 2.6 demonstrates the dependence of $g(R)$ on temperature, or equivalently on the parameter ϵ. The illustration corresponds to one density $\rho = 0.6$ (with $\sigma = 1.0$), and variations of ϵ/kT. The value of ϵ/kT is indicated next to each curve. One sees that as ϵ increases (or the temperature decreases) all the peaks become more pronounced. The general form of the curves, however, is not changed.

Before turning to mixtures, we should like to emphasize that the limiting behavior of $g(R)$ at large distances ($R \to \infty$) holds only for the limit of infinitely large systems. Failure to take the proper limit may lead to spurious conclusions. This may be demonstrated by a very simple example. Using the normalization conditions (2.5) and (2.8) for a system of finite number of particles, we obtain

$$\int_V [\rho(\mathbf{R}'/\mathbf{R}) - \rho(\mathbf{R}')] \, d\mathbf{R}' = N - 1 - N = -1 \qquad (2.17)$$

However, if one first takes the limit of $g(R)$ for an infinitely large system, or computes $g(R)$ in the Grand (open) ensemble, one obtains the well-known compressibility relation [6, 9]:

$$\int_V [\rho(\mathbf{R}'/\mathbf{R})-\rho]\ d\mathbf{R}' = \rho \int_0^\infty [g(R) - 1]4\pi R^2\ dR = -1 + \rho kT\kappa_T \qquad (2.18)$$

where κ_T is the isothermal compressibility

$$\kappa_T = -\frac{1}{V}\left(\frac{\partial V}{\partial P}\right)_T \qquad (2.19)$$

The discrepancy between (2.17) and (2.18) arises from the difference in the limiting behavior of $g(R)$ at $R \rightarrow \infty$ in the finite and in the infinite systems. In all the following sections we assume the limiting behavior (2.12). We define the quantity

$$G = \int_0^\infty [g(R) - 1]4\pi R^2\ dR \qquad (2.20)$$

which later on plays a crucial role in the theory of solutions. Clearly, following the assumption made above and observing the general form of the function $g(R)$, we expect that the integrand in (2.20) will contribute to the integral only for short distances, say, $0 \leqslant R \lesssim 5\sigma$. Therefore, for practical application, the upper limit of the integration may be taken to be of the order of magnitude of a few molecular diameters.

Next we elaborate on the meaning of the quantity G. We recall that the quantity $\rho g(R)4\pi R^2\ dR$ is the average number of particles in a spherical shell of width dR at a distance R from the center of a given particle. For instance, one may choose one of the spherical shells drawn on the left-hand side of Fig. 2.5. However, the quantity $\rho 4\pi R^2\ dR$ is simply the average number of particles in the same shell, the origin of which has been picked up at random. Therefore the quantity $\rho[g(R) - 1]4\pi R^2\ dR$ reflects the excess (or deficiency) in the number of particles found in a given shell, due to the fixing of a particle at the origin of the sphere. Figure 2.4 shows that there are regions where the quantity $g(R)$ – 1 is either positive or negative. We may say that the magnitude of $g(R)$ – 1 measures the *affinity* of one particle for another at a specific distance R. (This quantity is also referred to as the total correlation at distance R. However, in our discussion of mixtures we find the term affinity more appropriate. See also further discussion below.)

Thus the integral defined in (2.20) may be interpreted as the *overall affinity* of one particle for another. It should be emphasized, however, that the quantity G does not reflect only the molecular properties of the particles involved. For a given system of particles, G is dependent on density (as demonstrated in Table 2.1) as well as on temperature. Figure 2.7 illustrates the temperature

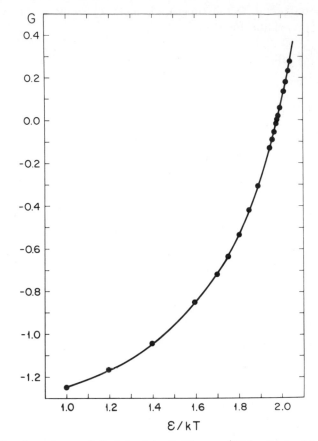

Fig. 2.7. The dependence of G, defined in (2.20), on ϵ/kT for Lennard-Jones particles with unit diameter and density $\rho = 0.6$. Note that at $\epsilon/kT \approx 1.98$ the value of G becomes zero.

dependence of G (for a given system of Lennard-Jones particles with fixed parameters σ and ϵ). One observes that, as the temperature decreases (i.e., ϵ/kT increases), the overall affinity becomes larger and may even change in sign. We shall later see how considering G as an overall affinity reveals its usefulness mainly for mixtures of different particles.

The generalization of all the concepts discussed above for mixtures is quite straightforward. For simplicity we consider only a two-component system of simple and spherical molecules. The two components are designated A and B. There are two singlet distribution functions which are defined as the average local density of A (or B) at a certain point **R**. Following similar arguments as for a one-component system, we obtain the generalization of (2.7) for mixtures

$$\rho_A(\mathbf{R}) = \frac{N_A}{V} = \rho_A \tag{2.21}$$

$$\rho_B(\mathbf{R}) = \frac{N_B}{V} = \rho_B \tag{2.22}$$

where N_A and N_B are the number of molecules of type A and B, respectively.

In a similar fashion one defines the average local density of A (or B) molecules at $\mathbf{R}'$, given an A (or B) molecule at $\mathbf{R}$. Hence we have four correlation functions generalizing relation (2.11):

$$\rho_{AA}(\mathbf{R}'/\mathbf{R}) = \rho_A g_{AA}(R) \tag{2.23}$$

$$\rho_{AB}(\mathbf{R}'/\mathbf{R}) = \rho_A g_{AB}(R) \tag{2.24}$$

$$\rho_{BA}(\mathbf{R}'/\mathbf{R}) = \rho_B g_{BA}(R) \tag{2.25}$$

$$\rho_{BB}(\mathbf{R}'/\mathbf{R}) = \rho_B g_{BB}(R) \tag{2.26}$$

where $R = |\mathbf{R}' - \mathbf{R}|$. For spherical particles we have the symmetry relation

$$g_{AB}(R) = g_{BA}(R) \tag{2.27}$$

Such a relation does not hold for nonspherical particles. However, for most of our purposes, and even for complex molecules, we need the orientational average pair correlation functions for which (2.27) is valid.

Figure 2.8 illustrates some of the salient features of various pair correlation functions for a two-component system. The most important new feature that enters into the case of solutions is the dependence of the spacing between the various peaks on the mole fraction. We recall that in a one-component system the spacing between successive peaks corresponds roughly to σ, 2σ, 3σ, and so on, which corresponds to filling the space between two particles by zero, one, two, and so on, particles of the *same* kind. In mixtures the filling of the space between the two particles under observation depends on the relative concentration of the two components. The illustration in Fig. 2.8 pertains to a mixture of two Lennard-Jones particles with the same energy parameters but with different diameters

$$\epsilon_{AA}/kT = \epsilon_{AB}/kT = \epsilon_{BB}/kT = 0.5 \tag{2.28}$$

$$\sigma_{AA} = 1.0 \qquad \sigma_{BB} = 1.5 \qquad \sigma_{AB} = \tfrac{1}{2}(\sigma_{AA} + \sigma_{BB}) \tag{2.29}$$

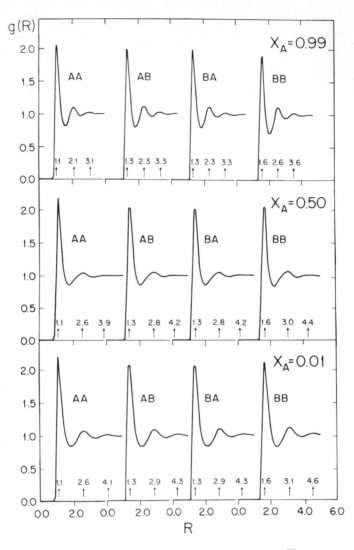

Fig. 2.8. Pair correlation functions for a two-component system. The two components are Lennard-Jones particles with parameters given in (2.28) and (2.29). The total volume density, defined in (2.30) is $\xi = 0.35$. The mole fractions, indicated in each row, are $x_A = 0.99, 0.5, 0.01$. The locations of the various maxima are indicated on the abscissa.

Computations have been carried out for three compositions: $x_A = 0.99$, $x_A = 0.5$, and $x_A = 0.01$. The total density ρ_T of the system has been determined by fixing the composition x_A, and the volume density ξ [14, 15] defined by

$$\xi = \frac{\pi}{6} \left(\rho_A \sigma_{AA}^3 + \rho_B \sigma_{BB}^3 \right) = \frac{\pi \rho_T}{6} \left[x_A \sigma_{AA}^3 + (1 - x_A) \sigma_{BB}^3 \right] \quad (2.30)$$

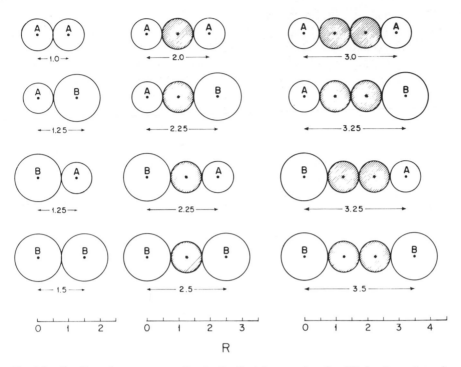

Fig. 2.9. Configurations corresponding to the first three peaks of $g_{ij}(R)$ for the system of B diluted in A (e.g., for x_A = 0.99 of Fig. 2.88). The two particles at the edge of each cluster are the "observed" particles, and their species (A or B) is indicated within the circle [the σ_{ij}'s are as in (2.29)]. The particles filling the space between the particles under observation are the darker circles, and for this particular composition are invariably A particles. Also indicated are the rounded distances at which maxima in the $g_{ij}(R)$ are expected.

The significance of the volume density has been stressed by Throop and Bearman [14, 15], who have also made a systematic study of the pair correlation functions for mixtures of two components.

Consider first the case of x_A = 0.99 in Fig. 2.8. Here, because of the preponderance of A molecules in the system, it is expected that the filling of space between any two observed particles will be done invariably by A molecules. This is demonstrated in Fig. 2.8 and 2.9. For simplicity we ignore the small difference between σ_{ij} and $2^{1/6}\sigma_{ij}$ (the latter being the location of the minimum of the Lennard-Jones potential). The first row in Fig. 2.9 shows the configurations corresponding to the first three peaks of $g_{AA}(R)$ in Fig. 2.8. (The locations of the actual peaks of $g_{AA}(R)$ are indicated in Fig. 2.8, and are seen to be slightly larger than the corresponding distances recorded in Fig. 2.9. This discrepancy is mainly due to the difference between σ_{AA} and $2^{1/6}\sigma_{AA}$, and also to the inaccuracies resulting from the approximate numerical computations. See also Appendix A.)

The second and third rows in Fig. 2.9 show the configurations corresponding to the peaks of $g_{AB}(R)$ and $g_{BA}(R)$, respectively. Note that, by virtue of the symmetry condition (2.27), we expect that these two functions will be the same, and indeed they are in Fig. 2.8; however, because of the approximate nature of the numerical computations, discrepancies in these two functions may be obtained. The fourth row in Fig. 2.9 corresponds to $g_{BB}(R)$.

In all the configurations depicted in Fig. 2.9, we have indicated by A and B the two particles under observation. The particles filling the space between the two observed particles are indicated by dark circles which, as explained above, are only of type A.

The whole situation is reversed if we consider the mixture with $x_A = 0.01$. Here the B molecules are in the majority, and therefore fill the space between any pair of observed particles. Figure 2.8 shows that at $x_A = 0.01$ the spacing between successive peaks of all the $g_{ij}(R)$'s is determined by the diameter of the B molecule, namely, $\sigma_{BB} = 1.5$. A chart similar to that depicted in Fig. 2.9 may be drawn for this case, the only difference being that all the dark circles would be replaced by B particles.

Thus far we have looked at the two extreme concentrations, one in which B is diluted in A, and the other in which A is diluted in B. If we change the composition from $x_A = 0.99$ to $x_A = 0.01$, we observe gradual changes in the locations of the peaks (except for the first one which is always at about σ_{ij}). Indeed, as one varies the mole fraction x_A, one reaches a kind of transition point at which the filling of space changes from A molecules to B molecules. The second and third peaks become very flat, indicating that there is no clear-cut preference as to which kind of particle should fill the intermediate space. One such intermediate composition is shown in Fig. 2.8.

Having computed the various pair correlation functions in mixtures, we proceed to define the quantities

$$G_{ij} = \int_0^\infty [g_{ij}(R) - 1] 4\pi R^2 \, dR \tag{2.31}$$

which, on account of the discussion following (2.20), may be interpreted as the overall affinity of species i for species j in the specified mixture. We note that G_{ij} always fulfilled the symmetry condition

$$G_{AB} = G_{BA} \tag{2.32}$$

This is true even for nonspherical particles, in which case $g_{ij}(R)$ appearing in the integrand is understood to be the orientational average pair correlation function. As we shall see in the following sections, the quantities G_{ij} play a very important role in characterizing the conditions for ideality and expressing small deviations from ideal solutions. The term overall affinity that we have affixed to the quantities G_{ij} should be used with some care. First, this term has already been used in connection with the thermodynamics of chemical reactions [10]. This should not create any source of error, however. Second, one must bear in mind that G_{ij} is an *overall* affinity of i for j. The function $[g_{ij}(R) - 1]$ is an indicator telling us whether at distance R we are to expect a surplus or deficiency of pairs of species i and j.

Once we proceed to the integral (2.31), we should recognize the occurence of positive and negative regions which may cancel each other. In fact, as we have seen in Fig. 2.7 and Table 2.1 (for a one-component system), one may find conditions for which G becomes exactly zero.

3 RELATIONS BETWEEN THE CHEMICAL POTENTIAL AND THE PAIR CORRELATION FUNCTION

In the previous sections we surveyed some of the most important aspects of the pair correlation function in one- and two-component systems. The ample illustrations presented should be sufficient for the nonexpert reader to gain a general feeling for the meaning and the properties of these functions. We now turn to some of the relations between pair correlation functions and thermodynamic quantities. More specifically, we present relations involving either the chemical potential or its density derivatives. For simplicity of notation, we confine our discussion to a two-component system of simple and spherical particles, but some of our results will be of much wider validity. Also, because of its relative simplicity, we develop the expression for the chemical potential in a constant-volume system. However, since a system at constant pressure is of greater practical importance, we include some of the relations for these systems in Appendix B.

The Chemical Potential in a One-Component System

Consider a system of N spherical molecules contained in a volume V at a temperature T. We further assume that the total potential energy of the system at each configuration $\mathbf{R}_1, \cdots, \mathbf{R}_N$ may be approximated as a sum over pair potentials:

$$U_N(\mathbf{R}_1, \cdots, \mathbf{R}_N) = \sum_{1 \leqslant i < j \leqslant N} U(\mathbf{R}_i, \mathbf{R}_j) \tag{2.33}$$

The chemical potential in this system is given by

$$\mu = \left(\frac{\partial A}{\partial N}\right)_{T,V} = A(T,V,N+1) - A(T,V,N) \tag{2.34}$$

where A is the Helmholtz free energy. The second equality on the right-hand side of (2.34) holds for macroscopic systems in which the addition of a single particle may be viewed as an infinitesimal change in the variable N.

Similarly, in the TPN ensemble the chemical potential is given by

$$\mu = \left(\frac{\partial G}{\partial N}\right)_{T,P} = G(T,P,N+1) - G(T,P,N) \tag{2.35}$$

where G is the Gibbs free energy. In the following discussion we develop some relations for the chemical potential in the TVN ensemble; the corresponding relations in the TPN ensemble are treated in Appendix B.

The statistical mechanical expression for the chemical potential is based on the fundamental relation connecting the Helmholtz free energy and the partition function of the system:

$$\exp\left[-\beta A(T,V,N)\right] = \frac{q^N}{N!\Lambda^{3N}} \int \cdots \int d\mathbf{R}_1 \cdots d\mathbf{R}_N \exp\left[-\beta U_N(\mathbf{R}_1,\cdots,\mathbf{R}_N)\right] \tag{2.36}$$

where $\beta = (kT)^{-1}$, k is the Boltzmann constant, and Λ^3 is the momentum partition function of a single moleucle:

$$\Lambda^3 = h^3 (2\pi mkT)^{-3/2} \tag{2.37}$$

h being the Planck constant and m the molecular mass of the molecule; q includes any internal (excluding the momentum) partition functions of a single molecule [we have assumed in (2.36) that internal degrees of freedom of a molecule are not affected by interactions with neighboring molecules].

Rewriting (2.36) for a system of N and $N + 1$ particles, we obtain an expression for the chemical potential

$$\exp\left(-\beta\mu\right) = \exp\left\{-\beta[A(T,V,N+1) - A(T,V,N)]\right\}$$

$$= \frac{q^{N+1}N!\Lambda^{3N} \int \cdots \int d\mathbf{R}_0\, d\mathbf{R}_1 \cdots d\mathbf{R}_N \exp\left[-\beta U_{N+1}(\mathbf{R}_0,\cdots,\mathbf{R}_N)\right]}{(N+1)!\Lambda^{3(N+1)}\, q^N \int \cdots \int d\mathbf{R}_1 \cdots d\mathbf{R}_N \exp\left[-\beta U_N(\mathbf{R}_1,\cdots,\mathbf{R}_N)\right]}$$

$$= \frac{q}{(N+1)\Lambda^3} \int \cdots \int d\mathbf{R}_0\, d\mathbf{R}_1 \cdots d\mathbf{R}_N\, P(\mathbf{R}_1,\cdots,\mathbf{R}_N) \exp\left[-\beta \sum_{i=1}^{N} U(\mathbf{R}_0,\mathbf{R}_i)\right] \tag{2.38}$$

In the last form of (2.38), we have introduced the probability density of finding a specific configuration $\mathbf{R}_1, \cdots, \mathbf{R}_N$ in the TVN system:

$$P(\mathbf{R}_1, \cdots, \mathbf{R}_N) = \frac{\exp \ [-\beta U_N(\mathbf{R}_1, \cdots, \mathbf{R}_N)]}{\int \cdots \int dR_1 \cdots dR_N \ \exp \ [-\beta U_N(\mathbf{R}_1, \cdots, \mathbf{R}_N)]} \quad (2.39)$$

and the total potential energy for the system of $N + 1$ particles has been written as

$$U_{N+1}(\mathbf{R}_0, \mathbf{R}_1, \cdots, \mathbf{R}_N) = U_N(\mathbf{R}_1, \cdots, \mathbf{R}_N) + \sum_{i=1}^{N} U(\mathbf{R}_0, \mathbf{R}_i) \quad (2.40)$$

Denote by

$$\phi(\mathbf{R}_0, \mathbf{R}_1, \cdots, \mathbf{R}_N) = \sum_{i=1}^{N} U(\mathbf{R}_0, \mathbf{R}_i) \quad (2.41)$$

and transform to the variables

$$\mathbf{R}_i' = \mathbf{R}_i - \mathbf{R}_0 \qquad i = 1, \ 2, \ 3, \cdots, N \quad (2.42)$$

We may write (2.38) in the form

$\exp \ (-\beta\mu)$

$= \frac{q}{(N+1)\Lambda^3} \int \cdots \int dR_0 \ dR_1 \cdots dR_N P(\mathbf{R}_1, \cdots, \mathbf{R}_N) \exp \ [-\beta\phi(\mathbf{R}_0, \mathbf{R}_1, \cdots, \mathbf{R}_N)]$

$= \frac{q}{(N+1)\Lambda^3} \int \cdots \int dR_0 \ dR_1' \cdots dR_N' P(\mathbf{R}_1', \cdots, \mathbf{R}_N') \exp \ [-\beta\phi(\mathbf{R}_1', \cdots, \mathbf{R}_N')]$

$= \frac{qV}{(N+1)\Lambda^3} \int \cdots \int dR_1' \cdots dR_N' \ P(\mathbf{R}_1' \cdots \mathbf{R}_N') \exp \ [-\beta\phi(\mathbf{R}_1', \cdots, \mathbf{R}_N')]$

$$= \frac{q}{\rho\Lambda^3} \left\langle \exp \ (-\beta\phi) \right\rangle \quad (2.43)$$

In the second form on the right-hand side of (2.43), we made the transformation of variables [see (2.42)]. Note that the function ϕ, as defined in (2.41), depends only on the *relative* coordinates $\mathbf{R}_i' = \mathbf{R}_i - \mathbf{R}_0$. Hence once we have made the transformation to these variables the resulting integrand is no longer a function of $\mathbf{R}_0$. In the third step in (2.43), we integrated over $\mathbf{R}_0$ to obtain the volume V. Also, we introduced the bulk density $\rho = N/V \cong (N+1)/V$ (macroscopic systems).

Finally, we used the notation $\langle \rangle$ for an average over all possible configurations of the system. The quantity ϕ [see (2.41)] may be viewed as an "external" field acting on a system of N particles, which is produced by fixing a particle at $\mathbf{R}_0$. Of course in a fluid all points are equivalent, and the effect of fixing a particle at $\mathbf{R}_0$ is independent of $\mathbf{R}_0$ (except for a small region near the surface of the system which we always neglect).

Let us rewrite (2.43) in a more convenient form:

$$\mu = kT \ln \rho\Lambda^3 q^{-1} - kT \ln \langle \exp(-\beta\phi) \rangle \qquad (2.44)$$

At this stage we cite an equivalent relation which employs the pair correlation function [6]:

$$\mu = kT \ln \rho\Lambda^3 q^{-1} + \rho \int_0^1 d\xi \int_0^\infty U(R)g(R,\xi)4\pi R^2 \, dR \qquad (2.45)$$

where ξ is a coupling parameter serving to introduce the new particle to the system in a continuous fashion, $U(R)$ is the pair potential, and $g(R,\xi)$ is the pair correlation function around the added particle, the latter being coupled to the extent ξ. We have mentioned (2.45) to illustrate the relation between the chemical potential and the pair correlation function.

We now interpret the second term on the right-hand side of (2.44). Consider the following process of adding one particle at a *fixed* position $\mathbf{R}_0$ to a system specified by the variables TVN. The corresponding change in the Helmholtz free energy is

$$\Delta A(\mathbf{R}_0) = A(T,V,N + 1;\mathbf{R}_0) - A(T,V,N) \qquad (2.46)$$

and the corresponding statistical mechanical expression [compare with (2.34) and (2.38)] is

$$\exp[-\beta\Delta A(\mathbf{R}_0)]$$

$$= \frac{q^{N+1}\Lambda^{3N} N! \int \cdots \int d\mathbf{R}_1 \cdots d\mathbf{R}_N \exp[-\beta U_{N+1}(\mathbf{R}_0, \cdots, \mathbf{R}_N)]}{\Lambda^{3N} N! q^N \int \cdots \int d\mathbf{R}_1 \cdots d\mathbf{R}_N \exp[-\beta U_N(\mathbf{R}_1, \cdots, \mathbf{R}_N)]}$$

$$= q \int \cdots \int d\mathbf{R}_1 \cdots d\mathbf{R}_N \, P(\mathbf{R}_1, \cdots, \mathbf{R}_N) \exp[-\beta\phi(\mathbf{R}_0,\mathbf{R}_1, \cdots, \mathbf{R}_N)]$$

$$= q \langle \exp[-\beta\phi] \rangle \qquad (2.47)$$

It is important to observe the difference between (2.38) and (2.47). Since the added particle in (2.47) is devoid of the translational degree of freedom, it does not carry a momentum partition function. Therefore we have Λ^{3N} instead of $\Lambda^{3(N+1)}$ as in (2.38). In addition, since the added particle was placed at a fixed position, it is distinguishable from the remaining N particles, hence the factor $(N + 1)!$ appearing in (2.38) is replaced by $N!$ in (2.47). Finally, the integration in both the numerator and the denominator extends over only the N locational vectors $\mathbf{R}_1, \cdot \cdot \cdot, \mathbf{R}_N$ and not over $\mathbf{R}_0$ as in (2.38). The external potential function ϕ may be identified in the two expressions. (Note that the choice of the point $\mathbf{R}_0$ is of no importance.)

Comparing the final result in (2.47) with (2.44), we obtain

$$\mu = \Delta A(\mathbf{R}_0) + kT \ln \rho \Lambda^3 \qquad (2.48)$$

This relation has a simple and important interpretation. The work (at T and V constant) required to add a particle to the system μ is split into two parts. First we add the particle to a fixed position $\mathbf{R}_0$, the corresponding work in this step being $\Delta A(\mathbf{R}_0)$. Next we remove the constraint imposed by fixing the position of the particle, the corresponding work being the second term on the right-hand side of (2.48). In the second step the particle "gains" the factor Λ^3 due to the release of translational kinetic energy. The factor V is gained, since the total volume is now accessible to the particle. Finally, once the particle has been released, it is no longer distinguishable from the other particle, and this gives the factor N (or more precisely $N + 1$) which together with the volume V forms the density ρ appearing in (2.48).

Another important property of $\Delta A(\mathbf{R}_0)$ is its "local character." We elaborate on this by using (2.45), although the result is quite general. From (2.45) and (2.48) we obtain

$$\Delta A(\mathbf{R}_0) = -kT \ln q + \rho \int_0^1 d\xi \int_0^\infty U(R) \, g(R,\xi) 4\pi R^2 \, dR \qquad (2.49)$$

Since $U(R)$ falls to zero as R^{-6} (we exclude here the case of ionic solutions), and since $g(R)$ decays to unity at distances of a few molecular diameters, the integration over R effectively extends from zero to some finite distance R_{max}, which is of a microscopic order of magnitude. Hence $\Delta A(\mathbf{R}_0)$ depends on the internal properties through q and on the *immediate* local environment of the particle at $\mathbf{R}_0$. The latter property will be of importance when we consider the solvation properties of various media in Section 9.

Before turning to mixtures, a few comments on the extent of validity of the relations given above are in order.

1. If the total potential cannot be split into pairwise terms as in (2.33), relation (2.48) still holds. However, relation (2.45) must be modified to include integrals over higher-order molecular distribution functions.

2. For complex particles having an orientational-dependent pair potential,

relation (2.48) still holds (in which case one may also require that the added particle be at a fixed position and orientation in the system). If only the position is held fixed, Λ^3 should also include the rotational partition function. If, however, both position and orientation are held fixed, the rotational partition function will be carried out by q. Relation (2.45) must also be modified to include the full angle-dependent pair potential and pair correlation function. In addition, proper modifications for nonadditivity of the potential may be needed, as discussed in the first comment above.

3. For systems in which internal properties of a molecule are appreciably affected by interaction with the environment, the very fundamental relation (2.36) is not valid. The proper treatment in such cases should start with the quantum mechanical partition function. In the following discussion we assume the validity of (2.36) and of (2.48).

Chemical Potentials in a Two-Component System

In a two-component system of A and B molecules, the chemical potential of, say, A, may be obtained from the definition

$$\mu_A = \left(\frac{\partial A}{\partial N_A}\right)_{T,V,N_B} = A(T,V,N_A+1,N_B) - A(T,V,N_A,N_B) \qquad (2.50)$$

One now repeats exactly the same procedure that we carried out in the one-component system and obtains the relation [compare with (2.48)]

$$\mu_A = \Delta A_A(T,\rho_A,\rho_B;\mathbf{R}_0) + kT \ln \rho_A \Lambda_A^3 \qquad (2.51)$$

where $\rho_A = N_A/V$, and Λ_A^3 is the momentum partition function of molecule A. The quantity ΔA_A has the same significance as in (2.48), but now we have explicitly introduced the variables on which this quantity depends. Again, $\mathbf{R}_0$ is introduced to remind us that the position of the added particle is fixed. The quantity ΔA_A does *not* depend on $\mathbf{R}_0$.

The generalization of relations (2.44) and (2.45) is straightforward:

$$\mu_A = kT \ln \rho_A \Lambda_A^3 q_A^{-1} - kT \ln \left\langle \exp\left(-\beta\phi_A\right)\right\rangle \qquad (2.52)$$

$$\mu_A = kT \ln \rho_A \Lambda_A^3 q_A^{-1} + \rho_A \int_0^1 d\xi \int_0^\infty U_{AA}(R)g_{AA}(R,\xi)4\pi R^2 \ dR$$

$$+ \rho_B \int_0^1 d\xi \int_0^\infty U_{BA}(R)g_{BA}(R,\xi)4\pi R^2 \ dR \qquad (2.53)$$

where ϕ_A is the "external" field produced by an A molecule at a fixed position, say, $\mathbf{R}_0$, and all the other symbols have obvious meanings. We note again the local character of ΔA_A $(T,\rho_A,\rho_B;\mathbf{R}_0)$ [see discussion following (2.49)].

We now introduce the notation of the standard chemical potential, defined for the general case by

$$\mu_A^0(T,\rho_A,\rho_B) = \Delta A_A(T,\rho_A,\rho_B;R_0) + kT \ln \Lambda_A^3$$

$$= -kT \ln \left\langle \exp (-\beta\phi_A) \right\rangle + kT \ln \Lambda_A^3 q_A^{-1} \qquad (2.54)$$

Hence relation (2.51) may be rewritten in a more conventional form as

$$\mu_A(T,\rho_A,\rho_B) = \mu_A^0(T,\rho_A,\rho_B) + kT \ln \rho_A \qquad (2.55)$$

[Note that one can formally define μ_A^0 through (2.55).]

The last relation, although it has the conventional *form*, is not being used here in the conventional *sense*. In fact, (2.55) is commonly used only for very dilute solution of A in B. In this case the standard chemical potential tends to a constant independent of ρ_A (this may conveniently be seen from (2.53)), that is,

$$\mu_A(T,\rho_A,\rho_B) = \mu_A^0(T,\rho_B) + kT \ln \rho_A, \quad \rho_A \to 0 \qquad (2.56)$$

Another special case of (2.55) is for pure A, that is, when $\rho_B = 0$. Then we have

$$\mu_A(T,\rho_A) = \mu_A^0(T,\rho_A) + kT \ln \rho_A \qquad (2.57)$$

which is essentially the relation for a one-component system (2.44).

In Section 9, we shall see that all the standard chemical potentials defined in (2.55) to (2.57) are useful for calculating thermodynamic quantities of transfer from one phase to another. It is important to recognize that in all these cases μ_A^0 is related to the work required to introduce a molecule A into a *fixed* position in the solution. We see in (2.54) that, in addition, it contains the term $kT \ln \Lambda_A^3 q_A^{-1}$. However, when we form differences of standard chemical potential, this term drops out and the only remaining difference is in the coupling work of A against its local environment. This aspect is further exploited in Section 9 in interpreting the standard free energies of transfer.

In conclusion, we note that from the thermodynamic point of view the only useful relation is (2.56), which originates from Henry's law. This law is an experimental one appended to the general framework of the laws of thermodynamics. The statistical mechanical treatment not only correctly predicts the form of Henry's law [i.e., relation (2.56)] but also shows that the general split of the chemical potential into two terms is a universal property, as we have demonstrated in relations (2.55) and (2.57).

Density Derivatives of the Chemical Potential

The Kirkwood-Buff [8] theory of solutions provides direct relations between various derivatives of the chemical potential and expressions involving the quantities G_{ij} defined in (2.31). We also here restrict our treatment to a two-component system of A and B molecules. More general results for a multicomponent system may be found in the original work of Kirkwood and Buff [8].

The most important derivatives that concern us are the following, [which are listed in the order of increasing complexity, not of importance]:

$$\left(\frac{\partial \mu_A}{\partial \rho_A}\right)_{T, \mu_B} = kT \left(\frac{1}{\rho_A} - \frac{G_{AA}}{1 + \rho_A G_{AA}}\right) \tag{2.58}$$

$$\left(\frac{\partial \mu_A}{\partial \rho_A}\right)_{T, P,} = kT \left[\frac{1}{\rho_A} - \frac{G_{AA} - G_{AB}}{1 + \rho_A (G_{AA} - G_{AB})}\right] \tag{2.59}$$

$$\left(\frac{\partial \mu_A}{\partial \rho_A}\right)_{T, \rho_B} = kT \left[\frac{1}{\rho_A} - \frac{G_{AA} + \rho_B(G_{AA}G_{BB} - G_{AB}^2)}{1 + \rho_A G_{AA} + \rho_B G_{BB} + \rho_A \rho_B (G_{AA}G_{BB} - G_{AB}^2)}\right]$$

$$\tag{2.60}$$

It is worthwhile noting that the response of μ_A to variations in ρ_A *strongly* depends on the independent variables chosen for the system. The three derivatives listed above correspond to three different processes which are schematically shown in Fig. 2.10. The three derivatives have one common feature, the ρ_A^{-1} divergence at $\rho_A \to 0$. In fact, as further discussed later, in the limit of very dilute solutions, the chemical potential has the same formal appearance. This is no longer true for finite concentrations of A.

The first derivative measures the change in μ_A on changing ρ_A, when μ_B is constant. (The temperature is constant in the following discussion.) This derivative is important in osmotic experiments in which the system is allowed to be in contact with a reservoir of constant μ_B (Fig. 2.10a). Clearly, the corresponding derivative is the simplest, and in fact it is also the one with exactly the same form as for a one-component system. Here we can merely drop the condition of constant μ_B to obtain the appropriate derivative for the chemical potential of a one-component system.

The second derivative is the most practical (Fig. 2.10b), since most experiments are carried out at constant pressure (and temperature). The third one, although quite complicated, may be of some use when working under constant-volume conditions. We explore some of the implications of these relations in Sections 6 and 7 when we deal with dilute solutions.

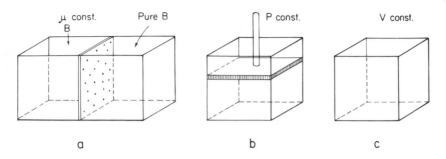

Fig. 2.10. Schematic illustration of the three different experimental setups corresponding to relations (2.58) to (2.60). In all cases the temperature is presumed to be constant. In addition, μ_B is constant in *a*, *P* in *b* and *v* in *c*.

For symmetric solutions (see next section), a somewhat more useful, though equivalent, relation to (2.59) may be obtained by replacing the density derivative by the mole fraction derivative:

$$\left(\frac{\partial \mu_A}{\partial x_A}\right)_{T,P} = kT \left[\frac{1}{x_A} - \frac{\rho_B(G_{AA} + G_{BB} - 2G_{AB})}{1 + \rho_B x_A(G_{AA} + G_{BB} - 2G_{AB})} \right] \quad (2.61)$$

where

$$x_A = \frac{N_A}{N_A + N_B} = \frac{\rho_A}{\rho_A + \rho_B} \quad (2.62)$$

It should be noted that relations (2.58) to (2.61) do not involve the pairwise additivity assumption. Relation (2.45), for instance, is valid only under the assumption of pairwise additivity of the total potential energy. Furthermore, for complex molecules where in general the pair correlation function depends on distance as well as on orientation, (2.58) to (2.61) are still valid, with the understanding that the $g_{ij}(R)$'s are the orientational averaged pair correlation functions. In this respect these latter relations are very general and may be applied to a very wide class of substances.

4 SYMMETRIC IDEAL SOLUTIONS, NECESSARY AND SUFFICIENT CONDITIONS

Most of this section is devoted to the study of the molecular origin of symmetric ideal (SI) behavior in a two-component system at a given temperature T and pressure P. These are the most common variables from the practical point of view. The analogous treatment of mixtures at constant temperature and volume is only briefly discussed in Appendix C.

One possible way to define a SI solution—the one we shall adopt here—is to require that the chemical potential of each component obey the relation

$$\mu_i = \mu_i^p + kT \ln x_i \qquad 0 \leqslant x_i \leqslant 1; \quad P \text{ and } T \text{ constant} \tag{2.63}$$

That is, we require that (2.63) be valid in the entire range of composition $0 \leqslant x_i \leqslant 1$, keeping P and T constant. Clearly, since we can put $x_i = 1$ in (2.63), the quantity μ_i^p may be assigned the meaning of the chemical potential of pure component i at the same P and T. The superscript p affixed to μ_i^p serves to stress the sharp difference between this quantity, which is a chemical potential of pure i, and the standard chemical potential defined for various cases in the previous section.

It is also useful to require that (2.63) be valid in a small neighborhood of P and T to allow differentiation with respect to these variables. It is superfluous to require the validity of (2.63) for any P and T. No real mixture is expected to fulfill such an exaggerated requirement. For the moment we assume however that (2.63) is given at fixed values of P and T. We come back to this point at the end of this section.

From here on we concentrate on a two-component system of species A and B. Relation (2.63) may be differentiated to yield

$$\left(\frac{\partial \mu_A}{\partial x_A}\right)_{T,P} = \frac{kT}{x_A} \qquad 0 \leqslant x_A \leqslant 1 \tag{2.64}$$

Conversely, if (2.64) holds in the entire range of composition, by integration we obtain

$$\mu_A = \text{constant} + kT \ln x_A \qquad 0 \leqslant x_A \leqslant 1 \tag{2.65}$$

The constant of integration in (2.65) is clearly identified with μ_A^p in (2.63). Hence the conditions (2.63) and (2.64) are equivalent, and either may serve to define the concept of a SI solution.

We now recall relation (2.61) of the previous section, and rewrite it in the form

$$\left(\frac{\partial \mu_A}{\partial x_A}\right)_{T,P} = kT \left(\frac{1}{x_A} - \frac{\rho x_B \Delta_{AB}}{1 + \rho x_A x_B \Delta_{AB}}\right) \tag{2.66}$$

where $\rho = \rho_A + \rho_B$ is the total density, and

$$\Delta_{AB} = G_{AA} + G_{BB} - 2G_{AB} \tag{2.67}$$

Equations (2.64) and (2.66) show that, for any nonzero density ρ, a necessary and sufficient condition for a SI solution (in a binary system at constant P and T) is

$$\Delta_{AB} = 0 \qquad 0 \leqslant x_A \leqslant 1 \qquad\qquad (2.68)$$

Note that, if $\rho \rightarrow 0$, then (2.66) reduces to (2.64), which is the case of an ideal gas mixture (see also Section 8). However, in (2.63) and (2.64) we have chosen P and T and, in general, this choice determines a nonzero total density. Furthermore, if $x_B \rightarrow 0$, then (2.66) reduces to an equation similar to (2.64), but because of the requirement $0 \leqslant x_A \leqslant 1$ it is not equivalent to (2.64). This case is treated in detail in Section 6.

That (2.68) is a sufficient condition of SI solutions follows directly from the nullification of the second term on the right-hand side of (2.66). Conversely, in order that (2.64) and (2.66) be identical, we must require that

$$x_B \rho \Delta_{AB} = 0 \qquad 0 \leqslant x_A \leqslant 1 \qquad\qquad (2.69)$$

Therefore, excluding the case of an ideal gas mixture, the necessary condition should be (2.68).

It is very important to stress that condition (2.68) for SI solutions does not depend on any modelistic assumptions on the nature of our system and, in this respect, is very general. It is appropriate to cite at this juncture a *sufficient* condition for a SI solution within the regime of lattice models of solutions (with particles of equal "size"):

$$W = W_{AA} + W_{BB} - 2W_{AB} = 0 \qquad\qquad (2.70)$$

where W_{ij} is the interaction energy between the species i and j, situated on two neighboring lattice points. Condition (2.70) is meaningful only within the context of the lattice model; it is a sufficient condition for a SI solution but not a necessary one.

Condition (2.68) certainly implies some degree of similarity between the two components. It is a far less stringent requirement than is usually implied in elementary textbooks. Let us further elaborate on this point.

Consider a system of two simple and spherical components. The *strongest* condition of similarity is to require that the pair potential operating between any pair of species is the same function for all pairs:

$$U_{ij}(R) = f(R) \qquad \text{for all species } i,j \qquad\qquad (2.71)$$

It is very easy to show that (2.71) is a sufficient condition for a SI solution. (This is shown in a direct manner in Appendix B, or may be inferred from the discussion following (2.73).) This condition may be expected for *almost identical* components, such as two molecules differing by isotopic substitution. Yet condition (2.68) is a far weaker one. To see this, consider the following successive series of sufficient conditions for SI solutions:

$$U_{AA}(R) = U_{AB}(R) = U_{BA}(R) = U_{BB}(R) \tag{2.72a}$$

$$g_{AA}(R) = g_{AB}(R) = g_{BA}(R) = g_{BB}(R) \tag{2.72b}$$

$$G_{AA} = G_{BB} = G_{AB} \qquad 0 \leqslant x_A \leqslant 1 \tag{2.72c}$$

$$G_{AA} + G_{BB} - 2G_{AB} = 0 \tag{2.72d}$$

[Note that for spherical particles $U_{AB} = U_{BA}$ and $g_{AB} = g_{BA}$; however, these equalities do not hold in the general case. On the other hand, the equality $G_{AB} = G_{BA}$ is always valid. All the forthcoming arguments hold for a specific temperature and pressure.]

Clearly, each of the conditions in (2.72) follows from its predecessor, and may be stated symbolically as

$$(2.72a) \quad \Rightarrow \quad (2.72b) \quad \Rightarrow \quad (2.72c) \quad \Rightarrow \quad (2.72d) \tag{2.73}$$

The first relation $(2.72a) \Rightarrow (2.72b)$ may be shown to follow directly from the definition of the pair correlation function for systems with a pairwise additive potential. Qualitatively, if the field of force produced by, say, an A molecule, is the same as that produced by a B molecule, the correlation function for any pair of molecules must be independent of their species. The second and third conditions in (2.73) follow directly from the definitions.

We have seen that (2.72d) is a sufficient condition for a SI solution, hence any condition that precedes (2.72d) in (2.72) must also be a sufficient condition for a SI solution. In particular, we see that (2.72a) is a sufficient condition, an assertion that is shown in a direct manner in Appendix B.

Note also that, in general, the conditions in (2.72) are in a decreasing order in the sense that each one is a weaker condition than its predecessor. For instance, (2.72c) requires equality between *integrals,* whereas (2.72b) requires equality of the *integrands*. Clearly, (2.72d) is much weaker than (2.72c). Note that, if (2.72d) is required to hold identically for *all T* and *P,* it is likely that this condition will approach (2.72a). However, as we have noted before, such a requirement usually makes the concept of ideality meaningless, since no real mixture is expected to fulfill such a condition.

Furthermore, since condition (2.72d) was shown to be a *necessary* condition for a SI solution, it must also be the weakest sufficient condition. To prove this, suppose we have been given a new condition (2.72e) as a sufficient condition which is weaker than (2.72d):

$(2.72d) \Rightarrow (2.72e)$, $(2.72e)$ weaker than $(2.72d)$

$(2.72e) \Rightarrow SI$, $(2.72e)$ is a sufficient condition for SI

Since $(2.72d)$ is a necessary condition, that is,

$$SI \Rightarrow (2.72d)$$

it follows, from the above three relations, that

$$(2.72e) \Rightarrow SI \Rightarrow (2.72d)$$

hence $(2.72e)$ and $(2.72d)$ must be equivalent.

We conclude by reemphasizing that, not only is condition $(2.72d)$ [i.e., (2.68)] the weakest for SI solutions, but it is also the condition that is applicable to the widest class of systems (i.e., systems of nonspherical molecules having nonpairwise additive potential).

For the purposes of this chapter, we refer to the two components fulfilling condition $(2.72d)$ in the entire range of compositions as *similar*. This term should contrast with the term *almost identical*, appropriate for two components fulfilling condition $(2.72a)$. We note, however, that the concept of similarity defined above depends on temperature and pressure. Thus in general a pair of components may be similar at one temperature and not similar at another temperature. Moreover, the concept of similarity may sometimes be contradictory to our intuition. The following example demonstrates this point.

Consider two components A and B interacting via a Lennard-Jones potential with parameters

$$\sigma_{AA} = 1.0 \quad \sigma_{BB} = 0.5 \quad \sigma_{AB} = \tfrac{1}{2}(\sigma_{AA} + \sigma_{BB}) \qquad (2.74)$$

$$\epsilon_{AA}/kT = 0.5 \quad \epsilon_{BB}/kT = 0.5 \quad \epsilon_{AB} = (\epsilon_{AA}\epsilon_{BB})^{\frac{1}{2}} \qquad (2.75)$$

and suppose that the total density is low enough, say, $\rho = 0.1$, so that relation (2.15) is approximately correct, that is,

$$G_{ij} = \int_0^\infty \left\{ \exp\left[-\beta U_{ij}(R)\right] - 1 \right\} 4\pi R^2 \; dR \qquad (2.76)$$

Clearly, if $\sigma_{BB} = 1.0$, the two components will be identical and $\Delta_{AB} = 0$. We have introduced a dissimilarity by choosing $\sigma_{BB} = 0.5$, but have kept the same energy parameters. The value of Δ_{AB} computed for this case was about 0.6. (Note that because of the particular choice of density all G_{ij}'s and Δ_{AB}

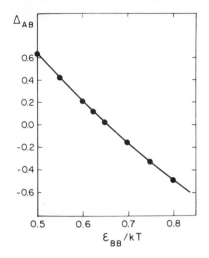

Fig. 2.11. Dependence of Δ_{AB} [defined in (2.67)] on ϵ_{BB}/kT for Lennard-Jones particles with parameters given in (2.74) and (2.75).

are independent of composition.) Starting with the parameters as in (2.74) and (2.75), we now change ϵ_{BB}/kT and therefore make the two components even more dissimilar. Figure 2.11 shows the dependence of Δ_{AB} on ϵ_{BB}/kT. One sees that for $\epsilon_{BB}/kT \approx 0.65$ we obtain $\Delta_{AB} = 0$. At this point the two components, although seemingly far from similar in the usual sense, form a SI solution. This illustrates the idea that the energy and distance parameters in the Lennard-Jones potential may be adjusted in such a way so that the resulting quantity Δ_{AB} may be zero, even though the two components do not seem to be very similar at first sight. In other words, dissimilarity in the energies may compensate for dissimilarity in the diameters. Of course for more complicated molecules there are more possibilities for compensation. Here, although condition (2.68) is still applicable, the pair potential is no longer dependent on two parameters as in the Lennard-Jones particles. It is therefore no wonder that real mixtures other than isotopic mixtures show the behavior of SI solutions. One well-known example is the mixture of ethylene bromide and propylene bromide [3].

One of the most important features of condition (2.68) is that it involves the local environment around the various species. As shown in Section 2, the integral defining G_{ij} gains a contribution only from a short range R_{max} of an order of magnitude of a few molecular diameters. Condition (2.68) states that the overall affinity of A for B is the arithmetic average of the affinity of A for A and B for B. Failure to recognize the local character of this condition may lead to some paradoxical applications of the idea of SI solutions. An outstanding example is the application of the mixture model approach to fluids and, in particular, to liquid water. Here one usually defines two species of water molecules which differ in local environment. Thus, although the two components are *identical* from the point of view of chemical constituents, by definition they markedly differ in local environment. Hence the condition for SI solutions for two such components is unlikely to be obeyed.

Finally, we should mention a phenomenological characterization of SI solutions in terms of their partial molar enthalpies and entropies. To do this we must assume that (2.63) may be differentiated with respect to temperature and pressure. Or equivalently we assume that condition (2.68) is fulfilled for all compositions $0 \leqslant x_A \leqslant 1$ and also that the *first* derivatives of Δ_{AB} with respect to P and T are zero. Clearly, we do not have to assume that *all* the derivatives of Δ_{AB} are zero (as noted before, such an exaggerated requirement renders the concept of a SI solution useless).

By differentiating (2.63) with respect to T and P, we, obtain

$$-\overline{S}_i = \left(\frac{\partial \mu_i}{\partial T}\right)_P = -S_i^p + k \ln x_i \qquad (2.77)$$

$$\overline{H}_i = \mu_i + T\overline{S}_i = \mu_i^p + kT \ln x_i + T(S_i^p - k \ln x_i)$$
$$= \mu_i^p + TS_i^p = H_i^p \qquad (2.78)$$

$$\overline{V}_i = \left(\frac{\partial \mu_i}{\partial P}\right)_T = \left(\frac{\partial \mu_i^p}{\partial P}\right)_T = V_i^p \qquad (2.79)$$

It is easily shown that, if (2.77) and (2.78) are obeyed for each component in the entire range of composition, the solution is a SI one.

An alternative way of describing SI solutions is to introduce excess functions, defined for the whole mixture as follows:

$$G^E = G - G^{\text{ideal}} = G - \left[\sum_i n_i(\mu_i^p + kT \ln x_i)\right] \qquad (2.80)$$

$$S^E = S - S^{\text{ideal}} = S - \left[\sum_i n_i(S_i^p - k \ln x_i)\right] \qquad (2.81)$$

$$V^E = V - V^{\text{ideal}} = V - \sum_i n_i V_i^p \qquad (2.82)$$

$$H^E = H - H^{\text{ideal}} = H - \sum_i n_i H_i^p \qquad (2.83)$$

The ideal solutions are characterized by having zero excess functions (2.80) to (2.83).

5 SMALL DEVIATIONS FROM SYMMETRIC IDEAL SOLUTIONS

We again consider mixtures of two components at a given temperature and pressure. As noted in the introduction, any mixture may be viewed as a SI solution with a correction term to account for the dissimilarity between the two components. Such a viewpoint is most useful, however, when the two components indeed deviate slightly from similarity. More precisely, the condition for a SI solution is $\Delta_{AB} = 0$; a small deviation from similarity is referred to whenever Δ_{AB} is very small in the entire range of compositions. Thus expansion in the first order with respect to the parameter Δ_{AB} is appropriate. Within the realm of the first-order deviations from SI solutions, we also include regular and athermal solutions. These may, however, be defined for a more general system using phenomenological rather than molecular criteria.

General Treatment

Consider again the exact relation (2.61) which we rewrite here as

$$\left(\frac{\partial \mu_A}{\partial x_A}\right)_{T,P} = kT\left(\frac{1}{x_A} - \frac{\rho x_B \Delta_{AB}}{1 + \rho x_A x_B \Delta_{AB}}\right) \tag{2.84}$$

with

$$\Delta_{AB} = G_{AA} + G_{BB} - 2G_{AB} \tag{2.85}$$

and $\rho = \rho_A + \rho_B$, the total number density.

Similar components were defined by the requirement that $\Delta_{AB} = 0$ for all compositions $0 \leqslant x_A \leqslant 1$ (P and T constant). Here we assume that Δ_{AB} is very small in the entire range of composition. More precisely, we require that

$$\rho x_A x_B \Delta_{AB} \ll 1 \qquad 0 \leqslant x_A \leqslant 1 \tag{2.86}$$

[Note again that we exclude in the present section the trivial case of ideal gas mixtures attained at $\rho \to 0$, which assures the validity of (2.86) independently of the value of Δ_{AB}. For nonzero total density, condition (2.86) is in fact a condition on Δ_{AB}.] Expanding (2.84) to first order in the dissimilarity parameter Δ_{AB}, we obtain

$$\left(\frac{\partial \mu_A}{\partial x_A}\right)_{T,P} = kT\left(\frac{1}{x_A} - \rho^0 x_B \Delta_{AB} + \cdots\right) \tag{2.87}$$

where ρ^0 is the limit of the total density of the same system but with $\Delta_{AB} = 0$, namely,

$$\rho^0 = \lim_{\Delta_{AB} \to 0} \rho \qquad\qquad [x_A, P, \text{ and } T \text{ constant}] \qquad (2.88)$$

Since (2.86) is presumed to be valid for all compositions, we may integrate (2.87) to obtain

$$\mu_A = \mu^p + kT \ln x_A + kT \int_0^{x_B} \rho^0 x_B \Delta_{AB} \, dx_B \qquad (2.89)$$

The constant of integration has been identified as μ_A^p by substituting $x_B = 0$. In general, both ρ^0 and Δ_{AB} may depend on temperature, pressure, and composition. The dependence of ρ^0 on composition is known, since at $\Delta_{AB} = 0$ we have a SI solution. Therefore

$$\rho^0 = \frac{1}{x_A V_A^p + x_B V_B^p} \qquad (2.90)$$

where V_A^p and V_B^p are the molar volumes of A and B and thus independent of composition.

A simple case occurs when $\rho^0 \Delta_{AB}$ is independent of composition. For instance, if $V_A^p \approx V_B^p$, then ρ^0 is independent of x_A. This case also occurs in the treatment of lattice models, where A and B may interchange on the same sites. (An example for which Δ_{AB} is independent of x_A is treated in Section 8). In this case the integration in (2.89) is immediate, and the result is

$$\mu_A = \mu_A^p + kT \ln (x_A) + \frac{kT\rho^0 \Delta_{AB} x_B^2}{2} + \cdots \qquad (2.91)$$

It is important to recognize that the expansion in terms of Δ_{AB} is carried out at a given P, T, and x_A; that is, we have exploited the total of the 3 degrees of freedom the system possesses. This means that the expansion (2.87) cannot be realized experimentally, since Δ_{AB} is not an experimentally controllable parameter. Nevertheless, the variation in Δ_{AB} may be conceived either by a theoretical model or by considering a series of mixtures with the same T, P, and x_B but having different (however small) values of Δ_{AB}. We have stressed this point in order to show the contrast with a different expansion, discussed in Section 7, which may be carried out experimentally.

Using (2.89) or (2.91), we may introduce an activity coefficient

$$kT \ln \gamma_A^S = kT \int_0^{x_B} \rho^0 x_B \Delta_{AB} \, dx_B \qquad (2.92)$$

and rewrite (2.89) as

$$\mu_A = \mu_A^p + kT \ln x_A \gamma_A^S \qquad (2.93)$$

Here, the superscript S in γ_A^S serves to stress the fact that this activity coefficient bears the deviations from SI solutions. The limiting behavior of γ_A^S is

$$\lim_{\Delta_{AB} \to 0} \gamma_A^S = 1 \qquad 0 \leqslant x_A \leqslant 1; \; P \text{ and } T \text{ constant} \qquad (2.94)$$

(Note that γ_A^S also tends to unity when either $\rho \to 0$ or $x_B \to 0$. These two cases are considered separately in Sections 7 and 8.) The usefulness of γ_A^S is revealed for systems of finite density and arbitrary composition, provided that the dissimilarity parameter Δ_{AB} is sufficiently small.

We now discuss two special cases of the first-order expansion given in (2.91). These are the so-called regular and athermal solutions. An extensive investigation has been carried out on these solutions within the regime of lattice models for solution [3]. In this model [3, 5] there is a natural distinction between the size of the molecules and the strength of the interaction parameter. Such a distinction is somewhat arbitrary for liquid solutions, especially for nonspherical molecules.

Regular Solutions [3, 5]

The general phenomenological definition of a regular solution states that the partial molar entropy of each component has the same form it has in an ideal solution (2.77):

$$\bar{S}_i = S_i^p - k \ln x_i \qquad 0 \leqslant x_i \leqslant 1 \qquad (2.95)$$

This definition is very general and does not imply any specific model for the solution. Within the context of this section, we obtain by differentiating (2.93) with respect to temperature

$$\bar{S}_A = S_A^p - k \ln x_A - \frac{\partial}{\partial T}(kT \ln \gamma_A^S) \qquad (2.96)$$

Hence the phenomenological definition of a regular solution may be equivalently stated by requiring that

$$\frac{\partial}{\partial T}(kT \ln \gamma_A^S) = 0 \qquad \text{for } 0 \leqslant x_A \leqslant 1 \qquad (2.97)$$

From (2.95) it follows that

$$\bar{H}_i = \mu_i + T\bar{S}_i = H_i^p + kT \ln \gamma_i^S \qquad (2.98)$$

That is, in regular solutions the deviations from ideality are borne solely by the partial molar enthalpy.

We now consider first-order deviation, expressed in (2.91), and examine the conditions for regular solutions from the molecular point of view. The general condition (2.97) transcribed to this case is

$$\frac{\partial}{\partial T}(T\rho^0 \Delta_{AB}) = 0 \tag{2.99}$$

At this juncture it is appropriate to cite the relation for the activity coefficient of regular solutions within the lattice model, namely,

$$kT \ln \gamma_A^S = \frac{-cWx_B^2}{2} \tag{2.100}$$

where c is the number of nearest neighbors, and W is the exchange energy defined by

$$W = W_{AA} + W_{BB} - 2W_{AB} \tag{2.101}$$

and W_{ij} is the interaction energy between two molecules of species i and j, situated on a pair of adjacent sites on the lattice. Clearly, (2.100) fulfills the condition (2.97), both c and W are presumed to be temperature-independent.

For liquid mixtures at relatively high density, the temperature dependence of the density is usually small. If this may be neglected, then (2.99) is fulfilled whenever

$$\Delta_{AB} \sim \frac{1}{T} \tag{2.102}$$

We now demonstrate an example of the behavior (2.102), although not necessarily an example of a regular solution. We recall the low-density behavior of the pair correlation functions (2.15):

$$g_{ij}(R) = \exp\ [-\beta U_{ij}(R)] \tag{2.103}$$

Hence in this case we have

$$\Delta_{AB} = \int_0^\infty \left\{ \exp\ [-\beta U_{AA}(R)] - 1 \right\}\ 4\pi R^2\ dR$$
$$+ \int_0^\infty \left\{ \exp\ [-\beta U_{BB}(R)] - 1 \right\}\ 4\pi R^2\ dR$$
$$-2 \int_0^\infty \left\{ \exp\ [-\beta U_{AB}(R)] - 1 \right\}\ 4\pi R^2\ dR \tag{2.104}$$

Consider, as a special case, Lennard-Jones particles having the same "diameter," and that all ϵ_{ij} are small relative to kT:

$$\sigma \equiv \sigma_{AA} = \sigma_{BB} = \sigma_{AB} \qquad (2.105)$$

$$\beta\epsilon_{AA} \ll 1 \qquad \beta\epsilon_{BB} \ll 1 \qquad \beta\epsilon_{AB} \ll 1 \qquad (2.106)$$

Since for $R < \sigma$ the potential function becomes large and positive, we may split each integral in (2.104):

$$\int_0^\infty \left\{ \exp\left[-\beta U_{ij}(R) - 1\right] \right\} 4\pi R^2 \, dR \approx \int_0^\sigma -4\pi R^2 \, dR$$
$$+ \int_\sigma^\infty -\beta U_{ij}(R) 4\pi R^2 \, dR \qquad (2.107)$$

That is, for $R < \sigma$ we assume that $\exp\left[-\beta U_{ij}(R)\right] \approx 0$, and for $R > \sigma$ we expand the exponent on account of assumption (2.106). Substituting the approximate split of (2.107) into (2.104), we obtain

$$\Delta_{AB} = -\frac{1}{kT} \int_\sigma^\infty [U_{AA}(R) + U_{BB}(R) - 2U_{AB}(R)] 4\pi R^2 \, dR \quad (2.108)$$

Clearly, since $U_{ij}(R)$ is presumed temperature-independent, we have here an example of behavior (2.102). The integral in (2.108) may be viewed as a generalization of the concept of exchange energy, given in (2.101), to the continuous case in which the whole pair potential must be considered, rather than interactions on lattice distances.

It should be stressed that the above example showing the behavior $\Delta_{AB} \sim T^{-1}$ is not a good example for regular solutions. This follows from the fact that, in the region where (2.103) is valid, the density depends strongly on temperature, hence the original condition for regular solutions (2.99) would be violated.

Athermal Solutions

The phenomenological definition of an athermal solution requires the partial molar enthalpy of each component to be the same as the molar enthalpy of the corresponding pure component (at the same P and T), that is,

$$\overline{H}_i = H_i^p \qquad 0 \leqslant x_i \leqslant 1 \qquad (2.109)$$

Hence from (2.93) and (2.109) it follows that

$$\overline{S}_i = \frac{\overline{H}_i - \overline{\mu}_i}{T} = S_i^p - k \ln x_i \gamma_i^S \qquad (2.110)$$

However, we have for the entropy:

$$\overline{S}_i = - \left(\frac{\partial \mu_i}{\partial T}\right)_P = S_i^p - k \ln x_i - \frac{\partial}{\partial T}(kT \ln \gamma_i^S) \qquad (2.111)$$

Comparing (2.110) and (2.111) we arrive at the following condition for athermal solutions:

$$\frac{\partial}{\partial T}(\ln \gamma_i^S) = 0 \qquad 0 \leqslant x_i \leqslant 1 \qquad (2.112)$$

(compare this with condition (2.97) for regular solutions).

We see that from the phenomenological point of view athermal solutions are characterized by having the same partial molar enthalpy as ideal solutions. The whole nonideality is borne solely by the partial molar entropy.

Confining ourselves again to the first-order deviation from SI solutions as expressed in (2.91), we may state the molecular condition for an athermal solution as

$$\frac{\partial}{\partial T}(\rho^0 \Delta_{AB}) = 0 \qquad 0 \leqslant x_i \leqslant 1 \qquad (2.113)$$

[Compare with (2.99) for regular solutions.] Athermal solutions have been studied mainly within the lattice models of solutions. It is easily shown that in this case the mixture of molecules of different sizes, but otherwise having similar interaction parameters, leads to condition (2.113). Again, for solutions at relatively high densities, the temperature dependence of ρ^0 is usually small. Hence condition (2.113) reduces to the requirement that Δ_{AB} be temperature-independent. The latter may be demonstrated again by mixtures obeying relation (2.103).

Consider a mixture of hard spheres at low densities. Each integral in (2.104) may be evaluated to yield

$$\int_0^\infty \left\{ \exp\left[-\beta U_{ij}(R)\right] - 1 \right\} 4\pi R^2 \, dR = \int_0^{\sigma_{ij}} -4\pi R^2 \, dR = -\frac{4\pi\sigma_{ij}^3}{3}$$

$$(2.114)$$

Hence

$$\Delta_{AB} = -\frac{4\pi}{3}\left(\sigma_{AA}^3 + \sigma_{BB}^3 - 2\sigma_{AB}^3 \right) \qquad (2.115)$$

We see that the temperature dependence is destroyed in (2.114), since for hard spheres $U_{ij}(R)$ and $\beta U_{ij}(R)$ are the same functions of R. Equation (2.115) is an example for which Δ_{AB} is temperature-independent. Note again that (2.115) by itself does not ensure the validity of (2.113). At higher densities of hard spheres, one expects that the mixture will be athermal, that is, that (2.113) will be obeyed.

We have surveyed in this section the conditions for regular and athermal solutions. The general phenomenological characterization may be applied to any kind of deviation from ideality. Yet the most useful characterization of these solutions, from the molecular point of view, seems to adhere to the small deviations from SI solutions in the special form expressed in (2.91).

6   DILUTE IDEAL SOLUTIONS

This section is devoted to a different kind of ideality which is very important in various applications in solution thermodynamics. The very fact that we use the term dilute ideal (DI) solution implies that we distinguish between the status of the various components. In the general case, one component, presumed to be diluted in the rest of the system, is referred to as the solute. We again restrict our treatment to a two-component system, say, A and B, and assume that A (the solute) is diluted in B (the solvent).

Because of the asymmetric role of the two components, one should discuss the behavior of the solute and the solvent separately. However, in practice the former has enjoyed more attention, thus most of this section is devoted to the behavior of the solute. We shall, however, briefly discuss the relevant relations for the solvent at the end of this section.

From relations (2.58) to (2.60), we obtain the limiting behavior as $\rho_A \to 0$ which, incidentally, is the same for all three cases:

$$\left(\frac{\partial \mu_A}{\partial \rho_A}\right)_{T,\mu_B} = \left(\frac{\partial \mu_A}{\partial \rho_A}\right)_{T,P} = \left(\frac{\partial \mu_A}{\partial \rho_A}\right)_{T,\rho_B} = \frac{kT}{\rho_A} \qquad \rho_A \to 0 \quad (2.116)$$

which upon integration yield

$$\left.\begin{array}{l} \mu_A(T,\mu_B,\rho_A) = \mu_A{}^0(T,\mu_B) + kT \ln \rho_A \\[2mm] \mu_A(T,P,\rho_A) = \mu_A{}^0(T,P) + kT \ln \rho_A \\[2mm] \mu_A(T,\rho_B,\rho_A) = \mu_A{}^0 \ (T,\rho_B) + kT \ln \rho_A \end{array}\right\} \qquad \rho_A \to 0 \quad (2.117)$$

In (2.117) we have explicitly indicated the independent variables employed in each case. It is remarkable that the limiting behavior of the chemical potential of the solute as $\rho_A \to 0$ is the same, although the experimental situations are quite different in the three cases. This uniformity in the formal appearence of the expressions for the chemical potential is already destroyed in the first-order deviation from the limiting behavior. The fact that the density dependence of the chemical potential is different for the three cases is quite apparent from relations (2.58) to (2.60). In the next section we consider in more detail the first-order deviations from DI solutions.

A solution is defined as DI (or ideal dilute) whenever the chemical potential of the solute A has the form (2.117), provided μ_A^0 is independent of the solute density ρ_A.

It is important to stress that none of the three constants of integration denoted by μ_A^0 in (2.117) is the chemical potential of the component A in a pure state. This is an essential difference between the behavior of DI and SI solutions. The quantity μ_A^0, although commonly referred to as a standard chemical potential, is in fact not a chemical potential at all. We shall see, however, in Section 9, that differences in standard chemical potential may be interpreted as differences in chemical potential of A in two real states. In the meantime μ_A^0 should be regarded merely as a constant of integration. It is useful, however, to recall the molecular interpretation of μ_A^0 as discussed in Section 3. There we elaborated on one set of thermodynamic variables. It turns out that μ_A^0 consists in part of the work required to introduce a single molecule A to a fixed position in the solvent in all three cases. This work is carried out under different experimental conditions, according to the choice of the independent variables.

Instead of starting from (2.58) to (2.60) to derive the limiting behavior (2.117), we could have started from relation (2.61) to obtain the limiting form

$$\mu_A(T,P,x_A) = \mu_A^*(T,P) + kT \ln x_A \qquad x_A \to 0 \qquad (2.118)$$

The relation between $\mu_A^*(T,P)$ and $\mu_A^0(T,P)$ in (2.117) is readily obtained by noting that $\rho_A = x_A \rho$, with ρ being the total density. Hence from (2.117) we obtain, after changing variables,

$$\mu_A(T,P,x_A) = \mu_A^0(T,P) + kT \ln \rho + kT \ln x_A \qquad x_A \to 0 \quad (2.119)$$

Comparing (2.118) with (2.119) yields

$$\mu_A^*(T,P) = \mu_A^0(T,P) + kT \ln \rho \qquad (2.120)$$

Also, since we are dealing only with the limiting behavior at $\rho_A \to 0$, we may replace ρ by $\rho_B{}^0$, the density of pure B at the specified P and T. Similarly, we may derive the limiting form of the chemical potential of A, expressed in terms of the molality m_A, by the transformation

$$m_A = \frac{1000}{M_B} x_A \qquad (2.121)$$

where M_B is the molecular weight of B. From (2.118) we obtain, after changing variables,

$$\mu_A(T,P,m_A) = \mu_A^*(T,P) + kT \ln \frac{M_B}{1000} + kT \ln m_A$$

$$= \mu_A^{**}(T,P) + kT \ln m_A \qquad m_A \to 0 \qquad (2.122)$$

The quantities $\mu_A^*(T,P)$ and $\mu_A^{**}(T,P)$, introduced in (2.120) and in (2.122), may be referred to as the standard chemical potential of A, using x_A and m_A as the respective concentration scales, and P and T as the independent thermodynamic variables.

From the formal point of view, it is possible to characterize the DI solutions by any one of the relations (2.117), (2.118), and (2.122). All these are equivalent and easily transformable into each other. In spite of this apparent equivalency, and because this question has been the subject of considerable controversy, it would be evasive to refrain from discussing the relative preferences of the various forms. We now present some arguments in favor of the application of the volume concentration ρ_A, as in (2.117), over the other concentration scales. These arguments apply only to the case of DI and not for SI solutions.

One very naive argument against the use of the mole fraction scale for SI solutions is the formal similarity between (2.118) and (2.63) for SI solutions. This similarity is sometimes even enhanced by the common practice of using $\mu_A{}^0$ for the standard chemical potential as well as for the chemical potential of pure A. However, since the origin of the two phenomena is completely different, it is advisable not to overstress the similarity in their formal appearance. Furthermore, the very fact that one writes relation (2.118) without appending the essential condition $x_A \to 0$ may lead to fallacious interpretations of the quantity $\mu_A^*(P,T)$. A very common one is to interpret μ_A^* as the *chemical potential* of pure A at a hypothetical state of $x_A = 1$. This is an unfortunate and a misleading interpretation for the following reasons:

1. The DI behavior holds only at $x_A \to 0$, therefore it is meaningless to substitute $x_A = 1$ in (2.118).

2. $\mu_A^*(T,P)$ in (2.118) cannot be interpreted as a chemical potential of A in

the pure state. We recall the general form of the chemical potential of a pure liquid in (2.57) (and in Appendix B). Specifically, we note that this quantity always contains the term $kT \ln \rho_A$, ρ_A being the density of *pure* A. This term is not included in $\mu_A^*(T,P)$ and, in fact, a glance at (2.120) shows that it contains the term $kT \ln \rho$, with ρ being essentially the density of *pure* B.

3. The interpretation of $\mu_A^*(T,P)$ as a chemical potential of pure A creates a physically paradoxical situation. We recall that $\mu_A{}^0(T,P)$ in (2.120) is essentially the coupling work of A against pure B (see Section 3). If $\mu_A^*(T,P)$ represented pure A, it should have contained the coupling work of A against pure A. Thus one is forced to envisage a hypothetical state of pure A, in which each A molecule "feels" a surrounding of pure B. Such a situation may be quite awkward, especially for a system of two very different components.

For all these reasons we believe that the use of such hypothetical states is hazardous, and in fact never really necessary. We shall see in Section 9 that in practice we always encounter *differences* in standard chemical potentials, and that these may be interpreted either on a molecular level or as a work of transfer, without resorting to hypothetical states.

The second argument favoring the volume concentration scale over that of the mole fraction or the molality is that ρ_A is a well-defined quantity in any solvent, whereas the other concentration scales are sometimes defined equivocally. The difficulty arises mainly when the solvent is itself a mixture of a few components. For instance, when dealing with solutions of argon in mixtures of water and alcohol, it is not clear how to define the molality of argon, particularly if we change the composition of the solvent. Similarly, some ambiguity may arise in the definition of the mole fraction when it is not clear how one should count the total number of molecules in the system. For instance, in a solution of n_a molecules of argon in n_w molecules of water and n_{KCl} molecules of KCl, two definitions of the mole fraction of argon are possible:

$$x_a = \frac{n_a}{n_a + n_w + n_{KCl}} \tag{2.123}$$

and

$$x_a' = \frac{n_a}{n_a + n_w + n_{K^+} + n_{Cl^-}} \tag{2.124}$$

In general, there can be many possibilities of defining the mole fraction, none of which may be regarded as the "natural" one, hence confusion is likely to arise.

The third argument and in our opinion the most important one, is concerned with the interpretation of differences in standard chemical potentials. Suppose we have two solvents α and β and we measure the difference in the standard chemical potential of A in the two solvents. Following our discussions in Section 3, we may interpret the quantity $\mu_A^{0\alpha}(T,P) - \mu_A^{0\beta}(T,P)$ as the work required to transfer a single A molecule from a fixed position in β to a fixed position in α. Because of its local character (for more details see Section 3), this quantity may serve as a useful probe to investigate the difference in the solvation properties of the two solvents. This is also true for the corresponding enthalpy and entropy changes. Such a simple and straightforward interpretation is lacking when we use differences in μ_A^* or μ_A^{**} in the two solvents. We shall return to this point in Section 9.

One argument against the density scale is often raised and should be mentioned. That is, ρ_A, in contrast to either x_A or m_A, is temperature-dependent. Hence the latter concentration scales are obviously more convenient. This argument indeed holds if the ultimate reported quantity is the concentration (or solubility) of A. However, once we proceed to compute standard thermodynamic quantities of transfer, the above advantage of x_A and m_A does not hold. Furthermore, we shall see in Section 9 that standard thermodynamic quantities based on ρ_A have distinct advantages for molecular interpretations.

We now turn to a brief discussion of the behavior of the solvent in a DI solution of a two-component system. One direct way of obtaining the chemical potential of the solvent B, when the chemical potential of the solute is known, is by means of the Gibbs-Duhem relation. For instance, applying the Gibbs-Duhem relation to (2.118), we obtain

$$\frac{\partial \mu_B}{\partial x_A} x_B + \frac{\partial \mu_A}{\partial x_A} x_A = -\frac{\partial \mu_B}{\partial x_B} x_B + kT = 0 \qquad (2.125)$$

which, upon integration, yields

$$\mu_B(T,P,x_B) = c(T,P) + kT \ln x_B \qquad x_A \to 0 \qquad (2.126)$$

But since the condition $x_A \to 0$ is the same as the condition $x_B \to 1$, one can substitute $x_B = 1$ in (2.126) to identify the integration constant $c(T,P)$ with the chemical potential of pure B, that is,

$$\mu_B(T,P,x_B) = \mu_B^P(T,P) + kT \ln x_B \qquad x_B \to 1 \qquad (2.127)$$

Note that (2.127) has the same form as in the case of SI solutions, except for the restriction $x_B \to 1$. In both cases μ_B^P has the same significance.

A different way of obtaining relation (2.127) is to use the general expression (2.61) which, when written for B in the limit of $x_A \to 0$, yields

$$\left(\frac{\partial \mu_B}{\partial x_B}\right)_{T,P} = kT \quad x_A \to 0 \tag{2.128}$$

$$\mu_B(T,P,x_B) = c(T,P) + kTx_B \quad x_A \to 0 \tag{2.129}$$

The constant of integration is identified by substituting $x_B = 1$, that is,

$$\mu_B^P(T,P) = c(T,P) + kT \tag{2.130}$$

Hence

$$\mu_B(T,P,x_B) = \mu_B^P(T,P) - kT + kTx_B$$

$$= \mu_B^P(T,P) - kTx_A \approx \mu_B^P(T,P) + kT \ln (1 - x_A)$$

$$= \mu_B^P(T,P) + kT \ln x_B \quad x_B \to 1 \tag{2.131}$$

which is the same as (2.127).

7 SMALL DEVIATIONS FROM DILUTE IDEAL SOLUTIONS

In the preceding section we saw that the limiting behavior of the chemical potential at $\rho_A \to 0$ has the same form, independent of the set of thermodynamic variables used. This uniformity is lost, however, when small deviations from DI solutions are considered. In the present section we are concerned with first-order deviations from DI behavior, which in fact are of foremost importance in their application to real systems. Nowadays, there exist well-known statistical mechanical methods of expressing higher-order corrections to the DI behavior. Their practical value is questionable, however, because their evaluation requires knowledge of the higher-order molecular distribution function, information that is presently unavailable. As in the previous section, we focus our interest mainly on the behavior of the solute and discuss the solvent only briefly at the end of the section.

We start by expanding, to first order in ρ_A, the nonsingular part of relations (2.58) to (2.60) to obtain

$$\left(\frac{\partial \mu_A}{\partial \rho_A}\right)_{T,\mu_B} = kT \left(\frac{1}{\rho_A} - G_{AA}^0 + \cdots\right) \tag{2.132}$$

$$\left(\frac{\partial \mu_A}{\partial \rho_A}\right)_{T,P} = kT \left(\frac{1}{\rho_A} - (G_{AA}^0 - G_{AB}^0) + \cdots\right) \tag{2.133}$$

$$\left(\frac{\partial \mu_A}{\partial \rho_A}\right)_{T,\rho_B} = kT\left(\frac{1}{\rho_A} - \frac{G_{AA}^0 + \rho_B^0(G_{AA}^0 G_{BB}^0 - G_{AB}^{0\,2})}{1 + \rho_B^0 G_{BB}^0} + \cdots\right) \quad (2.134)$$

The superscript zero in (2.132) to (2.134) stands for the limiting value of the corresponding quantity at $\rho_A \to 0$. Note that the limit should be taken under T and μ_B constant in (2.132), T and P constant in (2.133), and T and ρ_B constant in (2.134). These relations may be integrated, in the region of ρ_A, for which the first-order expansion is valid, to obtain

$$\mu_A(T,\mu_B,\rho_A) = \mu_A{}^0(T,\mu_B) + kT \ln \rho_A - kTG_{AA}^0 \rho_A + \cdots \quad (2.135)$$

$$\mu_A(T,P,\rho_A) = \mu_A{}^0(T,P) + kT \ln \rho_A - kT(G_{AA}^0 - G_{AB}^0)\rho_A + \cdots \quad (2.136)$$

$$\mu_A(T,\rho_B,\rho_A) = \mu_A{}^0(T,\rho_B) + kT \ln \rho_A - kT(G_{AA}^0$$
$$- \frac{\rho_B^0 G_{AB}^{0\,2}}{1 + \rho_B^0 G_{BB}^0})\rho_A + \cdots \quad (2.137)$$

Relations (2.135) to (2.137) clearly demonstrate the fact that, once we consider deviations from DI behavior, the correction terms will depend on the independent variables chosen to describe the system.

In a formal fashion we may now define the activity coefficients, corresponding to first-order deviations from a DI solution, as

$$kT \ln \gamma_A{}^D(T,\mu_B,\rho_A) = -kTG_{AA}^0 \rho_A \quad (2.138)$$

$$kT \ln \gamma_A{}^D(T,P,\rho_A) = - kT(G_{AA}^0 - G_{AB}^0)\rho_A \quad (2.139)$$

$$kT \ln \gamma_A{}^D(T,\rho_B,\rho_A) = -kT (G_{AA}^0 - \frac{\rho_B^0 G_{AB}^{0\,2}}{1 + \rho_B^0 G_{BB}^0})\rho_A \quad (2.140)$$

and rewrite relations (2.135) to (2.137) in the form

$$\mu_A(T,\mu_B,\rho_A) = \mu_A{}^0(T,\mu_B) + kT \ln \rho_A \gamma_A{}^D(T,\mu_B,\rho_A) \quad (2.141)$$

$$\mu_A(T,P,\rho_A) = \mu_A{}^0(T,P) + kT \ln \rho_A \gamma_A{}^D(T,P,\rho_A) \quad (2.142)$$

$$\mu_A(T,\rho_B,\rho_A) = \mu_A{}^0(T,\rho_B) + kT \ln \rho_A \gamma_A{}^D(T,\rho_B,\rho_A) \quad (2.143)$$

Here the superscript D serves to indicate that we are dealing with deviations from DI solutions and not from SI solutions. We have also included in our notation the set of thermodynamic variables selected in each case. In practical application, when these variables are presumed to be known, they may be deleted from the notation. The most useful, from the practical point of view, is the activity coefficient defined in (2.139). Its limiting behavior is

$$\lim_{\rho_A \to 0} \gamma_A^D(T,P,\rho_A) = 1 \qquad P \text{ and } T \text{ constant} \qquad (2.144)$$

which is completely different from the limiting behavior of γ_A^S discussed in Section 5.

We have thus far stressed the first-order deviations from DI solutions using the molar concentration ρ_A. As discussed in the previous section, we feel that this choice is the most advantageous for treating DI solutions. However, it is perfectly legitimate to use any other concentration scale for the solute. For instance, using the mole fraction x_A, we may expand the nonsingular term of (2.61) to first order in x_A to obtain

$$\left(\frac{\partial \mu_A}{\partial x_A}\right)_{T,P} = kT \left[\frac{1}{x_A} - (G_{AA}^0 + G_{BB}^0 - 2G_{AB}^0)\rho_B^0 + \cdots\right] \qquad (2.145)$$

Upon integration, in the region of x_A for which the expansion to first order is valid, we obtain

$$\mu_A(T,P,x_A) = \mu_A^*(T,P) + kT \ln x_A - kT(G_{AA}^0 + G_{BB}^0 - 2G_{AB}^0)\rho_B^0 x_A$$

$$= \mu_A^*(T,P) + kT \ln x_A \gamma_A^{Dx}(T,P,x_A) \qquad (2.146)$$

where D_x serves to indicate that we are dealing with deviations from a DI solution with respect to the choice of x_A as a concentration variable. Again, we suggest that one refrains from using (2.146) to express deviations from DI, if only to avoid possible confusion with the corresponding expression for deviations from SI solutions, as discussed in Section 5. Note that here the limiting value of Δ_{AB} (at $x_A \to 0$) appears, but there is no restriction on the magnitude of Δ_{AB}. However, relation (2.91) of Section 5 is restricted to small values of Δ_{AB} but is presumed valid for any value of the composition $0 \leqslant x_A \leqslant 1$.

The three expressions for the activity coefficients listed in (2.138) to (2.140) become nine once we also consider the corresponding expressions for the mole fraction and the molality scales. Such a multiplicity of cases certainly adds to the complexity of the situation and becomes a common source of confusion.

Regarding the content of the first-order activity coefficients in (2.138) to (2.140), we note that all of them contain the "new" term G_{AA}^0. It is new in the sense that DI solutions are characterized by the complete absence of solute-solute interaction, a quantity contained in G_{AA}^0. The other quantities, G_{AB}^0 and G_{BB}^0, convey the solute-solvent and solvent-solvent interaction, respectively, and must be operative in the determination of the standard chemical potential [although they are not simply expressible in terms of the G_{ij}'s; see section following relation (2.183) for an explicit example].

A comment on the terms solute-solute and solute-solvent interaction is also in order, as reference is made to the corresponding quantities of G_{ij}. A more precise description of G_{ij} should employ the concept of correlation between the species i and j and not just the *interaction* between them. For instance, two cavities in a solvent do not *interact* with each other, but there is a spatial correlation between them and the corresponding value of G may be nonzero.

The foregoing comment follows from the fact that the very term solute-solute interaction is sometimes a source of misinterpretation. Consider an imperfect gas, for which deviation from the ideal gas law is noticeable. In this case one may imagine a process of "switching off" all the interactions between the particles. Such a process immediately turns the system into an ideal gas. Care must be employed when extending the analogy to solutions. Suppose we have a dilute solution of A in B, in which deviations from DI behavior are significant. Here, if we "switch off" the *interaction* between the solute particles, we will *not* obtain a DI solution. The quantity one should switch off is the solute-solute correlation, which nullifies the right-hand side of (2.138). It is more difficult to point out the precise quantity to be switched off in the other examples given in (2.139) and (2.140). Therefore care must be exercised in identifying the absence of deviations from DI solutions with absences of solute-solute interaction.

In the context of this section, the term solute-solute interaction refers to the direct interaction between two solute particles. Of course one can use this term in a broader sense to mean also solute-solute correlation or, equivalently, the potential of average force [6] between two solute particles, in which case the above comment does not apply.

A different point that should be noted is that the expansions in (2.135) to (2.137) correspond to experimentally realizable procedures. This is in contrast to the expansion in the parameter Δ_{AB}, dealt with in Section 5. Here we can actually start with a pure solvent B and successively add solute A while keeping $T\mu_B$, TP, or $T\rho_B$ constant. The corresponding response of the chemical potential is represented by the expansions in ρ_A given in (2.135) to (2.137), respectively.

The most important set of thermodynamic variables is of course T, P, and ρ_A, employed in (2.136). However, relation (2.135) has received considerable attention in osmotic experiments, where μ_B is kept constant. This case is also of importance from the formal point of view, since it leads to a virial expansion of the osmotic pressure π, which is the analog of the virial expansion of the pressure. The latter is obtained from the thermodynamic relation

$$\left(\frac{\partial \pi}{\partial \rho_A}\right)_{T,\mu_B} = \rho_A \left(\frac{\partial \mu_A}{\partial \rho_A}\right)_{T,\mu_B} \tag{2.147}$$

The first-order expansion of π may be obtained either from (2.135) or from the more general expression (2.58)

$$\left(\frac{\partial \pi}{\partial \rho_A}\right)_{T,\mu_B} = \frac{kT}{1 + \rho_A G_{AA}} \xrightarrow{\rho_A \to 0} kT(1 - G^0_{AA}\rho_A + \cdots) \tag{2.148}$$

On integration we obtain

$$\frac{\pi}{kT} = \rho_A - \tfrac{1}{2} G^0_{AA}\rho_A^2 + \cdots \tag{2.149}$$

which is better known in the notation

$$\frac{\pi}{kT} = \rho_A + B_2^* \, \rho_A^2 + \cdots \tag{2.150}$$

B_2^* is the analog of the second virial coefficient in the density expansion of the pressure, that is,

$$\frac{P}{kT} = \rho + B_2\rho^2 + \cdots \tag{2.151}$$

We now turn to a brief treatment of the chemical potential of the solvent in the case of first-order deviations from a DI solution. The simplest way to arrive at the required relation is by using the Gibbs-Duhem relation to (2.146), that is,

$$\left(\frac{\partial \mu_B}{\partial x_A}\right)_{T,P} = -\frac{x_A}{x_B}\left(\frac{\partial \mu_A}{\partial x_A}\right)_{T,P} = -\frac{x_A}{x_B} kT \left(\frac{1}{x_A} - \Delta^0_{AB}\rho_B^0 + \cdots\right)$$

$$x_A \to 0 \tag{2.152}$$

which, on integration, yields

$$\mu_B = c(T,P) + kT \ln x_B + kT\rho_B{}^0 \Delta_{AB}^0 \int \frac{x_A}{1 - x_A} \, dx_A$$

$$= c(T,P) + kT \ln x_B + kT \frac{\rho_B{}^0 \Delta_{AB}^0}{2} x_A{}^2 + \cdots \tag{2.153}$$

The constant of integration is easily identified as $\mu_B{}^P(P,T)$ by substituting $x_A = 0$ in (2.153). Thus we obtain the final result

$$\mu_B = \mu_B{}^P(T,P) + kT \ln x_B + \frac{kT}{2} \rho_B{}^0 \Delta_{AB}^0 x_A{}^2 \qquad x_A \ll 1 \tag{2.154}$$

It is important to stress the difference between (2.154) and the similar relation (2.91). Here the expansion is valid only for small x_A, but otherwise the value of Δ_{AB}^0 is unrestricted. However, (2.91) was obtained only for small Δ_{AB} but was otherwise valid in the entire range of composition $0 \leqslant x_A \leqslant 1$.

8 A COMPLETELY SOLVABLE EXAMPLE

In Section 1 we made the distinction between three kinds of ideal mixtures. The first and simplest is an ideal gas mixture which, like an ideal gas, is characterized by a complete absence (or negligence) of intermolecular interactions. This is, however, the least important case for solution chemistry. The other two cases involve systems in which interactions among the molecules are appreciable. The second case, referred to as a SI solution, emerges whenever there is a kind of similarity between the two components—the precise requirement has been discussed in Section 4. The third case occurs whenever one component is very diluted in the other.

We have also noted that any given mixture may be viewed as a system deviating from any one of the ideal reference systems. The degree of deviation depends of course on the choice of the reference system. We now illustrate all three cases in a relatively simple case. The system under observation is a slightly imperfect gas mixture of two components A and B. More specifically, we assume that the density is sufficiently low so that the pair correlation functions for each pair of species is given by

$$g_{ij}(R) = \exp\left[-\beta U_{ij}(R)\right] \tag{2.155}$$

We now discuss separately a system at constant volume and a system at constant pressure.

Constant Volume

By applying relation (2.155) in (2.49), the integration over the parameter

ξ may be carried out to obtain an explicit expression for the chemical potential:

$$\mu_A = kT \ln \rho_A \Lambda_A^3 - kT\rho_A \int_0^\infty \left\{ \exp \left[-\beta U_{AA}(R) \right] - 1 \right\} 4\pi R^2 \; dR$$

$$- kT\rho_B \int_0^\infty \left\{ \exp \left[-\beta U_{AB}(R) \right] - 1 \right\} 4\pi R^2 \; dR \qquad (2.156)$$

(For the purpose of this section we assume, for simplicity of notation, that the internal partition function q_A has also been included in Λ_A^3.)

The integrals appearing in (2.156) are just the explicit form of G_{ij} in this system. A more common notation for these quantities is

$$B_{ij} = - \tfrac{1}{2} \int_0^\infty \left\{ \exp \left[-\beta U_{ij}(R) \right] - 1 \right\} 4\pi R^2 \; dR \qquad (2.157)$$

or

$$B_{ij} = \lim_{\rho \to 0} \left(- \tfrac{1}{2} G_{ij} \right) \qquad (2.158)$$

Hence

$$\mu_A = kT \ln \rho_A \Lambda_A^3 + 2kTB_{AA}\rho_A + 2kTB_{AB}\rho_B \qquad (2.159)$$

We now examine (2.159) from three different points of view.

An Ideal Gas Mixture as a Reference System

An ideal gas mixture is obtained when all $U_{ij}(R) \equiv 0$ (in the present example, it is sufficient to assume the weaker condition $B_{ij} = 0$), that is,

$$\mu_A^{\text{id. gas}} = kT \ln \rho_A \Lambda_A^3 = \mu_A^{0,g} + kT \ln \rho_A \qquad (2.160)$$

Thus the second and the third terms on the right-hand side of (2.159) may be viewed as corrections to ideal gas behavior.

We may define the corresponding activity coefficient as

$$kT \ln \gamma_A^{\text{id. gas}} = 2kTB_{AA}\rho_A + 2kTB_{AB}\rho_B \qquad (2.161)$$

and rewrite (2.159) as

$$\mu_A = \mu_A^{0,g} + kT \ln \rho_A \gamma_A^{\text{id. gas}} \qquad (2.162)$$

where $\mu_A^{0,g}$ is defined in (2.160). Using the relation $\rho_A = x_A\rho$, with ρ the total density, we may transform to mole fraction x_A which is also frequently used in this case.

A SI Solution as a Reference System

Suppose that the two components are similar in the sense that

$$B = B_{AA} = B_{AB} = B_{BB} \qquad (2.163)$$

(which is clearly a weaker requirement than the equality of all the pair potentials but stronger than the requirement $\Delta_{AB} = 0$; see (2.179)). Then, from (2.159) we obtain

$$\mu_A = kT \ln \rho_A \Lambda_A^3 + 2kTB\rho$$

$$= kT \ln x_A \rho \Lambda_A^3 + 2kTB\rho \qquad (2.164)$$

where $\rho = \rho_A + \rho_B$ is the total density of the system.

The chemical potential of pure A at the same temperature and total density ρ is

$$\mu_A^p = kT \ln \rho \Lambda_A^3 + 2kTB\rho \qquad (2.165)$$

Hence

$$\mu_A = \mu_A^p + kT \ln x_A \qquad (2.166)$$

which is SI behavior. If, however, (2.163) is not obeyed, we may view the general expression (2.159) in the form

$$\mu_A = kT \ln \rho_A \Lambda_A^3 + 2kTB_{AA}\rho_A + 2\,kTB_{AB}\rho_B$$

$$= kT \ln \rho \Lambda_A^3 + 2kTB_{AA}\rho + kT \ln x_A + 2kT\rho_B(B_{AB} - B_{AA}) \quad (2.167)$$

Denote

$$\mu_A^p = kT \ln \rho \Lambda_A^3 + 2kTB_{AA}\rho \qquad (2.168)$$

and

$$kT \ln \gamma_A^S = 2kT\rho_B(B_{AB} - B_{AA}) \qquad (2.169)$$

We rewrite (2.167) as

$$\mu_A = \mu_A^p + kT \ln x_A \gamma_A^S \qquad (2.170)$$

where now the activity coefficient accounts for the dissimilarity between the two components. (Note that here we use the total density, not the pressure, as the independent variable. Further discussion of SI solutions at constant volume rather than pressure is presented in Appendix C.)

A DI Solution as a Reference System

Finally, we view the same relation (2.159) in reference to a DI solution. Suppose A is very diluted in B; then instead of (2.159) we have

$$\mu_A = kT \ln \rho_A \Lambda_A{}^3 + 2kTB_{AB}\rho_B$$

$$= \mu_A{}^0 + kT \ln \rho_A \tag{2.171}$$

Note the essential difference between the standard chemical potential defined in (2.160), (2.168), and (2.171). Using the last-mentioned we can define the activity coefficient

$$kT \ln \gamma_A{}^D = 2kTB_{AA}\rho_A \tag{2.172}$$

and rewrite relation (2.159) as

$$\mu_A = \mu_A{}^0 + kT \ln \rho_A \gamma_A{}^D \tag{2.173}$$

where the activity coefficient in this case accounts for deviations from DI behavior. A summary of the three cases is presented in Table 2.2.

Constant Pressure

In order to examine the case of constant pressure, we have to express the chemical potential as a function of T, P, and the mole fraction x_A. To do this we use the well-known virial expansion for the pressure, which is:

$$\frac{P}{kT} = (\rho_A + \rho_B) + (\rho_A + \rho_B)^2 (x_A{}^2 B_{AA} + 2x_A x_B B_{AB} + x_B{}^2 B_{BB}) + \cdots \tag{2.174}$$

This may be inverted to obtain $\rho = \rho_A + \rho_B$ as a power series in P which, to second order in P, is

$$\rho = \rho_A + \rho_B = \frac{P}{kT} - \frac{x_A{}^2 B_{AA} + 2x_A x_B B_{AB} + x_B{}^2 B_{BB}}{(kT)^2} P^2 + \cdots \tag{2.175}$$

Substituting $x_A = \rho_A/\rho$ and ρ from (2.175) in (2.159), we obtain the required expression for the chemical potential:

Table 2.2 Various Ways of Splitting the Chemical Potential Given in (2.159)

Reference Ideal System	Standard Chemical Potential	Deviation from Ideality
Ideal gas mixture	$\mu_A^{0,g} = kT \ln \Lambda_A^3$	$kT \ln \gamma_A^{\text{id. gas}} = 2kT(\rho_A B_{AA} + \rho_B B_{AB})$
Symmetric ideal solution	$\mu_A^p = kT \ln \rho/\Lambda_A^3 + 2kTB_{AA}\rho$	$kT \ln \gamma_A^S = 2kT(B_{AB} - B_{AA})\rho_B$
Dilute ideal solution	$\mu_A^0 = kT \ln \Lambda_A^3 + 2kTB_{AB}\rho_B$	$kT \ln \gamma_A^D = 2kTB_{AA}\rho_A$

$$\mu_A = kT \ln x_A \Lambda_A^3 + kT \ln P/kT + PB_{AA} - Px_B^2(B_{AA} - 2B_{AB} + B_{BB}) \quad (2.176)$$

Here we have an explicit expression for the chemical potential. It is particularly convenient for illustrating the various ideal cases discussed in the previous sections. If no interactions are present (in the limit of $P \to 0$), we have an ideal gas mixture. If $\Delta_{AB} = B_{AA} + B_{BB} - 2B_{AB} = 0$, we have a SI solution and, when $x_A \to 0$ (or $x_B \to 1$), we obtain a DI solution.

We now view relation (2.176) in various ways, differing in the choice of the reference ideal system, and examine the appropriate correction terms, or activity coefficients.

An Ideal Gas Mixture as a Reference System

If no interactions exist (in the present case it is sufficient to assume that all $B_{ij} = 0$), we have

$$\mu_A^{\text{id. gas}} = kT \ln x_A \Lambda_A^3 + kT \ln P/kT$$

$$= \mu_A^{0,g,x} + kT \ln x_A \quad (2.177)$$

(the superscript x on μ_A^0 is used to indicate that here we use the mole fraction as a concentration variable).

Using the definition of $\mu_A^{0,g,x}$ in (2.177), we can rewrite (2.176) as

$$\mu_A = \mu_A^{0,g,x} + kT \ln x_A \gamma_A^{\text{id. gas}}$$

where

$$kT \ln \gamma_A^{\text{id. gas}} = PB_{AA} - Px_B^2(B_{AA} - 2B_{AB} + B_{BB}) \quad (2.178)$$

accounts for deviations due to the existence of interactions among the molecules. Using the relation $x_A = \rho_A/\rho$ and (2.175), we can easily express the chemical potential in terms of ρ_A instead of x_A.

A SI Solution as a Reference System

If the two components were similar in the sense of Section 4:

$$\Delta_{AB} = B_{AA} + B_{BB} - 2B_{AB} = 0 \quad (2.179)$$

then we obtain from (2.176)

$$\mu_A = kT \ln \Lambda_A^3 + kT \ln P/kT + PB_{AA} + kT \ln x_A$$

$$= \mu_A^P(T,P) + kT \ln x_A \quad (2.180)$$

where we have an explicit expression for $\mu_A^P(T,P)$, the chemical potential of pure A at T and P. Note that for the pure A we have a term PB_{AA} which expresses the fact that interactions among particles are taken into account. We can now rewrite relation (2.176) as

$$\mu_A = \mu_A^P(T,P) + kT \ln x_A \gamma_A^S \qquad (2.181)$$

where we now have an explicit expression for the activity coefficient γ_A^S accounting for the dissimilarity between the two components:

$$kT \ln \gamma_A^S = - Px_B^2 (B_{AA} + B_{BB} - 2B_{AB}) \qquad (2.182)$$

Compare this with the more general form in (2.91). Using the relation $P = \rho kT$ and the definition (2.158), we can identify (2.182) with the general expression (2.91).

A DI Solution as a Reference System

If $x_A \rightarrow 0$, then x_B is almost unity and the limiting form of (2.176) is

$$\mu_A = kT \ln \Lambda_A^3 + kT \ln P/kT + P(2B_{AB} - B_{BB}) + kT \ln x_A$$

$$= \mu_A^*(T,P) + kT \ln x_A \qquad x_A \rightarrow 0 \qquad (1.183)$$

Here we can clearly see that $\mu_A^*(T,P)$ is not the chemical potential of pure A at T and P [compare with (2.180)]. Another point relevant to the discussion following relation (2.146), is that $\mu_A^*(T,P)$ [as well as $\mu_A^0(T,P)$, which is introduced below] depends on B_{AB} and B_{BB} but not on B_{AA}. The latter may be interpreted as the solute-solute interaction. Substituting $\rho_A = x_A \rho$ and using (2.175), we may transform (2.183) into

$$\mu_A = kT \ln \Lambda_A^3 + 2PB_{AB} + kT \ln \rho_A$$

$$= \mu_A^0(T,P) + kT \ln \rho_A \qquad \rho_A \rightarrow 0 \qquad (2.184)$$

where we have an explicit expression for the standard chemical potential $\mu_A^0(T,P)$. Using the DI solution as a reference system, we may rewrite (2.176) in the form

$$\mu_A = \mu_A^0(T,P) + kT \ln \rho_A \gamma_A^D \qquad (2.185)$$

where γ_A^D accounts for deviations from DI solution, that is,

$$kT \ln \gamma_A^D = 2Px_A (B_{AA} - B_{AB}) \qquad (2.186)$$

Table 2.3 Various Ways of Splitting the Chemical Potential Given in (2.176)

Reference Ideal System	Standard Chemical Potential	Deviation from Ideality
Ideal gas mixture	$\mu_A^{0,g,x} = kT \ln \Lambda_A^3 + kT \ln (P/kT)$	$kT \ln \gamma_A^{\text{id. gas}} = PB_{AA} - Px_B^2 (B_{AA} - 2B_{AB} + B_{BB})$
Symmetric ideal solution	$\mu_A^p = kT \ln \Lambda_A^3 + kT \ln (P/kT) + PB_{AA}$	$kT \ln \gamma_A^S = -Px_B^2 (B_{AA} + B_{BB} - 2B_{AB})$
Dilute ideal solution	$\mu_A^0 = kT \ln \Lambda_A^3 + 2PB_{AB}$	$kT \ln \gamma_A^D = 2Px_A(B_{AA} - B_{AB})$

Using again the relation $x_A = \rho_A/\rho \sim \rho_A \, kT/P$, we may rewrite (2.185) as

$$kT \ln \gamma_A^D = kT 2\rho_A (B_{AA} - B_{AB}) \qquad (2.187)$$

which is the same as (2.139) [note the definition of B_{ij} in (2.158)]. The various cases are tabulated in Table 2.3.

9 STANDARD THERMODYNAMIC QUANTITIES OF TRANSFER

This section is devoted to an elaboration of thermodynamic quantities currently presented or tabulated in the literature as a final output of processing experimental data. These are the standard free energies, entropies, enthalpies, and so on, associated with the transfer of a solute from one phase to another. We are invariably concerned with transfer under constant pressure and tempera- ture. Other thermodynamic variables may be used whenever warranted, but these two are the most prevalent ones. We are mainly interested in very dilute solutions of a solute A in a solvent B. The latter may be a pure component or a mixture of any number of components. Later on we point out which of the relations are also valid for any concentration of solute A.

The general form of the chemical potential of A in a two-component system described by the variables T, P, and ρ_A (see Section 3 and Appendix B) is

$$\mu_A(T,P,\rho_A) = kT \ln \rho_A \Lambda_A^3 q_A^{-1} - kT \ln \left\langle \exp\left(-\beta\phi_A\right) \right\rangle$$

$$= kT \ln \rho_A \Lambda_A^3 + \Delta G_A(T,P,\rho_A;\mathbf{R}_0)$$

$$= kT \ln \rho_A + \mu_A^0(T,P,\rho_A) \qquad (2.188)$$

All the quantities in (2.188) have a similar significance to the analogous ones defined in Section 3, except for the modifications imposed by the use of the variables T, P, and ρ_A. Specifically, $\Delta G_A(T,P,\rho_A;\mathbf{R}_0)$ is the Gibbs free-energy change for the introduction of a single A particle into a fixed position in a system characterized by the variables T, P, and ρ_A. (For complex molecules having internal rotational degrees of freedom, a proper averaging over all conformations is needed [1].)

The quantity $\mu_A^0(T,P,\rho_A)$ may be referred to as the *nonconventional* standard chemical potential, having a meaning similar to the quantity defined in (2.54). It is nonconventional in the sense that it *depends* on the solute density ρ_A.

The conventional standard chemical potential is independent of ρ_A and is defined by

$$\mu_A^0(T,P) = \lim_{\rho_A \to 0} [\mu_A(T,P,\rho_A) - kT \ln \rho_A]$$

$$= kT \ln \Lambda_A^3 q_A^{-1} - \lim_{\rho_A \to 0} [kT \ln \langle \exp(-\beta\phi_A) \rangle]$$

$$= kT \ln \Lambda_A^3 + \Delta G_A(T,P;R_0) \tag{2.189}$$

where $\Delta G_A(T,P,;R_0)$ is the limiting value of $\Delta G_A(T,P,\rho_A;R_0)$ as $\rho_A \to 0$, that is, it contains the coupling work of A against pure B.

Let us now focus our attention on DI solutions. Let a and β denote two phases which may be either pure solvents or mixtures of solvents at the same T and P. We assume that A forms a DI solution in both phases, hence

$$\mu_A^a(T,P,\rho_A^a) = \mu_A^{0a}(T,P) + kT \ln \rho_A^a \qquad \rho_A^a \to 0 \tag{2.190}$$

$$\mu_A^\beta(T,P,\rho_A^\beta) = \mu_A^{0\beta}(T,P) + kT \ln \rho_A^\beta \qquad \rho_A^\beta \to 0 \tag{2.191}$$

We have stressed in Section 3 that the standard chemical potential $\mu_A^0(T,P)$ is not a chemical potential of A in any real system. However, a difference in standard chemical potentials may be assigned the meaning of a difference in chemical potentials in the two phases. To see this we substract (2.191) from (2.190) to obtain

$$\mu_A^a(T,P,\rho_A^a) - \mu_A^\beta(T,P,\rho_A^\beta) = \mu_A^{0a}(T,P) - \mu_A^{0\beta}(T,P) + kT \ln \rho_A^a/\rho_A^\beta \tag{2.192}$$

Clearly, the quantity written on the left-hand side of (2.192) is the free-energy change for transferring A from β to a at constant T and P. [Where the density of A is ρ_A^a in a and ρ_A^β in β, both are presumed to be very small in order to validate (2.190) and (2.191).]

A particular case occurs when $\rho_A^a = \rho_A^\beta = \rho_A$, for which (2.192) reduces to

$$\mu_A^a(T,P,\rho_A) - \mu_A^\beta(T,P,\rho_A) = \mu_A^{0a}(T,P) - \mu_A^{0\beta}(T,P) \equiv \Delta\mu_A^0(\beta \to a) \tag{2.193}$$

Thus we arrive at the following conclusion:

The *experimental interpretation* of $\Delta\mu_A^0(\beta \to a)$ is the change in Gibbs free energy for the transfer of A from β to a at constant T and P, provided that the density of A is the same in the two phases and that it is small enough to ensure the validity of (2.190) and (2.191). In order to distinguish this interpretation from a forthcoming one, we also use the shorthand notation

$$\Delta\mu_A^0(\beta \rightarrow a) = \Delta G_A \begin{bmatrix} \beta \rightarrow a \\ \rho_A^a = \rho_A^\beta \end{bmatrix} \qquad (2.194)$$

with the understanding that T and P are always constant.

The second and more useful interpretation of $\Delta G_A^0(\beta \rightarrow a)$ may be referred to as the *molecular interpretation*, and follows from (2.189) to (2.191):

$$\Delta\mu_A^0(\beta \rightarrow a) = \Delta G_A^a(T,P;\mathbf{R}_0) - \Delta G_A^\beta(T,P;\mathbf{R}_0) \equiv \Delta G_A \begin{bmatrix} \beta \rightarrow a \\ \mathbf{R}_0 \end{bmatrix} \quad (2.195)$$

which is the change in Gibbs free energy for the transfer of a single A molecule from a fixed position in β to a fixed position in a, the process being carried out at constant T and P. Within the context of the present discussion, the above statement strictly refers to the limiting case of a DI solution. However, this restriction is not a necessary one. We return to this point below.

Thus far we have dwelt on the interpretation of $\Delta\mu_A^0(\beta \rightarrow a)$. The next question concerns the method of measuring it experimentally. Usually, one does not measure free energies directly, but concentrations. From (2.192) we see that if A is in equilibrium between the two phases, that is, if

$$\mu_A^a(T,P,\rho_A^a) = \mu_A^\beta(T,P,\rho_A^\beta) \qquad (2.196)$$

we have

$$0 = \Delta\mu_A^0(\beta \rightarrow a) + kT \ln (\rho_A^a/\rho_A^\beta)_{eq} \qquad (2.197)$$

where the subscript eq has been appended to stress the fact that we are considering a specific case in which the ratio ρ_A^a/ρ_A^β is measured at equilibrium. Relation (2.197) furnishes a means of determining $\Delta\mu_A^0(\beta \rightarrow a)$, provided we know that A forms a DI solution in the two phases. The more general definition [noting (2.189)] is

$$\Delta\mu_A^0(\beta \rightarrow a) = \lim_{\substack{\rho_A^a \rightarrow 0 \\ \rho_A^\beta \rightarrow 0}} [\mu_A^a(T,P,\rho_A^a) - \mu_A^\beta(T,P,\rho_A^\beta) - kT \ln \rho_A^a/\rho_A^\beta]$$

$$= \lim_{\rho_A \rightarrow 0} [-kT \ln (\rho_A^a/\rho_A^\beta)_{eq}] \qquad (2.198)$$

where, in the second form on the right-hand side of (2.198), we have introduced the condition of equilibrium for A in the two phases, hence only one concentration (either ρ_A^a or ρ_A^β) may be taken as an independent variable.

If we use the mole fraction of A as a concentration variable. we may form the new difference in standard chemical potentials (see Section 6 for more details):

$$\Delta\mu_A^* \ (\beta \rightarrow a) = \mu_A^{*a}(T,P) - \mu_A^{*\beta}(T,P)$$

$$= \mu_A^{0a}(T,P) - \mu_A^{0\beta}(T,P) + kT \ \ln \ \rho^a/\rho^\beta \qquad (2.199)$$

where ρ^a and ρ^β are the number densities of the phases a and β, respectively. (As noted in Section 6, these quantities may be somewhat ambiguous if the solvent itself is a multicomponent system.)

Following an argument similar to that for $\Delta\mu_A^0(T,P)$, we can state the *experimental interpretation* of $\Delta\mu_A^*(T,P)$ as the change in Gibbs free energy for the transfer of A from β to a at constant T and P, provided that the mole fractions $x_A{}^a$ and $x_A{}^\beta$ are the same in the two phases and that they are small enough to ensure the validity of DI behavior. By analogy with (2.194), we may write

$$\Delta\mu_A^*(\beta \rightarrow a) = \Delta G_A \left[\begin{array}{c} \beta \ \rightarrow \ a \\ x_A{}^a \ = \ x_A{}^\beta \end{array} \right] \qquad (2.200)$$

The experimental evaluation of $\Delta\mu_A^*(\beta \rightarrow a)$ may be carried out either directly through the limiting process [analogous to (2.198)] :

$$\Delta\mu_A^*(\beta \rightarrow a) = \lim_{x_A \rightarrow 0} \ [- \ kT \ \ln \ (x_A{}^a/x_A{}^\beta)_{eq}] \qquad (2.201)$$

or indirectly through (2.199), once we have evaluated $\Delta\mu_A^0(T,P)$.

In contrast to the case of $\Delta\mu_A^0(\beta \rightarrow a)$, we have no simple molecular interpretation of the quantity $\Delta\mu_A^*(\beta \rightarrow a)$.

In a similar fashion one treats the difference in standard chemical potentials constructed on the molality scale. The corresponding connection is (see Section 6 for more details)

$$\Delta\mu_A^{**}(\beta \rightarrow a) = \mu_A^{**a}(T,P) - \mu_A^{**\beta}(T,P)$$

$$= \mu_A^{0a}(T,P) - \mu_A^{0\beta}(T,P) + kT \ \ln \ \rho^a M^a/\rho^\beta M^\beta \quad (2.202)$$

where M^a and M^β are the molecular weights of the solvents a and β, respectively, [As noted in Section 6, some difficulties may arise in the very definition of the molality of A if the solvent is itself a mixture of several components.]

The *experimental interpretation* of $\Delta\mu_A^{**}(\beta \to a)$ is simply the change in the Gibbs free energy for the transfer of A from β to a at constant T and P, provided that the molalities $m_A{}^a$ and $m_A{}^\beta$ are the same in the two phases and that they are small enough to ensure the validity of DI behavior. By analogy with (2.194) and (2.200), we write

$$\Delta\mu_A^{**}(\beta \to a) = \Delta G_A \begin{bmatrix} \beta \to a \\ m_A{}^a = m_A{}^\beta \end{bmatrix} \qquad (2.203)$$

Finally, we note that the experimental evaluation of $\Delta\mu_A^{**}(\beta \to a)$ may be carried out either directly through the limiting process:

$$\Delta\mu_A^{**}(\beta \to a) = \lim_{m_A \to 0} [-kT \ln (m_A{}^a/m_A{}^\beta)_{eq}] \qquad (2.204)$$

or indirectly through (2.202).

For each of the concentration scales, we have constructed the corresponding differences in standard chemical potential. These in turn were given the usual *experimental interpretation* in terms of changes in free energies on transferring A from β to a. In a manner similar to this, one may devise any other new concentration scales (as is indeed often done), and proceed to construct the corresponding standard chemical potentials. Since all these are convertible from one to another, they are basically equivalent; the question arises which of these is the most useful one.

In Section 6 we pointed out some advantages of the use of the molar concentration scale for this purpose. This can be reinforced by noting that only $\Delta\mu_A{}^0(\beta \to a)$ has a simple *molecular interpretation*, which makes it a more adequate quantity for investigating the difference in the solvation properties of A in the two phases. Moreover, while the *experimental interpretation* of $\Delta\mu_A{}^0(\beta \to a)$, $\Delta\mu_A^*(\beta \to a)$, and $\Delta\mu_A^{**}(\beta \to a)$, expressed symbolically in (2.194) (2.200), and (2.203), respectively, are valid only for the limiting case of DI solutions, the *molecular interpretation* of $\Delta\mu_A{}^0(\beta \to a)$ may be extended to any solution without imposing restrictions on ρ_A. This extension may be arrived at by using the nonconventional form of the standard chemical potential defined in (2.188). For any two phases a and β, in which the density of A is $\rho_A{}^a$ and $\rho_A{}^\beta$, respectively, we have

$$\Delta\mu_A{}^0 (\beta \to a) = \mu_A{}^{0a}(T,P,\rho_A{}^a) - \mu_A{}^{0\beta}(T,P,\rho^\beta)$$

$$= \Delta G_A{}^a(T,P,\rho_A{}^a;\mathbf{R}_0) - \Delta G_A{}^\beta(T,P,\rho_A{}^\beta;\mathbf{R}_0) \qquad (2.205)$$

$\Delta\mu_A{}^0(\beta \to a)$ is the change in free energy for the transfer of A from a fixed position in β to a fixed position in α; the process is carried out at T and P constant. No restrictions whatsoever are imposed on $\rho_A{}^\alpha$ and $\rho_A{}^\beta$.

Such an extension is not possible for the experimental interpretation of $\Delta\mu_A{}^0$, $\Delta\mu_A^*$, or $\Delta\mu_A^{**}$. For instance, using (2.192), but with the nonconventional $\mu_A{}^0$, we realize that each of the μ^0's depends on $\rho_A{}^\alpha$ and $\rho_A{}^\beta$. Hence, by substitution of $\rho_A = \rho_A{}^\beta = \rho_A{}^\alpha$ we also affect $\mu_A{}^{0\alpha}$ and $\mu_A{}^{0\beta}$, and therefore we obtain no interpretation for the original $\Delta\mu_A{}^0$ in (2.192).

In using the nonconventional standard chemical potential, we have lost the independence of $\Delta\mu_A{}^0(\beta \to \alpha)$ on the density of A [compare with (2.195)]. However, the molecular significance of $\Delta\mu_A{}^0(\beta \to \alpha)$ has been retained. Therefore we can discard the limiting definition (2.198), which is now replaced by

$$\Delta\mu_A{}^0(\beta \to \alpha) = -kT \ln \left(\rho_A{}^\alpha/\rho_A{}^\beta\right)_{eq} \qquad (2.206)$$

It is important to realize that the foregoing use of the standard chemical potential is based on statistical mechanical arguments [derived from (2.188)] and cannot be arrived at by using classical thermodynamic arguments alone. The form of the chemical potential, as presented in (2.188), and the corresponding interpretations of its various terms, are not revealed by experiment. It is only the limiting behavior, such as (2.190), that shows up in experiments with systems obeying Henry's law.

Let us turn to a specific example of the application of the nonconventional standard chemical potential, in which the advantage of this quantity is further manifested. Let A be benzene and B water. The two phases α and β are (almost) pure benzene and water, respectively. We know that A forms a dilute solution in B, and for simplicity we assume that A in B is a DI solution. It is customary in this case to introduce the standard free energy of transfer of benzene from the pure liquid to the infinitely dilute solution in water. In our notation this corresponds to the difference

$$\mu_A{}^{0\beta}(T,P) - \mu_A{}^P(T,P) \qquad (2.207)$$

where $\mu_A{}^P(T,P)$ is the chemical potential of A in its pure liquid at the same T and P as the solution.

It is important to emphasize that the quantity in (2.207) is *not* the free-energy change on transferring benzene from its pure liquid to its dilute solution in water (such a quantity requires the addition of the term $kT \ln \rho_A{}^\beta$ to (2.207)). A common practice is to envisage a hypothetical state of unit molarity for which an interpretation in terms of free energy of transfer may be assigned to (2.207). This interpretation is subject to the same criticism raised in Section 6. A more severe flaw in using (2.207) is the unevenness in the status of the two states, namely, pure liquid vis-à-vis an infinitely dilute solution. In order to obtain more useful information on the difference in the solvation properties of A in the two phases, we use relation (2.188) which, when applied to pure A, is

$$\mu_A^P(P,T) = \mu_A^{0a}(T,P,\rho_A = \rho_A^{0a}) + kT \ln \rho_A^{0a} \qquad (2.208)$$

Note that, since we now have a one-component system, there are only 2 degrees of freedom, hence ρ_A^{0a}, the density of pure A, is viewed as a function of T and P. We now form the nonconventional standard free energy of transfer:

$$\mu_A^{0\beta}(T,P) - \mu_A^{0a}(T,P,\rho_A = \rho_A^{0a}) \qquad (2.209)$$

which, on account of the discussion following (2.188) (and Section 3), may be interpreted as the change in the Gibbs free energy for the transfer of A from a fixed position in *pure* A to a fixed position in *pure* B (at T and P constant). This quantity treats the two phases on a more fair basis and conveys the difference in the local environment of A in the two phases.

The quantity given in (2.209) may be evaluated experimentally through

$$\Delta\mu_A^0(a \to \beta) = \mu_A^{0\beta} - \mu_A^{0a} = -kT \ln (\rho_A^\beta/\rho_A^{0a})_{eq} \qquad (2.210)$$

where, in contrast to (2.206), we have ρ_A^{0a} as the density of benzene in pure benzene.

In conclusion, the quantity $\Delta\mu_A^0(a \to \beta)$ may always be considered as the change in free energy on transferring A from a fixed position in a to a fixed position in β at T and P constant. There are no restrictions on the phases a or β which may be pure solvents different from A, mixed solvents, or even pure A itself. Therefore this quantity not only enjoys a simple molecular interpretation, but also one that is applicable to a wider class of systems. [For complex solutes having internal rotation, the requirement of a fixed position is not sufficient. In such cases a proper averaging over all conformations [1] of the molecule should be carried out in the definition of $\Delta\mu_A^0(a \to \beta)$.]

Before turning to a discussion of some other thermodynamic quantities, let us summarize the two different interpretations of $\Delta\mu_A^0(\beta \to a)$. The first was referred to as the *experimental interpretation*, which corresponds to the transfer process described prior to relation (2.194) and which is designated process I. The second, referred to as the *molecular interpretation*, is described immediately following relation (2.195), and henceforth is referred to as process II. The important finding is that the two processes I and II produce the *same* change in the Gibbs free energy:

$$\Delta\mu_A^{\,0}(\mathrm{I}) = \Delta G_A \begin{bmatrix} \beta \to a \\ \rho_A^{\,a} = \rho_A^{\,\beta} \end{bmatrix} = \Delta G_A \begin{bmatrix} \beta \to a \\ R_0 \end{bmatrix} = \Delta\mu_A^{\,0}(\mathrm{II}) \quad (2.211)$$

where the symbol $\Delta\mu_A^{\,0}(\mathrm{I})$ refers to the change in free energy associated with process I.

In discussing the standard entropy, enthalpy, and so on, of transfer, care must be exercised in distinguishing between the two processes which in general produce different values for these quantities. The reason for this discrepancy is that the entropy is related to the temperature derivative of the Gibbs free energy at constant P, N_A, and N_B. Hence, to obtain the standard entropy associated with process I, we first differentiate (2.190) and (2.191) with respect to temperature to obtain

$$-S_A^{\,a} = \left(\frac{\partial \mu_A^{\,a}}{\partial T}\right)_{P,N_A,N_B} = \left(\frac{\partial \mu_A^{\,0a}}{\partial T}\right)_P + \frac{\partial}{\partial T}(kT \ln \rho_A^{\,a}) \quad (2.212)$$

$$-S_A^{\,\beta} = \left(\frac{\partial \mu_A^{\,\beta}}{\partial T}\right)_{P,N_A,N_B} = \left(\frac{\partial \mu_A^{\,0\beta}}{\partial T}\right)_P + \frac{\partial}{\partial T}(kT \ln \rho_A^{\,\beta}) \quad (2.213)$$

The general expression for the entropy of transfer of A, at constant P and T, from a phase β in which the density of A is $\rho_A^{\,\beta}$ to a phase a in which the density A is $\rho_A^{\,a}$ is

$$S_A^{\,a} - S_A^{\,\beta} = - \frac{\partial}{\partial T}(\mu_A^{\,0a} - \mu_A^{\,0\beta}) - k \ln (\rho_A^{\,a}/\rho_A^{\,\beta}) - kT \frac{\partial}{\partial T} \ln (\rho_A^{\,a}/\rho_A^{\,\beta}) \quad (2.214)$$

[See also relation (2.192) for the change in free energy associated with this process.] Since the derivatives in (2.214) are at constant pressure and number of particles, we have

$$\frac{\partial}{\partial T} \ln \rho_A^{\,a} = \frac{\partial}{\partial T} \ln \frac{N_A^{\,a}}{V^a} = - \frac{\partial}{\partial T} \ln V^a \quad (2.215)$$

where V^a is the volume of the phase a. For a pure solvent this is simply the coefficient of thermal expansion of the solvent.

As in the passage from (2.192) to (2.193), we now substitute $\rho_A^{\,a} = \rho_A^{\,\beta} = \rho_A$ in (2.214) (note that this substitution is made *after* the differentiation with respect to T) and obtain

$$\Delta S_A^{\,\circ}(\mathrm{I}) = - \frac{\partial}{\partial T} [\Delta\mu_A^{\,0}(\mathrm{I})] + kT \frac{\partial}{\partial T} \ln V^\beta/V^a \quad (2.216)$$

where $\Delta S_A^{\circ}(\mathrm{I})$ refers to the standard entropy of transfer for the same process as described for (2.194), or simply process I. We see that in general the standard entropy of transfer associated with process I is not given simply by the temperature derivatives of the corresponding standard free energy. One requires, in addition, the coefficients of thermal expansion for the two solvents.

This fact is often quoted as an argument against the use of the density ρ_A. It should be stressed, however, that such an objection does not arise if we use the standard entropy of transfer for process II; see below.

The standard entropy of transfer associated with process II may be obtained either from (2.195) or, for the more general case, from (2.205). For simplicity we show this in the simple case of a DI solution in two-component systems. Recalling the definition of ΔG_A^{α} [see (2.46) and Appendix B], we have

$$\frac{\partial}{\partial T} \Delta G_A^{\alpha}(T,P;\mathbf{R}_0) = \frac{\partial}{\partial T} [G(T,P,N_A = 1,N_B;\mathbf{R}_0) - G(T,P,N_A = 0,N_B)]$$

$$= - [S(T,P,N_A = 1,N_B;\mathbf{R}_0) - S(T,P,N_A = 0,N_B)]$$

$$= - \Delta S_A^{\alpha}(T,P;\mathbf{R}_0) \tag{2.217}$$

which is just the entropy change for introducing a molecule A into a fixed position in α (at T and P constant). Applying (2.217) to the two phases α and β, we obtain from (2.195)

$$\Delta S_A^{\circ}(\mathrm{II}) = \Delta S_A^{\alpha}(T,P;\mathbf{R}_0) - \Delta S_A^{\beta}(T,P;\mathbf{R}_0)$$

$$= - \frac{\partial}{\partial T} [\Delta G_A^{\alpha}(T,P;\mathbf{R}_0) - \Delta G_A^{\beta}(T,P;\mathbf{R}_0)]$$

$$= - \frac{\partial}{\partial T} \Delta \mu_A^{\circ} \tag{2.218}$$

Note the difference between (2.218) and (2.216). The quantity $\Delta S_A^{\circ}(\mathrm{II})$ has the same advantage over $\Delta S_A^{\circ}(\mathrm{I})$ as that which we have discussed regarding the two interpretations of $\Delta \mu_A^{\circ}$. Here $\Delta S_A^{\circ}(\mathrm{II})$ refers to the process of transferring A from a fixed position in β to a fixed position in α (at T and P constant). Hence this quantity is more suitable for investigation of solvation entropies of A in various solvents. Furthermore, we see that $\Delta S_A^{\circ}(\mathrm{II})$ is computed from (2.218) using only the temperature derivative of $\Delta \mu_A^{\circ}$, whereas for calculating $\Delta S_A^{\circ}(\mathrm{I})$ through (2.216) we also need the thermal expansion coefficients of the two phases. Finally, we note that by using (2.205) instead of (2.195) we can readily generalize relation (2.218) to any system without requiring DI behavior with respect to A. Such a generalization is not possible for $\Delta S_A^o(I)$, since (2.216) is valid strictly for DI solutions.

The standard enthalpy change associated with processes I and II may be obtained either by differentiating $\Delta\mu_A^{\,0}$ with respect to temperature and applying a similar procedure as in the case of $\Delta S_A^o(I)$ and $\Delta S_A^o(II)$, or more simply by

$$\Delta H_A^o(I) = \Delta\mu_A^{\,0} + T\Delta S_A^o(I) \qquad (2.219)$$

$$\Delta H_A^o(II) = \Delta\mu_A^{\,0} + T\Delta S_A^o(II) \qquad (2.220)$$

The advantages of $\Delta H_A^o(II)$ are the same as the ones mentioned above for $\Delta S_A^o(II)$.

In a similar way we derive the standard volume change for processes I and II:

$$\Delta V_A^{\,0}(I) - \frac{\partial}{\partial P}\,\Delta\mu_A^{\,0} + kT\,\frac{\partial}{\partial P}\,\ln\,V^\beta/V^\alpha \qquad (2.221)$$

$$\Delta V_A^{\,0}(II) = \frac{\partial}{\partial P}\,\Delta\mu_A^{\,0} \qquad (2.222)$$

where for $\Delta V_A^{\,0}(I)$ we need the isothermal compressibilities of the two solvents. Again we note both the relative simplicity of the expression for $\Delta V_A^{\,0}(II)$ and its advantages which are similar to the ones mentioned above for $\Delta S_A^o(II)$.

It is now quite a straightforward matter to derive other thermodynamic quantities of transfer. The expressions for process I are usually more complex, involving various derivatives of the volume of the two phases, than are the corresponding expressions for process II.

10 SUMMARY AND CONCLUSIONS

The main purpose of this chapter has been to clarify some concepts currently accepted in solution chemistry. The need for such clarification stems from the fact that thermodynamics in itself does not provide any meaning, on a molecular level, to concepts such as standard free energy, activity coefficient, and so on. Thus, in order to gain a deeper insight into the meaning and significance of these notions, one must resort to statistical mechanics.

In the course of our analysis, and in particular following the Kirkwood-Buff theory of solutions, we found that there are essentially three categories of "ideal solutions": the ideal gas mixture, the symmetric ideal solution, and the dilute ideal solution. These were considered as fundamentally different concepts since they occur under very different conditions. The next question refers to the meaning of the various activity coefficients. Here it must be realized that the mere writing of

$$\mu_i = \mu_i{}^0 + kT \ln \rho_i \gamma_i$$

does not lend *any* meaning to γ_i. In the first place, its meaning depends on which class of ideality is under consideration. Secondly, it depends on the concentration scale one is using for the species i. And, finally, it depends on the independent variables that are being used to describe our system.

Referring to the various concentration scales, we found that the mole fraction is the more natural concentration unit to be used in the case of symmetric ideal solutions. On the other hand, in dilute solutions use of the molarity (or the number density) is preferable. The reasons are essentially the following: (1) To avoid confusion with the symmetric ideal case; (2) the number density is the more natural variable that comes out in the statistical mechanical expression for the chemical potential; (3) there is a straightforward molecular interpretation to standard theromdynamic quantities based on this concentration scale.

Of course, there is no "absolute" advantage to any of the concentration scales over the others, and the final choice is made on grounds of convenience, and perhaps, on matters of personal taste.

We hope, however, that this chapter will help to bring about a unified system of conventions and notation, and thereby facilitate the comparison of data from different sources.

Acknowledgment

The author is grateful to I. D. Kuntz, H. L. Friedman, Y. Marcus, J. S. Rowlinson, R. Tenne, W. Y. Wen, and M. Yaacobi for reading the manuscript and offering useful comments.

Appendix A: Further Details on the Method of Computation of the Pair Correlation Functions Shown in the Illustrations

All the illustrations of either the pair correlation functions or of derived quantities were computed by solving numerically the Percus-Yevick (PY) equation [9, 13]. For a one-component system, the actual form of the integral equation that was solved is

$$z(r) = z + 2\pi\rho \int_0^\infty z(u)f(u) \, du \int_{|r-u|}^{r+u} z(v)f(v) \, dv \quad +$$

$$+ 2\pi\rho \int_0^\infty z(u)f(u)\, du \int_{|r-u|}^{r+u} [z(v) - v]\, dv \qquad (2.223)$$

where

$$z(r) = rg(r) \exp [\beta U(r)] \qquad (2.224)$$

$$f(r) = \exp [-\beta U(r)] - 1 \qquad (2.225)$$

The Lennard-Jones pair potential was used in all the computations:

$$U(r) = 4\epsilon [(\sigma/r)^{12} - (\sigma/r)^6] \qquad (2.226)$$

with parameters ϵ and σ as indicated in each case. Once these parameters have been chosen, $f(r)$ is a known function and (2.223) is an integral equation for the function $z(r)$. Having a solution for $z(r)$, we may compute the pair correlation function $g(r)$ through (2.224).

The computations were carried out by direct iterational procedure using 51 division points for the range of distances $0 \leqslant r \leqslant 5.0$, with the molecular diameter being unity. The density ρ is the number of particles per unit volume. The iterations were performed until two successive solutions did not differ appreciably from each other. Since accuracy was not the aim of these computations, we terminated the procedure whenever we felt that the solution converged to a stable result. In spite of the limited accuracy of the method, all the illustrations represent faithfully the essential features of the pair correlation functions we intended to demonstrate.

For a two-component system, we used the four PY equations for the four pair correlation functions;

$$z_{\alpha\beta}(r) = r + \sum_{\gamma=1}^{2} [2\pi\rho_\gamma \int_0^\infty z_{\alpha\gamma}(u)f_{\alpha\gamma}(u)\, du \int_{|r-u|}^{r+u} [z_{\gamma\beta}(v)f_{\gamma\beta}(v)]\, dv$$

$$+ 2\pi\rho_\gamma \int_0^\infty z_{\alpha\gamma}(u)f_{\alpha\gamma}(u)\, du \int_{|r-u|}^{r+u} [z_{\gamma\beta}(v) - v]\, dv \qquad (2.227)$$

where ρ_1 and ρ_2 are the number densities of the two components, and all the pairwise functions in (2.227) are defined similarly to those in (2.224) and (2.225). The same procedure was employed for solving the set of equations (2.227). All the starting input functions were

$$z_{\alpha\beta}(r) = r \qquad (2.228)$$

and iterations were carried out simultaneously for the four functions $z_{\alpha\beta}(r)$ ($\alpha, \beta = 1,2$).

Appendix B: The Chemical Potential in the TPN Ensemble

Consider a two-component system specified by the thermodynamic variables T, P, N_A, and N_B. We first show, by direct argument, that a SI solution is obtained when the pair potentials for all the pairs of species are identical. (In Section 3 we have given an indirect argument for this contention.) Next we show that the general form of the chemical potential of, say, A, is the same as in the T, V, N_A, and N_B ensemble [see (2.44) and (2.48)].

The chemical potential of A is defined by

$$\mu_A = \left(\frac{\partial G}{\partial N_A}\right)_{T,P,N_B} = G(T,P,N_A+1,N_B) - G(T,P,N_A,N_B) \qquad (2.229)$$

and its connection with statistical mechanics is established through the appropriate partition functions:

$$\exp\,[-\beta\mu_A] = \frac{q_A}{\Lambda_A^3(N_A+1)}$$

$$\frac{\int dV \int \cdots \int d\mathbf{R}^{N_A+1} d\mathbf{R}^{N_B} \exp\,[-\beta PV - \beta U(\mathbf{R}^{N_A+1},\mathbf{R}^{N_B})]}{\int dV \int \cdots \int d\mathbf{R}^{N_A} d\mathbf{R}^{N_B} \exp\,[-\beta PV - \beta U(\mathbf{R}^{N_A},\mathbf{R}^{N_B})]} \qquad (2.230)$$

where an obvious shorthand notation has been used in (2.230). For instance, $U(\mathbf{R}^{N_A+1},\mathbf{R}^{N_B})$ stands for the total potential energy of the system specified by the configuration $\mathbf{R}^{N_A+1}$ for the A particles, and $\mathbf{R}^{N_B}$ for the B particles.

Next, consider a system of pure A with N particles at the same pressure P and temperature T. The chemical potential is given by

$$\exp\,[-\beta\mu_A^P] = \frac{q_A}{(N+1)\Lambda_A^3}$$

$$\frac{\int dV \int \cdots \int d\mathbf{R}^{N+1} \exp\,[-\beta PV - \beta U(\mathbf{R}^{N+1})]}{\int dV \int \cdots \int d\mathbf{R}^{N} \exp\,[-\beta PV - \beta U(\mathbf{R}^{N})]} \qquad (2.231)$$

A special case occurs when the two components considered in (2.230) have the same pair potential for all the pairs of species. In such a case the total potential energy denoted by $U(\mathbf{R}^{N_A}, \mathbf{R}^{N_B})$ in (2.230) depends only on the configuration of the $N_A + N_B$ particles and not on the type of species occupying the various sites. Therefore, if we choose N to be equal to $N = N_A + N_B$, we have the two equalities

$$U(\mathbf{R}^{N_A}, \mathbf{R}^{N_B}) = U(\mathbf{R}^N) \qquad (2.232)$$

$$U(\mathbf{R}^{N_A+1}, \mathbf{R}^{N_B}) = U(\mathbf{R}^{N+1}) \qquad (2.233)$$

which simply means that the total potential function of the system depends in this case only on the configuration of the N (or $N + 1$) particles, and *not* on the assignment of the species to each of the $\mathbf{R}_i$'s. Using the equalities (2.232) and (2.233) and comparing (2.231) with (2.230), we obtain

$$\exp\left[-\beta\mu_A(T,P,N_A,N_B)\right] = \frac{q_A}{\Lambda_A^3(N+1)} \frac{(N+1)\Lambda_A^3}{q_A} \exp\left[-\beta\mu_A^P(T,P,N)\right] \qquad (2.234)$$

that is,

$$\mu_A(T,P,N_A,N_B) = \mu_A^P(T,P,N) + kT \ln \frac{N_A + 1}{N + 1} \qquad (2.235)$$

with the condition $N = N_A + N_B$. Since for macroscopic systems

$$\frac{(N_A + 1)}{(N + 1)} = x_A$$

and since μ_A is an intensive quantity, we may write (2.235) as

$$\mu_A(T,P,x_A) = \mu_A^P(T,P) + kT \ln x_A \qquad (2.236)$$

which is valid for all compositions, hence we have a SI solution. This concludes the demonstration that equality of the pair potentials of the two components is a sufficient condition for a SI solution. We have shown this in the T, P, N_A, N_B ensemble, but similar arguments apply also to the case of the T, V, N_A, N_B ensemble. Furthermore, the arguments given above rest on the assumption of pairwise additivity for the total potential function. The result is still valid for more complex systems in which case one must assume the validity of (2.232) and (2.233) for all configurations.

We now turn to the general form of μ_A in the T,P,N_A,N_B ensemble. For simplicity, we use the assumption of pairwise additivity of the total potential energy, but no other assumption on the similarity of the two components is imposed. The expression for the chemical potential in such a system is given by (2.230). Let us denote by $P(\mathbf{R}^{N_A},\mathbf{R}^{N_B},V)$ the probability density that a system in the T,P,N_A,N_B ensemble will be found with a volume V and configuration $\mathbf{R}^{N_A},\mathbf{R}^{N_B}$.

$$P(\mathbf{R}^{N_A},\mathbf{R}^{N_B},\ V) =$$

$$\frac{\exp\ [-\beta PV -\beta U(\mathbf{R}^{N_A},\mathbf{R}^{N_B})]}{\int dV \int \cdot\cdot\cdot \int d\mathbf{R}^{N_A}d\mathbf{R}^{N_B}\ \exp\ [-\beta PV -\ \beta U(\mathbf{R}^{N_A},\mathbf{R}^{N_B})]} \tag{2.237}$$

We also introduce the probability density $P(V)$ of finding a system with volume V (independent of the configuration), and the conditional probability density $P(\mathbf{R}^{N_A},\mathbf{R}^{N_B}/V)$ of finding the configuration $\mathbf{R}^{N_A}$, $\mathbf{R}^{N_B}$, given that the system has volume V. These are connected by the relation

$$P(\mathbf{R}^{N_A},\ \mathbf{R}^{N_B}/V) = \frac{P(\mathbf{R}^{N_A},\mathbf{R}^{N_B},V)}{P(V)} \tag{2.238}$$

Using (2.237) and (2.238), we may rewrite (2.230) as

$$\exp\ [-\beta\mu_A] = \frac{q_A}{\Lambda_A^3(N_A+1)}\ \int dV \int \cdot\cdot\cdot \int d\mathbf{R}^{N_A+1}\ d\mathbf{R}^{N_B}\ P(\mathbf{R}^{N_A},\mathbf{R}^{N_B},V)\ \exp\ [-\beta\phi_A]$$

$$= \frac{q_A}{\Lambda_A^3(N_A+1)}\ \int dV\ P(V)\ \int \underset{V}{\cdot\cdot\cdot} \int d\mathbf{R}^{N_A+1}d\mathbf{R}^{N_B}\ P(\mathbf{R}^{N_A},\mathbf{R}^{N_B}/V)\ \exp\ [-\beta\phi_A] \tag{2.239}$$

where ϕ_A stands for the total interaction energy of the added A particles with the rest of the system, at a given configuration. We have also indicated that the integration of the $\mathbf{R}_i$'s extends over the entire volume V. In a macroscopic system we assume that $P(V)$ is concentrated, very sharply, at the average value $<V>$. For simplicity we take $P(V)$ to be a Dirac delta function

$$P(V) = \delta(V - <V>) \tag{2.240}$$

which permits simplifying (2.240) into

$$\exp\ [-\beta\mu_A]$$

$$= \frac{q_A}{\Lambda_A^3(N_A+1)}\ \int \underset{<V>}{\cdot\cdot\cdot} \int d\mathbf{R}^{N_A+1}\ d\mathbf{R}^{N_B}\ P(\mathbf{R}^{N_A},\ \mathbf{R}^{N_B}/<V>)\ \exp\ [-\beta\phi_A]$$

$$= \frac{q_A<V>}{\Lambda_A^3(N_A+1)}\ \int \underset{<V>}{\cdot\cdot\cdot} \int d\mathbf{R}^{N_A}\ d\mathbf{R}^{N_B}\ P(\mathbf{R}^{N_A},\mathbf{R}^{N_B}/<V>)\ \exp\ [-\beta\phi_A]$$

$$= \frac{q_A}{\Lambda_A^3 \rho_A} < \exp \; [-\beta\phi_A]> \qquad (2.241)$$

when, in the second form on the right-hand side of (2.241), we have integrated over the locations of the added new A particles and transformed to relative coordinates. In the second form we have replaced $(N_A + 1)/<V>$ with the average density of A, ρ_A, and $<\exp \; [-\beta\phi_A]>$ is a conditional average quantity given by the integral in the predecessor form. We now have the final form of the chemical potential

$$\mu_A = kT \ln \rho_A \Lambda_A^3 q_A^{-1} - kT \ln \left< \exp \; [-\beta\phi_A] \right> \qquad (2.242)$$

In a very similar fashion we may now repeat the same derivation as in (2.242) but, instead of starting with (2.229), we start with the process of adding a new particle to a fixed position. The change in the Gibbs free energy is

$$\Delta G_A(\mathbf{R}_0) = G(T,P,N_A+1,N_B;\mathbf{R}_0) - G(T,P,N_A,N_B) \qquad (2.243)$$

(Compare this with (2.46) and the corresponding derivation in Section 3). By analogy with (2.51) we obtain for the present case the relation

$$\mu_A = \Delta G_A(\mathbf{R}_0) + kT \ln \rho_A \Lambda_A^3 \qquad (2.244)$$

The significance of the split of μ_A into two terms in (2.244) is the same as in (2.51). First we introduce the A particle into a fixed position, and then release it to wander in the system. The two processes are carried out at constant pressure instead of constant volume as in Section 3.

Appendix C: Symmetric Ideal Solutions in a Constant-Volume Mixture

In Section 4, we deal with the conditions for SI solutions for mixtures at constant pressure. This is indeed the most important case from the practical point of view. Here we present the condition for SI solutions at constant volume rather than at constant pressure. If the pair potentials for all the pairs of species are equal, following similar arguments as in Appendix B, we obtain the SI behavior. However, as in Section 4, we now seek weaker conditions of SI solutions for systems at constant volume.

We start from the thermodynamic relations

$$\left(\frac{\partial \mu_A}{\partial N_A}\right)_{T,V,N,} = \frac{1}{N}\left(\frac{\partial \mu_A}{\partial x_A}\right)_{T,\rho} = \frac{1}{V}\left(\frac{\partial \mu_A}{\partial \rho_A}\right)_{T,\rho} \qquad (2.245)$$

where $N = N_A + N_B$ and $\rho = N/V$. From (2.245) we obtain

$$\left(\frac{\partial \mu_A}{\partial x_A}\right)_{T,\rho} = \rho\left(\frac{\partial \mu_A}{\partial \rho_A}\right)_{T,\rho} = \rho\left[\left(\frac{\partial \mu_A}{\partial \rho_A}\right)_{T,\rho_B} - \left(\frac{\partial \mu_A}{\partial \rho_B}\right)_{T,\rho_A}\right] \qquad (2.246)$$

where now we have on the right-hand side of (2.246) quantities expressible in terms of the G_{ij}'s through the Kirkwood-Buff theory (see Section 3).

$$\left(\frac{\partial \mu_A}{\partial x_A}\right)_{T,\rho} = \frac{\rho kT(\rho_B + \rho_B^2 G_{BB} + \rho_A \rho_B G_{AB})}{\rho_A \rho_B[\ 1 + \rho_A G_{AA} + \rho_B G_{BB} + \rho_A \rho_B(G_{AA}G_{BB} - G_{AB}^2)]} \qquad (2.247)$$

After rearranging,

$$\left(\frac{\partial \mu_A}{\partial x_A}\right)_{T,\rho} = kT\left[\frac{1}{x_A} + \frac{\rho(G_{AB} - G_{AA} - \rho_B \delta_{AB}}{1 + \rho_A G_{AA} + \rho_B G_{BB} + \rho_A \rho_B \delta_{AB}}\right] \qquad (2.248)$$

where

$$\delta_{AB} = G_{AA}G_{BB} - G_{AB}^2 \qquad (2.249)$$

Clearly, a sufficient condition for SI solutions (defined in the same manner as in Section 4 but requiring the condition for all $1 \leqslant x_A \leqslant 1$, at T and P constant) is that all G_{ij} are equal:

$$G_{AA} = G_{AB} = G_{BB} \qquad (2.250)$$

This condition is definitely weaker than the equality of the intermolecular potentials. As a concrete example, consider the example of Section 8, in which G_{ij} reduces to

$$G_{ij} = -\frac{1}{2}\int_0^\infty \left\{\exp\ [-\beta U_{ij}(R)] - 1\right\} 4\pi R^2\ dR \qquad (2.251)$$

Hence equality of all the U_{ij}'s leads to equality of all the G_{ij}'s. The latter does not follow, however, from the former.

References

1. A. Ben-Naim, *J. Chem. Phys.*, **54**, 3696 (1971).
2. K. Denbigh, *The Principles of Chemical Equilibrium*, 2nd ed., Cambridge University Press, 1966.
3. E. A. Guggenheim, *Mixtures*, Oxford, New York, 1952.
4. E. A. Guggenheim, *Thermodynamics*, 4th ed., North Holland, Amsterdam, 1959.
5. J. H. Hildebrand, J. M. Prausnitz, and R. L. Scott, *Regular and Related Solutions*, Van Nostrand Reinhold, New York, 1970.
6. T. L. Hill, *Statistical Mechanics*, McGraw-Hill, New York, 1956.
7. J. O. Hirschfelder, C. F. Curtiss, and R. B. Bird, *Molecular Theory of Gases and Liquids*, Wiley, New York, 1954.
8. J. G. Kirkwood and F. P. Buff, *J. Chem. Phys.*, **19**, 774 (1951).
9. A. Münster, *Statistical Thermodynamics*, Vol. I, Springer-Verlag, Berlin, 1969.
10. I. Prigogine and R. Defay, *Chemical Thermodynamics*, D. H. Everatt, Translator, Longmans Green, London, 1954.
11. S. A. Rice and P. Gray, *The Statistical Mechanics of Simple Liquids*, Interscience, Wiley, New York, 1965.
12. J. S. Rowlinson, *Liquids and Liquid Mixtures*, 2nd ed., Butterworths, London, 1969.
13. H. N. V. Temperley, J. S. Rowlinson, and G. S. Rushbrooke, *Physics of Simple Liquids*, North Holland, Amsterdam, 1968.
14. G. J. Throop and R. J. Bearman, *J. Chem. Phys.*, **42**, 2838 (1965).
15. G. J. Throop and R. J. Bearman, *J. Chem. Phys.*, **44**, 1423 (1966).

Chapter III

SOLVATION OF IONS

Edward S. Amis

1 INTRODUCTORY REMARKS

The phenomenon of the solvation of ions includes many concepts. It would be expected that small and/or highly charged ions, that is, ions with high surface charge density, in polar solvents would exhibit great ion-solvent interaction, hence the highest degree of solvation [178, 193, 255]. This is not always the case, however, since specific effects are often present which dominate the situation, and ions are selectively solvated by a component of a mixed solvent that is less polar than the other component [98, 99, 177, 221, 222, 317, 354, 379]. Sometimes the ion may react with the solvent in more than one manner, and a particular method of measurement may indicate the domination of one effect over another effect or other effects, hence not detect the other effect or effects. Thus an ion may break the structure of a hydrogen-bonded solvent and at the same time be solvated by the solvent. A method that measures proton frequency shifts, such as nuclear magnetic resonance spectroscopy (NMR), may predominantly show the upfield proton frequency shift due to solvent structure breaking, and leave undetected the downfield proton frequency shift arising from solvation of the particular type of ion. This gives the impression that the ion is not solvated, though other methods of measurement would show the ion to be highly solvated [179]. Again a particular type of measurement may show the molecules of solvent in the neighborhood of an ion to have greater mobility than molecules of solvent surrounded by other molecules of solvent in the same solution. The residence time of a molecule of solvent in the neighborhood of an ion is shorter than in a location where the molecule is surrounded by other solvent molecules. This phenomenon is termed *negative solvation,* in contrast to *positive solvation* which is defined as the case in which the mobility of a molecule of solvent is less and its residence time longer in the vicinity of an ion than when surrounded by other solvent molecules [227, 228, 230, 340, 342, 344, 348]. Also, if the intercalation of an ion into the tetrahedral structure of water reduces the orderliness of the water, the phenomenon is termed *negative hydration* [17].

Some methods of measuring ion solvation indicate only an inner shell of solvent molecules [168, 175, 209, 328, 385], while other methods detect both inner and outer (or several outer) shells of ion solvation [29, 157, 158, 199, 267, 335, 336, 416, 418, 419].

Some other factors that should be pointed out as influential in any general discussion of ion solvation are the concentration of the ion in solution and the temperature of the solution. It appears that, the more dilute the solution of electrolyte in an ionic solution, the greater the solvation number of any particular ion in the solution, since there are more solvent molecules per ion to solvate the ion [178, 396]. In several methods of measurement this is found to be the case, but the reverse effect has been observed, probably as a result of the structure-breaking effects of an ion or the ions of an electrolyte and a consequent increase in the number of simple solvent molecules required to solvate the ion of interest.

Increase in temperature would ordinarily be expected to decrease the solvation of ions [95, 97, 266], since the increased kinetic energy of molecules hinders their orientation by ions in the solvation shells of the ions. In general, solvation is found to increase with decreasing temperature. However, there are exceptions in which solvation remains constant [28, 215] or decreases with decreasing [28, 178] temperature, perhaps because of an increased strength of solvent structure at lower temperatures, which prevents disruption of the solvent structure by the ions.

Solvation does not involve a fixed, permanent molecule of solvent attached to a given ion. Rather, molecules and parts of molecules solvating ions constantly exchange with parts of molecules and molecules in the unbound state [199, 295].

There are often consistent trends in the degree of solvation of the ions from a given group of elements in the periodic table [79, 255, 399, 411, 414], though the trends according to one method of measurement may be the reverse of those of another measurement [13, 28]. However, the trends may not be evident [145], or the degree of solvation random [12, 392].

Thus ion solvation is a complex and relative phenomenon. For most methods a selected degree of solvation of a particular ion must be chosen, and the degree of solvation of all other ions must be referred to this standard. Thus the solvation number depends on the ion chosen and its assumed solvation number, because most methods measure the total solvation of the electrolyte and not that of the individual ions [178, 246, 255, 392]. The limitations of measuring ionic and molecular phenomena with macroscopic probes give rise to most of the seeming conflicts in ion solvation observations. These methods in general yield averages of contributions to the phenomena of various structures. They are therefore not definitive in determining absolute solution structure, which is necessary if the results of different methods are to be reconciled [178].

2 METHODS OF MEASUREMENT

Transference

The transference or transport number of an ion is the fraction of the total current carried by that ion. Since the different ions of an electrolyte have different velocities, different concentration changes occur in the anode and cathode compartments of a transference cell, and these changes may be used to calculate transference numbers [180, 181, 413].

The transport number t_i for the ith type of ion is given by the expression

$$t_i = \frac{n_i u_i e_i}{\Sigma \, n_i u_i e_i} \tag{3.1}$$

where n_i is the number of ith type of ions per cubic centimeter, and u_i and e_i are the velocity and charge, respectively, of the ith type of ion. Now $e_i = z_i \epsilon$, where z_i is the valence of the ion, and ϵ is the charge on the electron; hence

$$t_i = \frac{n_i u_i z_i}{\Sigma \, n_i u_i z_i} \tag{3.2}$$

where the summations in the denominators of (3.1) and (3.2) are taken over all the ions in solution.

In the Hittorf method the electrode and central portions of the solution in an electrolytic cell are analyzed, and the number of faradays flowing through the cell is determined using silver or other types of coulometers, or other current-measuring devices in series with the cell. Such devices are usually placed in the circuit immediately preceding and following the cell, to verify the amount of electricity passing through the cell and at the same time to detect any leakage of current in the cell. Figure 3.1 is a sketch of a transference cell. The coulometers A immediately precede and follow the cell in the electric circuit. The solution in the middle portion of the cell between the stopcocks B should not change in concentration during the electrolysis, or else the experiment will have to be discarded. The composition of electrodes C and D must be selected to give the best results with the system being studied. For example, silver electrodes can be used for solutions of alkali chlorides in water-ethanol solvents [29, 416, 419]. Spirals of no. 20 B and S grades of silver wire are suitable for electrodes. The silver wire anode should be etched with nitric acid, and both electrodes washed many times with distilled water and then three or four times with the solution to be used in the transference cell. The cell must be suspended firmly to prevent mixing of the solutions in the cell compartments, and must be immersed in an oil-filled constant-temperature bath with the thermostatic liquid being slowly but thoroughly and continuously stirred. Slow stirring prevents vibration of the cell, and thorough stirring makes possible the careful control of temperature necessary in the measurements. Compartments E and F permit contact of the cell solution with the source of current through platinum, silver or other noble metals fused through the ball joints, and they protect the leads from the bath fluid.

A method has been formulated [413] for the calculation of transference numbers. In this procedure N_0 and N_F are the respective initial and final number of equivalents of an ion associated with a given weight of solvent. N_E is the

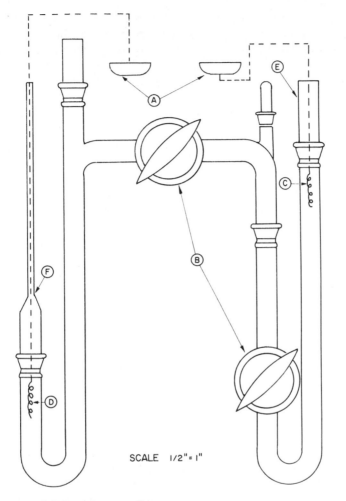

Fig. 3.1. Transference cell.

number of equivalents of ion added to the given weight of solvent by the electrode reaction, and tN_E is the number of equivalents of this ion lost to the given weight of solvent by migration. Hence

$$N_F - N_0 = N_E - N_E t \tag{3.3}$$

and

$$t = \frac{N_0 - N_F + N_E}{N_E} \tag{3.4}$$

A run is made with open stopcocks B in Fig. 3.1. At the end of the run the three compartments of the cell are isolated by closing the stopcocks. The analysis for the desired ions in the three compartments yields the required values of N_0 and N_F. The gain in weight during the run of a platinum crucible cathode of a silver coulometer due to deposited silver can be converted to N_E, and thus t can be calculated.

The moving-boundary method of transference numbers is based on the principle that, when a solution of one electrolyte is placed above a solution of another electrolyte in a vertical tube and a direct current passed from bottom to top, the boundary between the two solutions becomes sharp and moves up the tube.

If V is the volume in liters per faraday of current passed, VC is the equivalents per faraday of a selected ion constituent passing a fixed plane in the tube (where C is the equivalents per liter of the specified ion), and I is the constant current in amperes flowing for τ seconds, the transference number t is given by

$$t = \frac{VCF}{\tau I} \tag{3.5}$$

A current-measuring device in series with the transference tube can be used to determine I, and a timing device used to measure τ; or a coulometer in series with the tube can be used to ascertain the product τI. In this method of measurement, there must be no interdiffusion or mixing of the solutions meeting at the boundary; the motion of the boundary must be independent of the nature and concentration of the following or indicator ion solution; and there must be no volume changes in the tube to affect the motion of the boundary.

There is an adjusting effect whereby a faster-moving ion in a more concentrated solution, if mixed at the boundary with a slower-moving ion in a more dilute solution, will encounter a greater potential drop and will move forward more rapidly to overtake the boundary. The slower-moving ion, if mixed at the boundary with the more concentrated solution of faster-moving ions, will encounter a lower potential drop and will be slowed down so that the boundary overtakes it. Thus the boundary tends to adjust itself to a sharp condition. In spite of this adjusting effect, it is well to ensure a sharp boundary at the start of the experiment, rather than to depend on this restoring effect. One way to do this is to set both solutions in a jelly. Other ways are now described.

The autogenic method [131] in which the solution to be observed is placed in a tube the bottom of which is covered with a disc of metal which forms a soluble salt with the anion of the solution. If a current is passed from the disc upward through the solution, the boundary between the cations of the

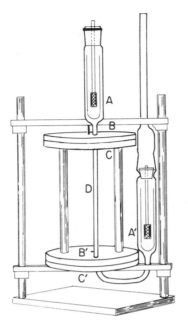

Fig. 3.2. Moving-boundary apparatus. A and A', electrode vessels; B and C, upper and lower sections of top disc; B' and C', upper and lower sections of bottom disc; D, calibrated tube.

solution and those formed from the metal will move up the tube, and the time required for the boundary to move a measured distance is observed. The method is applicable to alkali chlorides in solution and a cadmium disc. The method is simple but more limited than that of sheared boundaries [255, 256], to be described next.

The plane surfaces of each of two pairs of discs having a common central pivot are fitted together. The outlet tube of an electrode vessel containing the indicator solution extends downward through the top disc, and the electrode vessel is stoppered in such manner that a drop of liquid on the end of the outlet tube extends into a depression in the lower disc. A graduated tube filled with a slight excess of the solution to be studied extends upward through a hole in the bottom disc. Part of this slight excess of the solution forms a drop extending upward into a depression in the bottom of the top disc. The lower end of the calibrated tube extends downward through a hole in the top of a second pair of copivotal discs which also have conjoined plane surfaces and are copivotal with the upper pair of discs. The remainder of the excess solution extends as a drop into a depression of the lower of the second pair of discs. The properly bent outlet tube of a second electrode compartment extends upward through a hole in the lower of the second pair of discs, and the vertical electrode vessel is stoppered in such a manner that a drop extends upward into a depression in the upper of the lower pair of discs. The upper of the upper pair of discs and the lower of the lower pair of discs are fastened to a frame, as is the lower electrode vessel. All tube fittings into the discs are snug. Rods connecting the two inside discs of the two pairs can be used to rotate the calibrated tube so that its upper end fits snugly with the outlet tube of the upper electrode portion and its lower end over the upward-extending outlet tube of the vertical lower electrode vessel. The excess drops of solutions on the ends of the various tubes are sheared off smoothly by the rotating discs, and the surfaces of the solutions in the various tubes are brought together with very little mixing or diffusion at the boundaries. In Fig. 3.2 is a representation of the moving-boundary apparatus with tubes in place for an experiment.

Denison and Steel [103] used parchment paper to separate the solution of the ion whose transference number was to be determined in a lower tube from the solution of indicator in an upper tube, the bottom of which was flanged into a cone. Impressing a potential in the proper direction caused the current to move around the edges and through the paper, and the boundary moved down the tube. The top of the lower tube opened into a flared cup containing the same solution as the tube. After the motion of the boundary was established, the parchment-covered lower end of the upper tube was cautiously adjusted in the cup to a height above the upper end of the lower tube to give the best results. This device was simplified by replacing the cone and parchment paper with a flattened glass rod and a disc of soft rubber [258]. After the current that leaked around the edge of the disc started to cause the boundary to move downward, appropriate adjustment of the position of the rubber disc above the top of the lower tube was made. With the two last-mentioned methods it was difficult to obtain adequate separation of the solutions of the leading and indicator ion constituents. The sheared boundary method is applicable to systems of various leading and indicator ion constituents and to rising and falling boundaries, and gives a sharp boundary between solutions. The autogenic boundary is recommended because of its simplicity, when it can be applied. However, it is more restricted in its application than the method of sheared boundaries.

Boundaries can be made to move up or down. In upward-moving boundaries the indicator solution should be the denser, and in downward-moving boundaries the indicator should be less dense than the solution being observed. Also, the mobility of the indicator ion constituent should be less than that of the leading ion. The difference in refractive index, the difference in color of the two ions or, in the case of hydrogen or hydroxide ions, the change in color of an indicator may be used to follow the boundary.

It is necessary that solutions of observed and indicator ions do not mix, hence that the volume swept out by the boundary per coulomb of electricity passed be constant [see (3.5)]. Provided these conditions prevail, transference numbers can be determined without measuring current or time by employing two boundaries, one for the cation and one for the anion constituents. For the cation,

$$t_+ = \frac{v_+ cF}{\tau I} \tag{3.6}$$

where t_+ and v_+ are, respectively, the transference number and volume swept out by the cation boundary. For the anion an identical type of equation,

$$t_- = \frac{v_- cF}{\tau I} \tag{3.7}$$

can be written in terms of its transference number t_- and the volume v_- swept out by its boundary in τ seconds.

Since

$$t_+ + t_- = 1 \tag{3.8}$$

and the coulombs of current passed τI is the same in both cases,

$$t_+ = \frac{t_+}{t_+ + t_-} = \frac{\dfrac{v_+ cF}{\tau I}}{\dfrac{v_+ cF}{\tau I} + \dfrac{v_- cF}{\tau I}} = \frac{v_+}{v_+ + v_-} \tag{3.9}$$

and

$$t_- = \frac{v_-}{v_+ + v_-} \tag{3.10}$$

Most early transference measurements were made using this method. The one-boundary method prevailed, because it was always necessary to determine that the volume swept out by a boundary per coulomb of current passed was constant before mixing at the boundary could be ruled out.

Solvation of Ions: "True" Transference Numbers

Solvation or, in the case of water, hydration numbers of ions may be determined using the transference number of ions.

If a method uninfluenced by the movement of the solvent of solvation is used, true transference numbers are obtained. Nernst [297], Buckbock [72], Washburn [412, 414], and their associates were among the early investigators of solvation numbers who used the Hittorf method. A second solute, for example, sucrose, raffinose, and recently fructose [267] or a-methyl-D-glucoside [29, 416, 418, 419], was added to the solution in the Hittorf cell. The changes in salt and solvent concentrations were referred to this second solute which, being composed of neutral molecules, was not affected by the passage of current. When, however, solvent was carried along by the moving ions, the concentration of the reference substance was changed in the electrode regions of the cell. By accurately determining the change in concentration of the reference substance in an electrode region, the true transference number of the ion constituent and Δn (the increase or decrease in the number of moles of water) in a given electrode portion per faraday of electricity passed could be calculated. A polarimeter was used [29, 164, 267, 406, 412, 414, 417, 418, 419] to determine the concentrations of the optically active reference substance.

If Δn is the net effect of solvent carrying by all the ions present, for the cathode portion in the case of a binary electrolyte Δn becomes

$$\Delta n = \tau_c n_s^c - \tau_a n_s^a \tag{3.11}$$

where τ_c and τ_a are the true transference numbers of the cation and anion, respectively, and n_s^c and n_s^a are the number of moles of solvent carried per mole by cation and anion, respectively. Thus the number of moles of solvent carried by either the cation or the anion can be determined by using (3.11). For n_s^c, for example, the solution is

$$n_s^c = \frac{\Delta n}{\tau_c} + \frac{\tau_a}{\tau_c} n_s^a \tag{3.12}$$

The number of moles of solvent carried per mole of cation is found in terms of the number of moles of solvent carried per mole of anion, the net number of moles of solvent carried into the cathode portion of the cell, and the true transference numbers of the two ions. If τ_c, τ_a, and Δn are measured, and a value of n_s^a is assumed, n_s^c can be calculated. Thus a solvation number for one ion must be selected in order to determine the solvation numbers of a series of ions experimentally.

Methods have been devised for obtaining true transference numbers and the solvation numbers of ions in mixed solvents [416], using the Washburn [412] equation for the difference between the true transference number τ and the Hittorf transference number t. The Washburn equation is

$$\tau - t = \Delta n_{sol}^F \frac{n_s}{n_{sol}} \tag{3.13}$$

where Δn_{sol}^{F} is moles of solvent transferred per faraday, n_s is equivalents of solute per n_{sol} moles of solvent. By using the relationships

$$\Delta n_{sol}^{F} = \frac{\Delta g_{sol}^{F}}{M_{sol}} \tag{3.14}$$

and

$$n_{sol} = \frac{g_{sol}}{M_{sol}} \tag{3.15}$$

the following equation is obtained:

$$\tau - t = \Delta g_{sol}^{F} \frac{n_s}{g_{sol}} \tag{3.16}$$

whereby the true transference numbers can be obtained from the Hittorf transference numbers without knowing the molecular weight of the solvent mixture. In (3.15) and (3.16), Δg_{sol}^{F} is grams of solvent transferred per faraday, g_{sol} is grams of solvent, and M_{sol} is the molecular weight of the solvent.

By using true transference numbers, the following relations can be written for a solvent containing two components A and B:

$$\Delta g_{sol(A)}^{F} = M_{sol(A)} n_{sol(A)}^{c} \tau^{c} - M_{sol(A)} n_{sol(A)}^{a} \tau^{a} \tag{3.17}$$

$$\Delta g_{sol}^{F} = [M_{sol(A)} n_{sol(A)}^{c} + M_{sol(B)} n_{sol(B)}^{c}] \tau^{c} -$$

$$[M_{sol(A)} n_{sol(A)}^{a} + M_{sol(B)} n_{sol(B)}^{a}] \tau^{a} \tag{3.18}$$

In these equations $\Delta g_{sol(A)}^{F}$ and Δg_{sol}^{F} are grams of solvent A and grams of total solvent, respectively, transferred from the anode to the cathode per faraday; $M_{sol(A)}$ and $M_{sol(B)}$ are the molecular weights of solvent components A and B, respectively; $n_{sol(A)}^{c}$ and $n_{sol(B)}^{c}$ are moles of components A and B, respectively, solvating a mole of cation; $n_{sol(A)}^{a}$ and $n_{sol(B)}^{a}$ are moles of components A and B, respectively, solvating a mole of anion; and τ^{c} and τ^{a} are the respective true transference numbers of cation and anion.

Dividing (3.17) by $M_{sol(A)}$ yields (3.11) for a pure solvent. Equation (3.11) has an infinite number of solutions, but these can be reduced with the assumption that $n_{sol(A)}^{c}$ and $n_{sol(A)}^{a}$ must be integers and have upper and lower limits.

For unknown systems a reasonable lower limit is zero. The maximum upper limit for $n_{sol(A)}^{c}$ is the maximum possible number of moles of solvent A that can solvate one mole of cation in the solution assuming the anion is not solvated, and a corresponding upper limit for $n^a_{sol(A)}$ is the number of moles of solvent A per mole of anion in the solution assuming the cation is not solvated, and similarly for $n^c_{sol(B)}$ and $n^a_{sol(B)}$. The relations

$$n^c_{sol(A)} + n^a_{sol(A)} \leq \frac{n_{sol(A)}}{n_s} \qquad (3.19)$$

and

$$n^c_{sol(B)} + n^a_{sol(B)} \leq \frac{n_{sol(B)}}{n_s} \qquad (3.20)$$

where $n_{sol(A)}$ and $n_{sol(B)}$ are the total number of moles, respectively, of solvent A and of solvent B in the solution, and n_s is equivalents of solute. Equation (3.17) subtracted from (3.18) yields

$$\frac{\Delta g^F_{sol(B)} - \Delta g^F_{sol(A)}}{M_{sol(B)}} = n^c_{sol(B)} \tau^c - n^a_{sol(B)} \tau^a \qquad (3.21)$$

Equations (3.19) to (3.21) can be solved similarly, since they are of like form.

Differences in ionic hydration have been determined [24, 323] by measuring the volume change in the two halves of a solution separated from each other by a parchment paper diaphragm in a cell through which current was passed. Electroosmosis of the solvent alone caused negligible error in a $1.0\,N$ electrolyte solution at which concentration measurements were made [323].

Table 3.1 lists some examples of solvation numbers observed using transference methods. The number acutally obtained depends on the solvent, the reference ion selected, the extent of solvation assigned to the reference, whether the method used involved an inert reference substance or a moving boundary (e.g., parchment paper), temperature, and so on. Where possible, these conditions have been specified.

In general, transference measurements gave the orders of solvation: $Li^+ > Na^+ > K^+ > Rb^+ > Cs^+$; $Mg^{2+} > Ca^{2+} > Sr^{2+} > Ba^{2+}$; $F^- > Cl^- > Br^- > I^-$. These are the orders of surface charge densities.

Transference methods measure the total solvent carried by ions through the solvent, and not just an inner layer. There is, no doubt, exchange of solvent molecules between adjacent layers of solvent in solvation, and between the outer layer and the pure solvent, but the total entrainment of solvent is perhaps nearly constant as the ion moves through the solvent. This more-or-less constant entrainment of solvent by the ion is comparable to the retinue of people attracted to a famous personage parading along a street in a metropolis: some people pressing inward to approach or even touch the central personage;

Table 3.1 Solvation Numbers of Ions by Different Electrolytic Transference Methods

Solvent	Reference ion	Assumed solvation number of reference ion, n	Solvated ion	Solvation number of ion, n	Concentration of electrolyte (N)	Temp. $(°C)$	Ref.
Water	Cl^-	4	Li^+	14.0	1.2	25	255[a]
Water	Cl^-	4	Na^+	8.4	1.2	25	255[a]
Water	Cl^-	4	K^+	5.4	1.2	25	255[a]
Water	Cl^-	4	Cs^+	4.7	1.2	25	255[a]
Water	Cl^-	4	Li^+	14.3	1.0	25	255[b]
Water	Cl^-	4	Na^+	9.8	1.0	25	255[b]
Water	Cl^-	4	K^+	5.0	1.0	25	255[b]
Water	Cl^-	4	Li^+	14.0	1.2	25	414
Water	Cl^-	4	Na^+	8.4	1.2	25	414
Water	Cl^-	4	K^+	5.4	1.2	25	414
Water	Cl^-	4	Cs^+	4.7	1.2	25	414
NMA[c]	R_4N^+	0	Li^+	5.1	—	40	161
NMA	R_4N^+	0	Na^+	3.5	—	40	161
NMA	R_4N^+	0	K^+	3.3	—	40	161
NMA	R_4N^+	0	Cs^+	2.6	—	40	161
NMA	R_4N^+	0	Cl^-	2.1	—	40	161
NMA	R_4N^+	0	Br^-	1.7	—	40	161
NMA	R_4N^+	0	I^-	1.5	—	40	161
Water	Cl^-	4	Br^-	3-4	1	—	25
Water	Cl^-	4	I^-	2	1	—	25

[a]Inert reference substance.
[b]Parchment paper.
[c]N-methylacetamide.

117

some having attained this goal being willing to be displaced outward; some falling out of the stream as their place of residence, business, or means of transportation becomes more remote; but others joining the stream as their convenient locations for observation are approached. The net result is a fairly large, constant number of people in the immediate vicinity of the person of interest. While ions and molecules are not endowed with these personal motivations, the overall effect is much the same and results in the central ion carrying along a fairly constant number of molecules of solvent of solvation as it moves through the solution.

Solvation numbers of ions obtained from transference methods are usually relatively high compared with other methods of measurement. The moving-boundary method is more accurate than the procedure using an inert substance, since inert substances are not completely inert under all circumstances. However, Kortüm and Bockris [224] and Gordon [162] state that this is the best method for the simultaneous determination of transference numbers and total solvation. The method indicates [29, 267, 416, 418, 419] an alternate displacement of different layers of solvation by a first solvent on the cation and anion by a second solvent as larger and larger proportions of a second solvent are added to the solution.

Of the moving-boundary methods, that of sheared boundary is the most accurate of the general methods. The autogenic boundary method is the simplest, when applicable.

Conductance

The resistance R of a solution is measured, and the reciprocal of this resistance is termed the conductance K of the solution, that is,

$$K = 1/R \tag{3.22}$$

The specific resistance Γ can be calculated from the measured resistance R using the equation

$$R = \Gamma \frac{l}{A} \tag{3.23}$$

where l is the length, and A is the cross section of a uniform substance. The ohm is the unit of resistance. The international ohm is the resistance at $0°C$ to an unvarying electric current by a column of mercury 14.452 g in mass, having a constant cross-sectional area and a length of 106,300 cm. Now,

$$K = \frac{1}{R} = \frac{A}{\Gamma l} = L\frac{A}{l} \tag{3.24}$$

where L is the specific conductance. Specific conductance depends on the ions present and on their concentration. Equivalent conductance is used in most computations. It is the conductance of a solution of 1 g equivalent weight of an electrolyte placed between parallel electrodes 1 cm apart and extensive enough in area to contain the necessary volume of solution. It is calculated from the specific conductance using the equation

$$\Lambda = \frac{1000L}{c} \tag{3.25}$$

in which c is the concentration of the solution in gram equivalents per liter. The equivalent conductance at infinite dilution Λ_0 is an important constant for an electrolyte at a given temperature and is obtained by extrapolation of Λ as a function of $\sqrt{c}$ to $c = 0$, using the Kohlrausch [219] equation,

$$\Lambda = \Lambda_0 - K_1 \sqrt{c} \tag{3.26}$$

the Onsager [205] equation,

$$\Lambda = \Lambda_0 - (\theta \Lambda_0 + \sigma)\sqrt{c} \tag{3.27}$$

or some more sophisticated equation. In (3.26), K_1 is an empirical constant equal to the parenthetical term $\theta \Lambda_0 + \sigma$ of (3.27), where θ depends on the properties of the electrolyte, on the dielectric constant of the solvent, and on the temperature; and σ depends on the properties of the electrolyte, on the dielectric constant and viscosity of the solvent, and on the temperature. For a given electrolyte solvent and temperature, θ and σ can be obtained, hence Λ_0 found from measured values of Λ as a function of concentration under these conditions. A plot of Λ versus $\sqrt{c}$ extrapolated to $c = 0$ should, according to either (3.26) or (3.27), yield a straight line which intercepts the Λ axis at Λ_0. At this point the similarity between the two equations ceases. According to (3.26) the line has an empirical slope K_1, but according to (3.27) the line has a theoretical slope $\theta \Lambda_0 + \sigma$ which can be calculated from the value of Λ_0 and the valances of the ions composing the electrolyte, the transference number of the anion of the electrolyte at infinite dilution in case of an unsymmetric electrolyte, the temperature at which the measurements are made, and the dielectric constant and viscosity of the solvent.

Once Λ_0 values for electrolytes are found, λ_0^+ and λ_0^- values for ions can be evaluated by assuming a value of λ_0 for some standard reference ion such as ClO_4^-, or by using the relationship involving the transference number at infinite dilution t_0 of the ion,

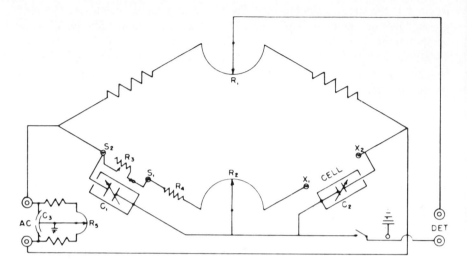

Fig. 3.3. Schematic diagram of conductivity bridge as applied to ac measurements. Courtesy of Leeds and Northrup Company, Sumneytown Pike, North Wales, Pennsylvania 19454.

$$\lambda_0 = t_0 \Lambda_0 \qquad\qquad (3.28)$$

provided t_0 for the ion is known from other measurements.

Figure 3.3 shows a schematic diagram of the Jones and Joseph bridge built by Leeds and Northrup [205]. The resistance of the sample in the cell connected across the positions x_1 and x_2 is measured by adjusting the known resistance R_3 until the balance point is attained. A Dumont model 304 H oscilloscope is a convenient instrument for locating the balance point where the instrument shows a minimum in the sine wave signal. The capacitance is regulated by two condensers C_1 and C_2, and the construction of the bridge ensures minimum inductance effects. The bridge is grounded externally, and all components are connected in such a manner that a closed circuit is formed with the ground. Measurements are generally taken with an alternating current of 1000-Hz frequency to prevent irreversible electrode reactions, hence change in the concentration, and in some cases the nature, of the electrolyte during resistance measurements. A U.S. Bureau of Standards (NBS) certified resistor is used to calibrate the resistors of the bridge [105]. Resistances up to 60,000 Ω can be made in series, and resistances up to 2×10^7 Ω made by placing the bridge in parallel with three of the 10,000-Ω resistors.

The Jones and Bollinger [203] type of conductivity cell generally used in conductance measurements is depicted in Fig. 3.4. The figure includes the flask which contains the solution to be measured, and the transfer system for

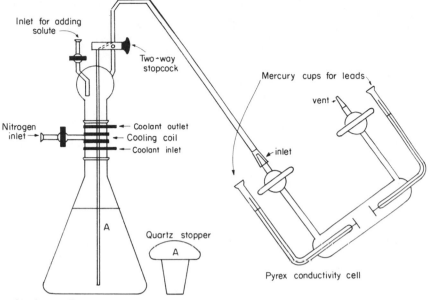

Fig. 3.4. Transfer system for solutions—quartz bottle to cell.

transferring the solution from the flask to the cell. Oil-containing thermostatic baths must be used to prevent leakage between the leads to the cell. Temperature control must be accurately maintained within ±0.01°C or better, and thermometers must be standardized against NBS thermometers or thermocouples. Conductivity solvents having specific conductances of 1 to 5 × 10⁻⁷ Ω or less must be used, and viscosities and dielectric constants of the solvents must be determined accurately if data are to be compared with theory.

Solvation numbers of ions can be calculated from conductance data on electrolytes, using the Stokes ionic radii r_s. Use is made in these calculations of the limiting ionic conductances at infinite dilution λ_0 which, as mentioned before, can be obtained from the limiting conductances of electrolytes and the limiting conductance of the standard reference perchlorate ion [287].

In calculating solvation numbers in sulfolane and benzene solvents, the following considerations were made. For tetraalkylammonium salts in benzene and sulfolane [287, 394], the λ_0 increase from ethyl to methyl substituent was less than anticipated. The small increase in the limiting increase in sulfolane, analogous to the behavior of nitrobenzene solutions, was explained as an interaction between the tetramethylammonium cation and the dipoles of the solvent molecules.

The Stokes ionic radii r_s were calculated from λ_0 values of unsolvated Et_4N^+, Pr_4N^+, and Bu_4N^+ by means of the equation

$$r_s = 0.82|z| \, / \, \lambda_0 \eta_0 \qquad (3.29)$$

and the Stokes-Robinson correction factors were evaluated [298] using the crystallographic radii r_0. Since the r_s values of the alkali metal ions were between or near those of the tetraalkylammonium ions, the correction factors for the former ions were obtained from linear plots of r_0/r_s versus r_s. These corrected radii r_{corr} of the solvated ions were used to calculate the volume V of the solvation shells of the ions from the equation

$$V = \tfrac{4}{3}\pi(r_{corr}^3 - r_0^3) \qquad (3.30)$$

Assuming electrostriction caused negligible contraction of the solvent sulfolane, the number of molecules in the solvodynamic unit was calculated from the molecular volume of sulfolane (158 Å³) and the volumes of the solvation shells. Other solvents were treated in a similar manner.

Solvents with the greatest molecular volume, such as sulfolane and nitrobenzene, gave the lowest values for solvation number of cations.

For ions with a Stokes radius greater than 2.5 Å, the above procedure yielded an effective hydrated radius [298]. The procedure was applied to ions of smaller Stokes radii, assuming all tetraalkylammonium ions except the tetramethylammonium ion were unhydrated. The effective hydrated radius r_H was defined by a calibration curve which possessed a finite limit as the Stokes radii r_s went to zero. The curve was a plot of the crystal radii r_0 versus r_s for the tetraalkylammonium ions, except for the tetramethylammonium ion. From this curve the deviations from, and the necessary corrections to, the Stokes'-law radius can be made. The method was applied to the hydration numbers of the lanthanide ions using conductance data [43, 84]. Solvation numbers of ions were calculated by Ulich [399], using the Stokes'-law radii from conductance data and the radii proper of the ions. Water added to the acetone-rich region of acetone-water solutions of LiBr caused a rapid decrease in the ion-pair association constant K_A of LiBr [42]. This was attributed to stronger solvation of the ions by water than by acetone; especially of the bromide ion as a result of hydrogen-bond formation.

In general the solvation numbers for the same ions from conductance data were similar to the values from transference data at like temperatures in the same solvent. For different solvents the same order of degree of solvation of ions did not always prevail. In some solvents the degree of solvation of ions from a group of the periodic table were the same as observed in transference measurements, in spite of the fact that the reference ion selected and its assumed extent of solvation were different for the two procedures.

Table 3.2 contains the solvation numbers as determined from conductance

Table 3.2 Solvation Number of Ions from Conductivity Measurements

Solvent	Temp. (°C)	Ion	Solvation Number	Ref.
Sulfolane	30	Li^+	1.4	287
Sulfolane	30	Na^+	2.0	287
Sulfolane	30	K^+	1.5	287
Sulfolane	30	Cs^+	1.3	287
Sulfolane	30	Cl^-	~0	287
Sulfolane	30	Br^-	~0	287
Sulfolane	30	I^-	~0	287
Ethanol	25	Li^+	6	399
Ethanol	25	Na^+	4-5	399
Ethanol	25	K^+	3-4	399
Ethanol	25	Cl^-	4-5	399
Ethanol	25	Br^-	4	399
Ethanol	25	I^-	2-3	399
Acetone	25	Li^+	4	399
Acetone	25	Na^+	4-5	399
Acetone	25	K^+	4	399
Acetone	25	Cl^-	2	399
Acetone	25	Br^-	1	399
Acetone	25	I^-	0-1	399
Water	—	Li^+	14	26
Water	—	Na^+	9	26
Water	—	K^+	5	26
Acetonitrile	25	Li^+	9	411
Acetonitrile	25	Na^+	6	411
Acetonitrile	25	K^+	3	411
Acetonitrile	25	Cl^-	2	411
Acetonitrile	25	Br^-	1-2	411
Acetonitrile	25	I^-	0-1	411
Methanol	25	Li^+	7	399
Methanol	25	Na^+	5-6	399
Methanol	25	K^+	4	399
Methanol	25	Cl^-	4	399
Methanol	25	Br^-	2-3	399
Methanol	25	I^-	0-3	399
Methanol	25	Li^+	7	411
Methanol	25	Na^+	5-6	411
Methanol	25	K^+	4	411
Methanol	25	Cl^-	4	411
Methanol	25	Br^-	3	411
Methanol	25	I^-	1	411

data for some selected ions in the specified solvents at the specified temperatures.

The variety of solvents and ions investigated with respect to solvation has been expanded by conductance measurements. The variation in solvation number of a particular ion with solvent is evident, and the order of the degree of solvation of the ions in a group of the periodic table is not the same in all solvents. The conductance method yields limiting solvation numbers, since limiting conductances of ions are used in the calculations. Since the equivalent conductances of ions are relative to some standard reference ion, so too are the solvation numbers of ions relative to some chosen ion. By using this approach the results of different workers under similar conditions seem consistent. The data for the solvent methanol at the bottom of Table 3.2 illustrate this point.

Electromotive Force

Electromotive force measurements can be used to give the solvation numbers of electrolytes and, provided the solvation number is assumed or taken from another source for one ion, the solvation numbers of other ions may be estimated. This was done for HCl [117], using the Hudson-Saville [186] approach, which is similar to that of Stokes and Robinson [377] and of Glueckauf [155] in the case of concentrated aqueous solutions.

Since ions are comparable in dimensions to solvent molecules, the Born equation, which depends on the potential of a charged particle in a uniform dielectric, cannot be applied accurately to an ion in solution. Close to an ion near-dielectric saturation prevails. The first layer of molecules can be regarded as completely oriented and treated as a firmly bound ionic solvation shell, the formation of which causes a loss of free energy by the coordinated water molecules as the gaseous ion enters the solution. The remaining free-energy change is assumed to be relatively small, especially in solutions of high dielectric constant, and can be calculated using a Born-type equation. Generally, in mixed solvents the ions are assumed to be preferentially solvated by more polar water molecules, especially when the water content is high in comparison to a less polar organic solvent.

The coordination of n water molecules, when 1 mol of HCl as ions goes from the gaseous state to a standard state of ions in aqueous solution, is written [117]:

$$(H + Cl)_{gas} + nW_1 \rightleftharpoons (H + Cl), nW \qquad (3.31)$$

where H^+ and Cl^- ions are written without their charges, and W stands for water. The change in free energy ΔG^W for the process is

$$\Delta G^W = \mu^W_{(H + Cl),nW} - \mu_{(H + Cl)} - n\mu_w^W \qquad (3.32)$$

where $\mu^w{}_{(H + Cl),nW}$ is the partial molal free energy of 1 mol of HCl solvated with n moles of water in aqueous solvation, $\mu_{(H + Cl)}$ is the partial molal free energy of the mole of pure HCl ions in the gaseous state, and $n\mu_w{}^w$ is the partial molal free energy of n moles of water in the pure state.

For the same process in an organic solvent-water mixture, the free-energy change ΔG^S is

$$\Delta G^S = \mu^S{}_{(H + Cl),nW} - \mu_{(H + Cl)} - n\mu_w{}^S \tag{3.33}$$

where $\mu^S{}_{(H + Cl),nW}$ is the partial molal free energy of 1 mol of HCl solvated with n moles of water in the organic substance-water solvent, $\mu_{(H + Cl)}$ is the same as above, and $n\mu_w{}^S$ is the partial molal volume of n moles of water in the organic substance-water solvent containing no HCl.

Now, if there is negligible interaction of the solvated ions with the solvent, the difference in the partial molal free energies of the solvated ions in the two solvents depends only on the difference in the concentrations of the ions in the two. We set

$$\mu^S_{(H + Cl),nW} = \mu^w_{(H + Cl),nW} \tag{3.34}$$

The terms are equal, whether the standard state is taken either as the mole fraction or the molar one. The equality is not exactly true for either scale. By using the assumption, the difference in (3.32) and (3.33) becomes

$$\Delta G^S - \Delta G^W = n(\mu_w{}^w - \mu_w{}^S) = -F(^SE^0 - {}^WE^0) \tag{3.35}$$

where F is the faraday, and $^SE^0$ and $^WE^0$ are the respective standard potentials of galvanic cells containing HCl as the electrolyte in the organic substance-water solvent and in the pure water solvent.

The partial molal volume for water in the mixture was expressed in terms of mole or of volume fraction statistics, and for simplicity expressions were developed for ideal mixtures. If N_w and ϕ_w are, respectively, the mole and volume fractions of water,

$$\mu_w{}^w - \mu_w{}^S = -RT \ln N_w = -RT \ln \phi_w \tag{3.36}$$

If $E_N{}^0$ and $E_\phi{}^0$ are used for the mole and volume fraction of water models, respectively, the following equations result:

$$^SE_N{}^0 = {}^WE_N{}^0 + n\frac{RT}{F} \ln N_w \tag{3.37}$$

on the mole fraction N scale, and

$$^{s}E_c^0 = {}^{w}E_c^0 + n\frac{RT}{F} \ln \phi_w \tag{3.38}$$

on the molar c scale.

Equations (3.37) and (3.38) can be put in the form

$$^{s}E_m^0 = {}^{w}E_m^0 + n\frac{RT}{F} \log w + \frac{RT}{F}(n-2) \log \frac{M_{xy}}{M_y} \tag{3.39}$$

$$^{s}E_m^0 = {}^{w}E_m^0 + n\frac{RT}{F} \log w + \frac{RT}{F}(n-2) \log \rho \tag{3.40}$$

where w is the weight fraction of water, and ρ is the density. For solutions rich in water, $\log \rho$ and $\log \dfrac{M_{xy}}{M_y}$ are small; and if $n \sim 2$, both (3.39) and (3.40) approximate to

$$^{s}E_m^0 = {}^{w}E_m^0 + n\frac{RT}{F} \ln w \tag{3.41}$$

where $^{s}E_m^0$ and $^{w}E_m^0$ are, respectively, the standard potential of the cell on the molal scale m when the solvent is mixed organic component-water and when the solvent is pure water. In the above equations M_{xy} is the mean molecular weight of the solvent calculated from the respective molecular weight M_x of the organic component and M_y of the water component of the solvent, and the percent by weight x of the organic component of the solvent, using the equation

$$M_{xy} = \frac{100}{\dfrac{x}{M_x} + \dfrac{(100 - x)}{M_y}} \tag{3.42}$$

Equations (3.37) and (3.38) can be used to obtain solvation numbers of electrolytes and, by assuming a solvation number for a standard reference ion, of ions by plotting $^{s}E_c^0$ versus $\ln \phi_w$ and by plotting $^{s}E_c^0$ versus $\ln N_w$. The plot of $^{s}E_c^0$ versus $\ln N_w$ showed no striking correlation and was not used [174].

In measuring the potentials of cells, an accurate potentiometer such as the Leeds and Northrup K-3 Universal potentiometer and a sensitive galvanometer to detect null points must be employed. Since the cell under study must have its emf balanced against an accurately known standard of emf, the cell used for standardization of the voltage for the potentiometer must be carefully calibrated. A NBS-certified Eppley standard cell is used in standardizing a saturated Eppley cell to be used as the reference source of potential in the potentio-meter circuit. The saturated cell is more precise than the unsaturated type. The saturated cell has an emf of 1.01864 V absolute at 20°C, which is constant over a long period of time but has a temperature coefficient sufficiently great to require the use of a temperature-controlled oil bath. Between 10 and 40°C the emf decreases about 40 μV per degree rise in temperature. The unsaturated cell is more portable and does not have as high a temperature coefficient of emf as the saturated cell. The former has an emf at 20°C of 1.01884 to 1.01964 V absolute.

The working cell must provide a current drain of 24.44 mA through the potentiometer, and a voltage of 1.8 V. The Leeds and Northrup No. 7597 lead storage cell, which has a rating of 2 V, 200 Ah, and such a low self-discharge as to provide 4000-hr uninterrupted service, is recommended.

For high precision, research-type potentiometers incorporate both guarding and shielding. Shielding is normally grounded and provides freedom from electrostatic effects, particularly in dry atmospheres. The panel and metal case of a research potentiometer are grounded and serve as an electrostatic shield. Guarding uses special circuitry insulated from ground to prevent adverse effects of leakage current paths between critical circuit points. For example, the battery and galvanometer circuits in research potentiometers are individually guarded.

The galvanic cell to be studied must be carefully thermostated in an oil bath which maintains the temperature to within ±0.01°C. The electrodes must be carefully prepared from extremely pure chemicals, and the solvents must be of conductivity grade and be free of oxidizing or reducing components. For example, alcohols as solvents must be of conductivity grade and free of aldehydes [45, 259, 277]. The cell vessel used depends on the type of cell being designed (e.g., whether the cell is with or without transference), but in many cases an H-type cell [277] or some modification thereof [196, 259] is suitable.

Figure 3.5 is a schematic diagram of a Leeds and Northrup K-3 potentiometer. The connection posts for the working battery $BA+$ and $BA-$, the standard cell $SC+$ and $SC-$, and the emf to be measured $EMF+$ and $EMF\ G$ are shown, as are many other details, for example, the three tapping keys for coarse, medium, and fine adjustment, and the guarding.

By using the above procedures, the cells containing the desired electrolyte ranging in concentration from 0.001 to a few hundredths molal in the selected solvent, and having properly chosen and designed electrodes, are placed in the thermostatic bath at the required carefully controlled temperature and allowed to equilibrate usually 30 to 60 min before their emfs are measured. Several emf measurements are made on each cell, to ensure that equilibrium has been reached and that the emf is steady. From the emfs of the cells at different concentrations of the chosen electrolyte and the selected solvent and tempera-

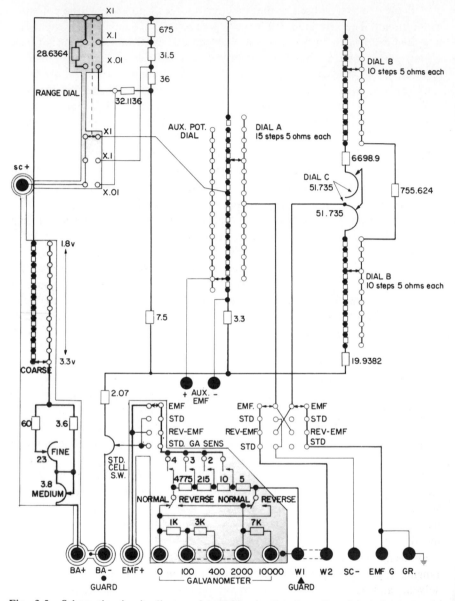

Fig. 3.5. Schematic circuit diagram for K-3 potentiometer. Guarding is indicated by shaded areas. Courtesy of Leeds and Northrup Company, Sumneytown Pike, North Wales, Pennsylvania 19454.

ture, the standard potential E^0 of the cell is calculated using the Harned and Owen [172, 174] equation or some modification thereof. The equation is complex and is employed most advantageously using computer techniques. It involves a least-squares statistical analysis of the experimental data. For example, in the cell

$$\text{Pt, } H_2(1 \text{ atm})|HI(m), \, X\% \, CH_3OH\text{-}Y\% \, H_2O|AgI, \, Ag \qquad (3.43)$$

the equation used [259] to calculate the standard potential of the silver-silver iodide electrode was

$$E_m{}^0 = E + 2k \log m + 2k \log \gamma_\pm \qquad (3.44)$$

where E is the measured emf corrected for hydrogen pressure by adding the value $760/(P - p_w)$; P is the barometric pressure and p_w is the vapor pressure of the solvent [174]. The molality is m, k is $2.303 \, RT/F$, and

$$\log \gamma_\pm = - \frac{A(d_0 m)^{1/2}}{1 + Ba(d_0 m)^{1/2}} - \log (1 + \frac{2M_{xy}}{1000}) + E_{ext} + bm \qquad (3.45)$$

Here A and B are constants of the Debye-Hückel theory, depending on the temperature and dielectric constant of the solvent, and d_0 is the measured density of the solvent. The symbols a and b are the ion size parameter and an adjustable parameter. M_{xy} is defined by (3.42), and E_{ext} represents the Debye-Hückel extended terms [174]. These standard potentials in different solvents are used to calculate solvation numbers when applying (3.39) to (3.41).

By taking Cl^- ion as the standard reference ion with zero solvation number in a water-methanol solvent, primary solvation numbers for Li^+, Na^+, K^+, and H^+ have been calculated [118] to be respectively, 4, 4, 4, and 1. Transsolvation ($H_2O \rightarrow MeOH$) equilibria for Li^+, Na^+, and K^+ ions were also determined. Equations have been given [397] for the dependence of the normal potentials $E_m{}^0$, $E_N{}^0$, and $F_C{}^0$ for cells without transference on the dielectric properties of a mixed solvent, primary hydration of an electrolyte, and cation-anion equilibria.

Galvanic cell potential studies were made with molten salt solutions containing water, in which water molecules were ligands competing with bromide ions in displacing nitrate ions from the coordination spheres of cadmium ions [236]. Bromide ions displaced nitrate ions more readily than water molecules from the coordination spheres, since the association constant for the reaction $Cd^{2+} + Br^- \rightleftharpoons CdBr^+$ increased with decreasing water content. The increase represented the tendency of the Cd^{2+} ions to hydrate.

The emf determination of solvation numbers apparently represents primary shell solvation. From the data Li^+, Na^+, and K^+ are equally hydrated in their primary shells.

The value of $n = 2.2$ for HCl from cell potential measurements is lower than that for other methods, with the exception of diffusion ($n = 2.1$). From activity coefficients the value of $n = 8$ was found [377]. Glueckauf [155] found $n = 4.7$. The ${}^{s}E_{N}{}^{0}$ versus $\ln N_{w}$ plots gave an n value higher than 2.2, namely, 2.7 to 5.0.

Thermodynamics

Theoretical and Experimental Data

As in other methods, solvation data for a reference ion or reference ions must be known from some source in order to determine solvation of ions from thermodynamic data, for example, solvation energies. Knowledge of solvation energies of standard reference ions is based on nonthermodynamic principles, generally theoretical, of the division of the observed thermodynamic function for a binary electrolyte. The principles are based on the dependence of the energy on some function of the reciprocal ionic radii.

One attempt to find individual ionic energies splits the total heat of solvation of $K^{+}F^{-}$ into equal parts, $\Delta H^{\circ}_{K^{+}} = \Delta H^{\circ}_{F^{-}} = -95.5$ kcal mol^{-1}, later adjusted to $\Delta H^{\circ}_{K^{+}} = -94$ kcal mol^{-1} and $\Delta H^{\circ}_{F^{-}} = -97$ kcal mol^{-1}, to account for the different spatial distribution of water molecules about the K^{+} and F^{-} ions arising from the noncentral location of the water molecule dipole. By applying these values, $\Delta H^{\circ}_{H^{+}}$ was determined [41] to be -276 kcal mol^{-1}. It was thought [87, 88, 292] that this was an oversimplification, and not in harmony with Bernal and Fowler's enthalpy and free energy of hydration.

The free energy of solvation of ions has been represented by the Born equation with empirically corrected radii [88], but this is believed to be a less satisfactory method than that of Bernal and Fowler [41].

Eley and Evans [108] made a comparative study of the methods up to 1953, and their calculations were selected [87] as the most acceptable for the determination of individual heats. $\Delta H^{\circ}_{K^{+}}$ and $\Delta H^{\circ}_{F^{-}}$ were given as -90 and -91 kcal mol^{-1}, respectively. Verwey [405] split ΔH°_{KF} into $\Delta H^{\circ}_{K^{+}} = -75$ and $\Delta H^{\circ}_{F^{-}} = -122$ kcal mol^{-1}.

A more complete calculation of hydration enthalpy has been formulated [73] which includes ion-dipole, dipole-dipole, and ion-quadrupole interactions, and the effects of induced moments.

$\Delta H^{\circ}_{H^{+}}$ has been obtained [169] using the difference in conventional energies of oppositely charged pairs of ions of the same radii. Absolute and conventional standard enthalpies of hydration for a cation M of valence z^{+} and of an anion A of valence z^{-} are related by

$$\Delta \overline{H}^{\circ}_{M^{z^{+}}} = \Delta H^{\circ}_{M^{z^{+}}} - z^{+} \Delta H^{\circ}_{H^{+}} \tag{3.46}$$

$$\Delta \overline{H}^{\circ}_{A^{z^{-}}} = \Delta H^{\circ}_{A^{z^{-}}} - z^{-} \Delta H^{\circ}_{H^{+}} \tag{3.47}$$

where the $\Delta \bar{H}^{\circ}$'s are the conventional relative enthalpies, and the ΔH°'s are the absolute ionic enthalpies of the indicated species. For uni-univalent electrolytes subtraction of (3.47) from (3.46) gives

$$\Delta \bar{H}^{\circ}_{M^{z^+}} - \Delta \bar{H}^{\circ}_{A^{z^-}} = [\Delta H^{\circ}_{M^{z^+}} - \Delta H^{\circ}_{A^{z^-}}] - 2\Delta H^{\circ}_{H^+} \qquad (3.48)$$

Experimental values of conventional heats of hydration have been compiled [37]. These values of $\Delta \bar{H}^{\circ}_{M^{z^+}}$ and $\Delta \bar{H}^{\circ}_{A^{z^-}}$ can be plotted against some function of the radii [53, 184, 431] of cations and anions, respectively, and from these curves a single curve can be plotted for the left-hand side of (3.48) against the same function of the ionic radius. In the latter plot r is not necessarily the radius of any particular ion, since identical radii are found in few cation-anion pairs. The extrapolation of the left-hand side of (3.48) to infinite radius should ideally yield $2\Delta H^{\circ}_{H^+}$ as the intercept on the $\Delta H^{\circ}_{M^{z^+}} - \Delta H^{\circ}_{A^{z^-}}$ axis, provided the theoretical relation between the conventional enthalpies of hydration and the inverse function of the radius against which they are plotted is known.

The reciprocal of the sum of the radius of the ion and of the radius of the water molecule becomes important, for usually no single-powered function of the hydrated radius represents the heat of hydration of ions [88], because of different dependencies on the radius of the primary shell [47, 87, 292] and outer regions. The primary shell depends mainly on the reciprocal of the square of the hydrated radius r_h^{-2}, and the outer region depends on the reciprocal of the first power of the sum of the radius of the ion r_i and of twice that of the water molecule $2r_{H_2O}$; that is, on $(r_i + 2r_{H_2O})^{-1}$. To surmount this difficulty, the best and most self-consistent of the conventional relative enthalpies $\Delta \bar{H}^{\circ}$ for cations and anions were plotted [169] against $(r_i + r_{H_2O})^{-3}$ taking r_{H_2O} as 1.38 Å. The r_i's were taken from Pauling [311] and from Ahrens [3].

After considering both "hard-sphere" and "soft-sphere" ion-contact models and allowing for a possible coordination number of 4 or 6 in the primary shell, the best value of $\Delta H^{\circ}_{H^+}$ was selected as -260.7 ± 2.5 kcal mol^{-1}.

The heat and standard free energy of formation in the gaseous state (g) of $H_3O^+(g)$ from $H^+(g)$ and $H_2O(g)$ have been investigated [87, 115, 165, 360]. Thermodynamic cycles involving H_3O^+, ClO_4^-, and NH_4^+, ClO_4^- were assigned the same crystal lattice energies, because the ions were of the same charge. Proton affinity and enthalpy changes for the protonation of water were assumed to be equal at room temperature, and the validity of the assumption was discussed [88]. At 298°C the values of $\Delta H^{\circ}_{H_2O}$ ($-P_{H_2O}$), $\Delta S^{\circ}_{H_2O}$, and $\Delta F^{\circ}_{H_2O}$ were given as -170 kcal mol^{-1}, -27 eu, and -160 kcal mol^{-1}, respectively.

High values [108] of -280 or -292 kcal mol^{-1} for free energy [92, 235] and -302 kcal mol^{-1} for enthalpy of hydration of H^+ are considered doubtful [88]. In the literature other data are -259 kcal mol^{-1} for free energy [300], and

-265 [404] and -263 kcal mol^{-1} [64] for enthalpy. The limits of uncertainty [47] in $\Delta H^\circ_{H^+}$ obtained by the method discussed above are perhaps no more than 7 kcal mol^{-1}.

The form of the function for large r_h has been criticized, but the plot of the left-hand side of (3.48) versus $(r_i + r_{H_2O})^{-3}$ has been approved [87].

Equilibrium Methods

Gaseous ion hydration, including that of hydrogen ions [214], was studied [182, 183, 211, 213] in irradiated water vapor from 0.1 to 6 torr and from 5 to 600°C. Mass spectrometric determinations were made of the relative concentration of the species $A^+ \cdot nS$, $A^+ \cdot (n-1)S$, $B^- \cdot nS$, and $B^-(n-1)S$ where A^+ and B^- are positive and negative ions produced by some ionizing radiation or thermal means, and S is a solvent molecule. From these relative concentrations equilibrium constants were calculated. In the equilibrium expression

$$K_{(n-1),n} = \frac{P_{A^+ \cdot nS}}{P_{A^+ \cdot (n-1)S} P_S} \qquad (3.49)$$

the equilibrium pressures were replaced by the ratio of the mass spectrometer intensities of the corresponding ions. From the equilibrium constants the stepwise free energy $\Delta G_{n-1,n}$, enthalpy $\Delta H_{n-1,n}$, and entropy $\Delta S_{n-1,n}$ were determined for the processes

$$A^+ + S \rightleftharpoons A^+ \cdot S \quad (0,1) \qquad (3.50)$$

$$A^+ \cdot S + S \rightleftharpoons A^+ \cdot 2S \quad (1,2) \qquad (3.51)$$

$$A^+(n-1)S + S \rightleftharpoons A^+ \cdot nS \quad [(n-1),n] \qquad (3.52)$$

For hydrogen-ion studies the H_2O^+ and OH^+ ions were formed by using electrons, protons, and α particles to irradiate water vapor. H_3O^+ was formed rapidly by water vapor reacting with H_2O^+ and OH^+, and H_3O^+ in turn reacted with water vapor to produce other hydrates [212, 237, 264], $H_5O_2^+$, $H_7O_3^+$, $H_9O_4^+$, $H_{11}O_5^+$, $H_{13}O_6^+$, $H_{15}O_7^+$, and $H_{17}O_8^+$.

Determination of the thermodynamic functions listed above shows the shell structure, since discontinuities in the change in $\Delta H_{n-1,n}$ and $\Delta S_{n-1,n}$ values occur whenever a shell is completed. The total heat of solvation ΔH_{solv} of the ion can also be found from the equation [211]

$$\Delta H_{solv} = \sum_{n=0}^{\infty} [\Delta H_{n-1,n} - \Delta H_{evap}(S)] \qquad (3.53)$$

and similar equations can be used for the free energy and entropy of solvation of the ions.

Comparative solvation of two different ions by the same solvent was studied, including the type in which the ions were both positively charged, as in the system H_3O^+, NH_4^+, and Na^+ in water vapor; the type in which one ion was positively charged and one negatively charged, as in the system K^+ and Cl^- in water vapor; and the type in which the ions were both negatively charged, as in the system Cl^-, BCl^-, and B_2Cl^- in water vapor.

Competitive solvation of either positively or negatively charged ions by two different solvents was also discussed [211], as in the systems NH_4^+, water vapor, and ammonia vapor, and $CH_3OH_2^+$, methanol vapor, and water vapor.

When water vapor containing several parts per million of ammonia vapor was irradiated, H_3O^+ and NH_4^+ ions were formed, the latter by proton transfer from hydrated hydronium ions to ammonia:

$$H_3O^+ \cdot wH_2O + NH_3 \rightarrow NH_4^+ \cdot n^*H_2O + (w + 1-n^*)H_2O \qquad (3.54)$$

This occurs because the proton affinity for ammonia is ca. 35 kcal higher than that for water. The ammonium hydrate $NH_4^+ \cdot n^*H_2O$ must make further collisions with water molecules before reaching the equilibrium composition $NH_4^+ \cdot nH_2O$. For higher concentrations of ammonia, complete disappearance of $H_3O^+ \cdot nH_2O$ occurred. Like studies were made with H_3O^+ and Na^+ in water vapor. The Na^+ ion solvates $Na^+ \cdot 2H_2O$, $Na^+ \cdot 3H_2O$, and $Na^+ \cdot 4H_2O$ were observed. The water molecules in the ammonium tri- and tetrahydrate are believed to be "inner-shell" molecules.

The types of mass spectrometers used were the a-particle mass spectrometer, electron-beam mass spectrometer, and proton-beam mass spectrometer. Figure 3.6 illustrates the ion source and the electrode system for the a-particle mass spectrometer. The other two types are discussed in some detail elsewhere [211].

The enthalpies $\Delta H_{n-1,n}$, that is, the enthalpies of hydration of $H^+ \cdot (n-1)H_2O$ to $H^+ \cdot nH_2O$, were compared with $Na^+ \cdot nH_2O$ in the literature [147]. By assuming the radius of the central H_3O^+ ion to be similar to that of the Na^+ ion, it was found that values of $\Delta H_{0,4}$ and $\Delta H_{0,6}$ for the Na^+ ion, -104 and -114 kcal, respectively, compared favorably with values of $\Delta H_{1,5}$ and $\Delta H_{1,7}$ for H_3O^+ which were -91 and -115 kcal, respectively. From the heat of solvation of the proton in liquid water [169, 300], -261 kcal mol^{-1}, and the enthalpy for the process $H^+(g) + 8H_2O(l) = H^+(H_2O)_8(g)$, ΔH_1 was found to be

$$\Delta H_1 = \Delta H_{0,8} + 8\Delta H_{evap}(H_2O) = -213 \text{ kcal mol}^{-1} \qquad (3.55)$$

and the heat of solvation of $H^+(H_2O)_8(g)$ into liquid water (-48 kcal mol^{-1}) was found from the difference. The Born equation, using a radius of 4 Å for the eight cluster, yields -40 kcal mol^{-1}, while a radius of 5 Å gives -35 kcal mol^{-1}. Thus the values found for $\Delta H_{n-1,n}$ are of the correct order of magnitude.

Data on solubility suggest [303] that NaCl solutions consist of NaCl·2H$_2$O aggregates separated by water. Ten molecules of water are removed per added molecule of HCl, and this reduces the solubility of NaCl. If the number of molecules of water is reduced below that required to separate the aggregates, the latter combine and precipitate.

If it is assumed that 10 water molecules, made up of two layers composed of 5

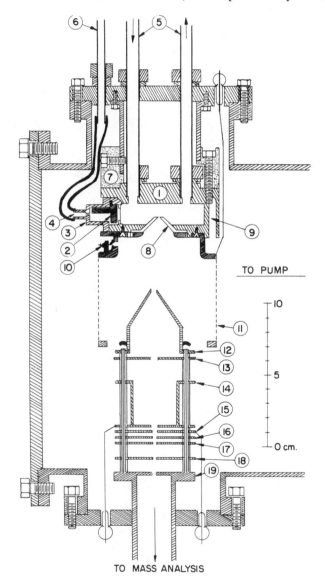

TO PUMP

TO MASS ANALYSIS

molecules each, surround each HCl molecule, the heat of neutralization of HCl by NaOH can be calculated by permitting two NaOH molecules to replace two water molecules in the outside layer. Thus, of the 10 water molecules, 2 adjacent molecules out of 5 are replaced. There are 10 ways of accomplishing this replacement, giving an entropy change:

$$\Delta S = 10R \ln 10 = 47 \text{ eu} \qquad (3.56)$$

This agrees well with that calculated from the heat of neutralization at $18°C$, which is 13.8 kcal mol^{-1} or 47 eu mol^{-1}, neglecting the change in heat content and the corresponding entropy change.

Solvation numbers of the electrolyte as a whole are dealt with in the two preceding paragraphs. When gaseous ions are dissolved in infinitely dilute solution, solvation numbers can be found [223] from the entropy decrease. The entropy decrease is caused by unbound water molecules entering tightly held solvation shells, and can be calculated from heat effects. If it is assumed that the decrease is the entropy change when free water becomes bound as water of crystallization, the entropy of dissolution of an ion can be used to calculate its solvation number.

Small, singly charged ions or multiply charged ions increase the viscosity of water and decrease the translational motion of water nearest them, while large, singly charged ions decrease the viscosity of water and increase the translational motion of water nearest them [344]. The decrease in viscosity is termed *negative hydration*, since the water molecules near the ions are more mobile than in pure water, and since these molecules exchange more frequently than those near other molecules in water. The structural temperature of water is lowered by ions that exhibit negative hydration.

Some data indicate [344] that negatively hydrated cations are associated with activity coefficients of water greater than unity over a considerable

Fig. 3.6. Ion source and electrode system for mass spectrometer. (1) Stainless-steel block forming ion source. (2) Alpha source, consisting of polonium deposited on a metal disc. Metal disc enclosed in container with stainless foil window and stainless porous plug allowing pressure equalization across foil. (3) Outer alpha source container with foil window and porous plug. Double container prevents spreading of polonium into pressure equalization system. (4) Porous stainless plug allowing pump-out of alpha source and pressure equalization across foils. (5) Gas supply to ion source and flow system (in the direction of the arrows). (6) Tube leading to vacuum system of alpha source container. (7) Insulating material allowing voltages different from ground to be applied to ion source. (8) Cone-carrying metal foil at its truncated apex. Foil has one or several leaks through which the gas and ions enter the pumping and electrode chamber. (9) Heater and thermocouple wells for temperature control of ion source. (10) Auxiliary electron gun for gas purity determinations. (11-19) Electrodes focusing ion beam into magnetic mass analyzer. [P. Kebarle, Advances in Chemistry Series, No. 72, p. 28 (1968).

concentration of electrolyte. In general, the activity coefficients of water in hydrated cation solutions are less than unity, except for a relatively small region at lower concentrations.

Thermochemical Method for Coordination Numbers

The coordination number of an ion is the average number of constantly exchanging solvent molecules that form the immediate surroundings of that ion. Values were estimated by using a thermochemical method [206-208, 289, 338, 339] in which the integral heats of solution of constant concentrations of salts in water and in aqueous solutions of acids of various concentrations were measured and used in calculating the coordination numbers of ions in dilute solutions. The heats of hydration of anhydrous salts were taken as the differences in the heats of solution of anhydrous and hydrated salts.

The protons from the acid and the proton charges were considered to be statistically distributed over the water molecules, so that each water molecule became an ion with the average electronic charge and the average proton per ion of ϵ/e, where ϵ is the average positive charge per ion, and e is the electronic charge. The ions produced from water had the formula $H_{2+(\epsilon/e)}O^{(\epsilon/e)^+}$. The coulombic interaction energies between these ions and cations and anions are, respectively,

$$E_k = -n_k \epsilon e/R_k \qquad (3.57)$$

and

$$E_a = n_a \epsilon e/R_a \qquad (3.58)$$

where n_k and n_a are the coordination numbers of the cations and anions, respectively. Repulsion (reduction of exothermic heat of solution) between positive ions is indicated by the minus sign. If $r_k + 1.38$ Å $= R_k$, and $r_a + 1.38$ Å $= R_a$, where r is the crystal radius of an ion, and $\epsilon = km$, (3.57) and (3.58) can be written

$$E_k = -\frac{n_k kme}{r_k + 1.38} \times 10^8 \qquad (3.59)$$

and

$$E_a = \frac{n_a kme}{r_a + 1.38} \times 10^8 \qquad (3.60)$$

The r's are in centimeters, and the energies in ergs per ion. Summing the two energies multiplied by Avogadro's number and by a number for converting ergs to calories (2.389×10^{-8}), and setting $K = 2.389 Nek$ yields

$$E = -K(\frac{n_k}{r_k + 1.38} - \frac{n_a}{r_a + 1.38})m \qquad (3.61)$$

where E in calories per mole of salt is greater than zero.

The integral heat of solution of the salt in m molal HCl is

$$L = L_0 - K(\frac{n_k}{r_k + 1.38} - \frac{n_a}{r_a + 1.38})m \qquad (3.62)$$

where L_0 is the integral heat of solution to the same concentration in pure water. Hence

$$\beta = -\frac{\Delta L}{\Delta m} = K(\frac{n_k}{r_k + 1.38} - \frac{n_a}{r_a + 1.38}) \qquad (3.63)$$

From (3.63) the difference between two salts with the same anion but different cation is

$$\beta_1 - \beta_2 = K(\frac{n_{k_1}}{r_{k_1} + 1.38} - \frac{n_{k_2}}{r_{k_2} + 1.38}) \qquad (3.64)$$

A factor of 2 is introduced into (3.63) and (3.64), and a factor of $\beta' = \beta/2$ is used for a biunivalent salt. The values of $\beta = \delta_k - \delta_a$, where

$$\delta_k = K \frac{n_k}{r_k + 1.38} \qquad\qquad \delta_a = K \frac{n_a}{r_a + 1.38} \qquad (3.65)$$

do not change with valence types of salts if β' is used for β. From (3.63) and (3.64), the coordination numbers can be obtained if K can be found. Thus coordination number was related to hydration.

There is no direct relation between coordination number and individual ion hydration numbers. Hydration is directly related to the surface density ρ' of the packing of water molecules in the first coordination layer of the ion, and ρ' must be determined. If n is the coordination number of the ion, r its radius, and $4\pi(r + 1.38)^2$ $Å^2$ the area of the first coordination sphere,

$$\rho' = \frac{n}{4\pi(r + 1.38)^2} \qquad (3.66)$$

and for the water molecule,

$$\rho'_{H_2O} = \frac{n_{H_2O}}{4\pi R^2} \tag{3.67}$$

where n_{H_2O} is the average coordination number of water molecules, and R is the distance between centers of neighboring molecules in water. From x-rays [289], $n_{H_2O} = 4.6$ and $R = 2.90$ Å. Therefore $\rho'_{H_2O} = 0.044$ Å^{-2}.

Hydrated ions such as Na$^+$, for which $\Delta E > 0$, reduce the frequency of the activated jumps of nearest water molecules compared to that in pure water by weakening their translational motion. For these ions,

$$\rho' > \rho'_{H_2O} \tag{3.68}$$

Ions showing negative hydration, such as K$^+$, Cl$^-$, Br$^-$, and I$^-$, have $\Delta E < 0$, and increase the frequency of activated jumps of neighboring molecules compared to pure water. These ions have

$$\rho' < \rho'_{H_2O} \tag{3.69}$$

Estimations of the lower limits for the coordination numbers of hydrated ions and the upper limits for negatively hydrated ions are possible from inequalities (3.68) and (3.69). Thus $n_{Na^+} > 3.05$, $n_{K^+} > 3.05$, $n_{K^+} < 4.02$, and $n_{Cl^-} < 5.57$. Limiting values for coordination numbers can be estimated by distinguishing between hydrated and nonhydrated (negatively hydrated) ions.

In (3.63) and (3.64) the lower limit for K was found [344] from β values for NaCl and NaBr (NaBr·2H$_2$O). From (3.63) for NaCl the corresponding equation for NaBr was subtracted, giving

$$K\left(\frac{n_{Br^-}}{3.34} - \frac{n_{Cl^-}}{3.19}\right) = 68 \tag{3.70}$$

where $r_{Br^-} + 1.38$ Å $= 3.34$, $r_{Cl^-} + 1.38$ Å $= 3.19$, and $\beta_{NaCl} - \beta_{NaBr} = 210 - 142 = 68$. From (3.70), since $K > 0$, $n_{Br^-} > n_{Cl^-}$. From certain properties, including heats of hydration of Cl$^-$, Br$^-$, and I$^-$ ions, presumably negative hydration increases in this order, hence $\rho'_{Br^-} < \rho'_{Cl^-}$. Therefore

$$\frac{n_{Br^-}}{4\pi(3.34)^2} < \frac{n_{Cl^-}}{4\pi(3.19)^2} \tag{3.71}$$

Thus $n_{Br^-} < 1.096\, n_{Cl^-}$. Substituting this value of n_{Br^-} into (3.70) gives

$$K(\frac{1.096n_{Cl^-}}{3.34} - \frac{n_{Cl^-}}{3.19}) > 68 \tag{3.72}$$

or

$$K > 68/0.0147 \; n_{Cl^-} \tag{3.73}$$

The upper limit for n_{Cl^-} substituted into (3.73) yields the lower limit for K. The equation for KCl,

$$K \left(\frac{n_{K^+}}{2.71} + \frac{n_{Cl^-}}{3.19}\right) = 11 \tag{3.74}$$

was used. In (3.74), $r_{K^+} + 1.38 = 2.71$, $r_{Cl^-} + 1.38 = 2.19$, and 11 is the experimental value for KCl. Now β and K are positive, and

$$\left(\frac{n_{K^+}}{2.71} - \frac{n_{Cl^-}}{3.19}\right) > 0 \tag{3.75}$$

Since $n_{K^+} < 4.02$, then $n_{Cl^-} < 4.74$. This is smaller than $n_{Cl^-} < 5.57$ obtained from the inequality $\rho'_{Cl^-} < \rho'_{H_2O}$. Substituting $n_{Cl^-} = 4.74$ into (3.73) gives $K > 976$.

The lower limit of n_{K^+} can be estimated by knowing the lower limit of K. Equation (3.64) for NaCl and KCl is

$$K\left(\frac{n_{Na^+}}{2.36} - \frac{n_{K^+}}{2.71}\right) = 199 \tag{3.76}$$

Here $r_{Na^+} + 1.38 \; Å = 2.36$. The quantity 2.71 was defined in (3.74), and $\beta_{NaCl} - \beta_{KCl} = 210 - 11 = 199$. Solving (3.76) for n_{K^+} yields

$$n_{K^+} = 2.71 \left(\frac{n_{Na^+}}{2.36} - \frac{199}{K}\right) \tag{3.77}$$

Since $n_{Ka^+} > 3.05$ and $K > 976$, $n_{K^+} > 2.95$ from (3.77). From (3.64) the upper limit of n_{Li^+} can be estimated as 4.23.

Plotting the differential heat of solution ΔH for HCl against concentration yields a slope $\Delta H/\Delta m$ or β of 233 cal mol^{-1} $[m]^{-1}$ which can be used to estimate K. The difference in the heat content of a mole of HCl in an infinitely dilute and in an m molal solution is determined by the interactions of $H_{2 + (e/e)}O^{(e/e)^+}$ with each other and with Cl$^-$ ions. The ΔH-versus-m plot indicated that an equation similar to (3.63) results. Then

$$K(\frac{n_{H^+}}{R} - \frac{n_{Cl^-}}{319}) = 233 \tag{3.78}$$

where n_{H^+} = 4.6 is the coordination number of water molecules in water. From x-rays R = 2.90 Å and is the average distance of separation of neighboring molecules in water. n_{H^+} is also the coordination number of $H_{2+(\epsilon/e)}O^{(\epsilon/e)^+}$ ions not in the vicinity of Cl⁻ ions.

Subtracting (3.63) from (3.78) gives

$$K(\frac{4.6}{2.90} - \frac{n_{K^+}}{2.71}) = 222 \tag{3.79}$$

where $222 = \beta_{HCl} - \beta_{KCl} = 233 - 11$. But $K > 976$, hence from (3.79), $n_{K^+} > 3.68$. Letting $n_{K^+} = \frac{1}{2}(3.68 + 4.02) = 3.8$ and substituting this for n_{K^+} in (3.79) gives $K = 1.17 \times 10^3$. By substituting these values for n_{K^+} and K in (3.60) and (3.64), those for the coordination numbers n for various ions were obtained and tabulated.

It was found that the coordination numbers of monatomic ions in dilute aqueous solutions nearly equal the coordination number (4.6) at 25°C of the water molecule in water. The structure of water is thought [201, 202, 344] to control the structure of dilute aqueous solutions. In dilute solutions the state of the ions corresponds to envelopment of the ions by water with minimum change in the structure of water [201].

From heats of solution, temperature coefficients of heats of solution, and other thermochemical data, the coordination numbers of ions and their change with temperature were determined [76, 77, 104, 148, 149, 194, 247, 248, 283, 285, 338, 340, 342, 345, 346, 381].

The methods used in solvation studies listed in this section are many and varied. Individual articles must be consulted to obtain the details. Calorimetric measurements of heats of solution and their temperature coefficients were among the techniques employed. In many cases calculations were made from existing thermochemical data. Many data on solvation and coordination numbers using the thermochemical approach have been tabulated [178].

Activity Coefficient Methods

An equation for osmotic and activity coefficients has been formulated [155] by applying volume statistics to an approximate Stokes-Robinson model [377], but the hydration numbers obtained do not show the anomalies of the Stokes-Robinson hydration parameters. Neglecting the covolume effect, these investigators obtained $n-$ values greater than any acceptable values for ionic hydration numbers, which vary widely with the anion and in a direction opposite that expected.

Hydration and covolume effects can be combined in a single theory if volume fraction rather than mole fraction statistics are used. A sharp difference between free and bound water is not necessarily implied by use of a mean hydration number h or the Stokes-Robinson n. The weak association of a large number of water molecules with an ion [159] is equivalent to the strong adsorption of a small fixed number for statistical purposes (377). These assumptions, coupled with the Gibbs' free energy for the solution of a hydrated electrolyte in volume fraction statistics, give the equation for the mean electrolyte activity coefficient $\gamma_{\pm}$ as

$$\log \gamma_{\pm} = \log \gamma_{\pm}^{el} + \frac{0.018mr(r + h - v)}{2.3v(1 + 0.018mr)} + \frac{h - v}{v} \log (1 + 0.018mr)$$

$$- \frac{h}{v} \log (1 - 0.018mh) \tag{3.80}$$

where h was defined above, v is the number of ions per electrolyte molecule, $r = \phi_v/v_w^\circ$, ϕ_v is the apparent molar volume of unhydrated electrolyte, v_w° is the mole volume of pure water, m is the molality, and $\gamma_{\pm}^{el}$ is the electrostatic contribution to the mean activity coefficient.

From data on activity coefficients, (3.80) was used to calculate hydration numbers [327, 374]. The value of r was constant in the calculation [377] and applied to concentrations $m = 1$ for 1:1 and $m = 0.7$ for 2:1 electrolytes, since at these concentrations (3.80) applied to all electrolytes. The halogen ions were assigned a value of 0.9 and the Cs^+ ion a value of zero in calculating the hydration numbers of individual ions.

Large ions were assumed to be unhydrated by Bernal and Fowler [41], and Stokes and Robinson [377] assumed that cations but not anions were hydrated. From apparent molal volumes the latter investigators obtained larger hydration numbers for cations than the former, or Glueckauf [155]. For example, Li^+ in water at $25°C$ was found by Stokes and Robinson to have a solvation number of 7.1 in 0.1 to 1.0 m LiCl solution, taking the solvation number of Cl^- ion as 0. Glueckauf, using Cs^+ as a standard of reference in 1.0 m solutions, found the Cl^- solvation number to be 3.4.

It was found [328] that the anomalies of large hydration numbers of cations and of their increase with increasing anion size (Li^+ from 7.1 to 7.6 to 9.0 for Cl^-, Br^-, and I^- ions, respectively) could be removed by adequate consideration of the volumes of hydrated ions.

From measured [398] solubilities and activities of electrolytes in mixed solvents, solvation numbers of ions and transsolvation constants were calculated.

The acidity function H_R in 44 to 64% H_2SO_4 was established [389], using seven aryl methanols. The acidity function H_0 was obtained in 60 to 75%

H_2SO_4, using amines. From the data the number of water molecules solvating the cations was found.

Colligative Properties

To determine their deviations from ideality, accurate freezing-point depressions of metal hydrogen sulfates in H_2SO_4 were measured [32], and osmotic coefficients ϕ calculated:

$$\phi = \theta(1 + 0.002\theta)/6.12 \; \Sigma m_{ij} \qquad (3.81)$$

where θ is the molal freezing-point depression measured from the freezing point of the hypothetical undissociated H_2SO_4, and Σm_{ij} is the molality summed over the various ion types. Pure H_2SO_4 has a minimum ionic strength μ of 0.0357 and a ϕ of 0.98. At the freezing point its dielectric constant is 120, which does not mean that solutions in H_2SO_4 are ideal.

The osmotic coefficient varies with molality according to the equation

$$\phi = 1 + \phi_{el} + b \; \Sigma m_i \qquad (3.82)$$

where ϕ_{el} is the electrostatic ionic contribution to the osmotic coefficient, Σm_i is the sum of the molalities of the ionic species, and b is a parameter related to the solvation numbers s by

$$b = (r + s)^2/40.8 - r/20.4 \qquad (3.83)$$

where r is the ratio of the apparent molar volume of the electrolyte to the molar volume of the pure solvent, 54 cm^3. This equation was used in calculating solvation numbers from freezing-point data [244]. The Debye-Hückel theory is used to calculate ϕ_{el}, using an appropriate distance for the distance of closest approach a for the dielectric constant of 120. Both a and b are adjustable parameters.

The freezing points were measured using [31] a cryoscope made from a 10 × 30 cm Dewar flask fitted with two lids made of ½-in polythene sheet and separated by about 1 cm. A slight positive pressure of dry nitrogen was maintained between the two lids to prevent leakage of moisture. The lids were perforated with four holes each, and the pairs of holes in the two lids vertically aligned. The four pairs of holes carried tightly fitting glass tubes with ground-glass joints for a thermometer, stirrer, withdrawal tube, and a tube for adding solutions for changing the composition. The Dewar flask was supported inside a larger Dewar flask containing alcohol and a coil through which alcohol, cooled by solid CO_2 to any desired temperature down to $-20°C$, could be pumped. The stirrer was operated at 120 rpm and below, at which speed the heat of stirring was negligibly small. After addition of solutions to alter the composition, rapid mixing was ensured by increasing the rate of stirring up to 800 rpm.

Freezing points were determined using a National Physical Laboratory calibrated platinum resistance thermometer and a Cambridge Smith's difference bridge in conjunction with a galvanometer of 150 mm/μA, on a scale of 1 m, and a Linsley photoelectric galvanometer amplifier.

Analyses of the solutions were made by directly delivering samples into a conductivity cell through the withdrawal tube, and measuring the conductivity of the solution at 25°C. The cell temperature was maintained constant by immersing the cell in an oil-filled thermostat maintained at 25.000 ± 0.002°C. Suction was applied through drying tubes to transfer solution from the Dewar flask through the withdrawal tube to the cell. The withdrawal tube terminated in a sintered-glass disc to exclude solid crystals. The cell constant was determined by measuring its resistance when filled with H_2SO_4 of minimum conductivity ($\mathcal{K}$ = 0.010433 Ω^{-1} cm^{-1}) [150]. The details of the apparatus used in the conductivity measurements have been described [150].

Using HSO_4^- as the reference ion with a solvation number of 0 in H_2SO_4, the solvation numbers of Li$^+$, Na$^+$, K$^+$, NH$_4^+$, Ag$^+$, H$_3$O$^+$, and Ba^{2+} were found to be, respectively, 2.3, 4.0, 2.1, 1.2, 2.1, 1.8, and 6.5.

Ebullioscopic and cryoscopic measurements of the total hydration of ions and of salts have been extensively used [54-61, 68, 152, 178, 187-189, 197, 331, 332, 364].

A graphical approach was devised [187] for obtaining solvation and association data from colligative properties applied to the total solvation of GaCl$_3$ in CH$_3$Cl. It proved the existence [197] of CH$_3$Cl·GaCl$_3$, and when applied to AlI$_3$-CH$_3$I and AlBr$_3$-C$_6$H$_6$ systems confirmed [69, 70] a 1:1 complex in the AlI-CH$_3$I but not in the AlBr$_3$-C$_6$H$_6$ system.

Entropy of Hydration

Entropy of hydration has been discussed [81, 239, 241, 242] in conjunction with entropies of aqueous ions. In aqueous systems a 1 m solution of ions, obeying the perfect solution laws and the ions possessing the same partial molal heat content that they have at infinite dilution, was chosen as the standard state for the calculation of entropies. These relative ionic entropies are referred to $S^\circ_{298.1}$ of H$^+$ taken as zero.

The summation of ΔS of solution and the entropy of the solid salt is the most direct method of obtaining ionic entropies. ΔS of solution is obtained from the free energy ΔG and the enthalpy ΔH of solution, using the expression

$$\Delta G = \Delta H - T\Delta S \qquad (3.84)$$

Relative ionic entropies referred to H$^+$ taken as zero, rather than electrolyte entropies, are often recorded.

The so-called entropy of hydration is a function of the radius r and charge e of the ion [241, 242]:

$$\Delta S_{\text{hydration}} = f\ e^2/r \tag{3.85}$$

The difference between the partial molal entropy of the ion in solution and its molal entropy in the gaseous state is defined as the entropy of hydration of the ion. This entropy of the ion in the aqueous state, if translation can be calculated from the Sackur equation, is

$$S_{298.1} = {}^3\!/_2 R\ \ln M + 26.03 \tag{3.86}$$

where R is the gas constant in calories per mole per degree, M is the molecular weight, and 26.03 is the constant for the gas at 1 atm and 298.15 K.

That specific hydration effects are small compared to the electrostatic action of the charged water dipoles is suggested [239] by linear plots of ionic entropies versus the reciprocal of the ionic radii [310]. The chemical properties of solutions of ions are those of a charged sphere of a given size in a medium of a certain dielectric constant, as was concluded earlier [241]; specific hydration numbers of H^+ and other ions [89, 223] from ionic entropy [400, 402] have been listed.

For several metal ions, using heats and free energies, the entropies of ammoniation were calculated and discussed [144] in relation to hydration numbers. The idea [207] that the average entropy change when a water molecule binds to a cation approximately equals the entropy of fusion of ice at 298 K (6 kcal mol^{-1}), and that the hydration number is obtained by dividing the absolute entropy of the gaseous ion by 6, was believed [144] to be an oversimplification, since it does not consider the different strengths of binding water molecules to a cation. Also, the average contribution of 9.4 kcal mol^{-1} to the entropy of a hydrate per water molecule is insufficient [240]. Instead, consideration must be given to ion-dipole forces causing orientation and decreasing the translational freedom of water molecules, and restricting true covalent bonding between a donating center of solvent and the ion. Consideration was given to which effect predominates and to when that effect occurs. That an ion carries no more water than ammonia covalently bound was indicated by the results. By assuming the entropy change per molecule of water replaced by one molecule of ammonia equals the difference in the entropies of fusion of ammonia and of water at 298 K (about $6 - 9 = -3$ kcal mol^{-1}), and by also assuming the number of water molecules in the aqua cation equal to the number of ammonia molecules in the amine, the number of ligands associated with H^+, Ag^+, Cu^{2+}, Ni^{2+}, Zn^{2+}, Cd^{2+}, Hg^{2+}, Li^+, and Mg^{2+} were found to be respectively, 1, 2, 4, 6, 4, 6, 4, 3, 4.

If the ion carried more water than ammonia, the calculated entropy change would have been more positive whereas, except for the H^+ ion, the observed results were more negative [144]. The larger decrease on ammoniation could arise partly from the shortening of the linkages in the amines accompanied by correspondingly more configurational restrictions.

The values of enthalpy, energy, and entropy of hydration of 15 lanthanides were calculated by different methods, and were in close agreement with experimental values [432]. Electrostatic volume and entropies of solvation were discussed [424]. Compression of the ion was found to have little effect on the enthalpy, energy, and entropy of solvation.

It was found that the structural-change entropy of water is the principal contribution to the entropy of exchange of negatively hydrated ions [369].

Heat Capacity

Na^+ ion was found [320] from heat capacity measurements and calculations to be more highly hydrated than K^+ ion, and the H^+ ion was found to be H_3O^+.

It was shown [36] that the third law of thermodynamics applied to the entropies of both LiOH and $LiOH \cdot H_2O$ from heat capacity data on LiOH and $LiOH \cdot H_2O$, and for the entropy change at $25°C$ for the reaction

$$LiOH \cdot H_2O(cryst) \rightarrow LiOH(cryst) + H_2O(gas)$$

It was concluded [111, 112] from specific heat and hydration studies on ions that hydrated ions contain 10 to 15 molecules of water in two energy layers, the second of which is dissociated at a temperature of $140°C$.

From heat capacity $\overline{C}_0$ and entropy S_{soln}^{298} data of ions in solution, the temperature dependence of the entropy ΔS_h, energy ΔE_h, and enthalpy of hydration ΔH_h for lithium, sodium, ammonium, and magnesium halides and hydroxides was calculated [243]. From measured [367] specific heats of solutions of $LaCl_3$, $NdCl_3$, $DyCl_3$, $ErCl_3$, and $YbCl_3$ at $25°C$ from $0.1\ M$ to saturation, apparent molal heat capacities and the concentration dependence of each of these for each solution were calculated; and from these were obtained the partial molal heat capacities of solute and solvent. The data indicate that heavy rare earths have a lower coordination number than light rare earths. As in crystals, it was assumed that ions of the type $[Cl_2Gd(OH_2)_6]$ also existed in saturated solutions [265].

Specific heats indicated that chlorides, bromides, and hydroxides of alkali metals were hydrated, as well as HCl and HBr [38]. A calorimeter is used in heat capacity measurements. Figure 3.7 shows a cross section of such a calorimeter [320]. A is a brass cylinder bolted to the main plate B, and C represents various extension tubes reaching above the oil in the bath in which

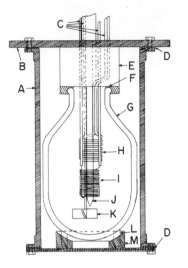

Fig. 3.7. Calorimeter used in measuring heat capacity.

the calorimeter is immersed. The gaskets D are made of Garlock sheet-rubber packing and shellacked to the cylinder. G is a vacuum flask supported by rubber pieces L held in a circular piece of balsa wood M. The vacuum flask G is tightly held against the rubber gasket F, shellacked to the balsa wood block E through which are inserted the three-vane all-glass stirrer K, the copper-constantan thermocouple protected by the Pyrex glass tube J, an entry opening, the leads of the cooling coil H, and the leads to the heating coil I made by drawing fine silver tubing over double silk-covered manganin wire coated thinly with Bakelite. The leads to the heating coil are of 0.32 German silver tubing.

Solubility and Dilution

By using the difference in the heats of solution in dilute and concentrated solutions of KF and of $KF \cdot 2H_2O$ at $25°C$, the heat of solvation of KF to $KF \cdot 2H_2O$ was found [238] to be 5912 cal mol^{-1}.

Solubilities of CO_2 measured in methyl and ethyl alcohols, in solutions of lithium and sodium chlorides, bromides, and iodides in these alcohols at 15 and $20°C$, permitted calculation of the solvation of these salts by the alcohols [225].

The solvation number S for the salts in the alcohols was calculated by letting the solubilities in moles of CO_2 per 100 mol of alcohol in the salt solutions and in the pure alcohols be, respectively, M''_{CO_2} and M'_{CO_2}, and by letting the moles of salt per 100 mol of alcohol be M_{salt}; then

$$S = 100 \frac{1 - (M''_{CO_2}/M'_{CO_2})}{M_{salt}} \tag{3.87}$$

and defining ΔM_{CO_2} as

$$\Delta M_{CO_2} = 100 \, \frac{M'_{CO_2} - M''_{CO_2}}{M'_{CO_2}} \tag{3.88}$$

then

$$S = \Delta M_{CO_2}/M_{salt} \tag{3.89}$$

In concentrated solutions 1 mol of the salts combined with 3 to 5 mol of alcohol, but in infinitely dilute solutions 1 mol of the salts combined with 7 to 9 mol of alcohol. These data compared fairly well with like data from Ulich [399], Ulich's data tending to give somewhat higher solvation numbers. If the solvation number of some reference ion is known, the solvation numbers of individual ions can be estimated.

By using the solubilities of some strong electrolytes in water, their solubilities in alcohols were calculated [125] from a formula involving only the charges and radii of the ions and the dielectric constant of the solvent. It was found that K^+, Rb^+, Cs^+, Cl^-, Br^-, and ClO_4^- ions were not solvated, since they had the same volume in solution and in the crystalline form, hence there was no enrichment of either water or alcohols around the ions. From the abnormal solubility curve of NaCl, the Na^+ ion was taken to be solvated.

Other solvation studies using solubilities have been reported [2, 17, 51, 52, 76, 82, 83, 107, 113, 220, 286, 315, 340, 356, 408] for pure and mixed solvents. These articles discussed solvation of salts rather than solvation of ions, but in each case the selection of the solvation of a reference ion makes possible the calculation of the solvation numbers of individual ions.

Structure and Ionic Heats and Entropies

By using a theoretical approach involving the structure of divalent ionic solutions, the heats of hydration of these ions were found [23] at 25°C to depend on four terms, the cationic sheath being the main contribution. The most probable permanent solvation numbers were found to be 4 for metals of group IIa, 6 for metals of group IIb, and 4 and 6 for transition metal divalent ions. The conflict of opinion on the nature of solvation bonding was discussed [109].

Heats of hydration and hydration numbers of the hydrogen and hydroxide ions at 25°C were investigated. In the ideal case the formulas of H^+ and OH^- ions in water were given as $(H_3O^+)(H_2O)_2 \cdot (H_2O)_9$ and $(OH^-)(H_2O)_{4.7}(H_2O)_{17}$.

Summaries of the hydration numbers of H^+ and OH^- ions determined by various methods have been tabulated [178, 425]. The hydration number of H^+ ion has been found to range from 0.3 to 10 and that for the OH^- ion to range from 0 to 8, depending on the method used. The number that occurs most frequently for either ion is 4.

Structural changes in water and positive and negative entropies of hydration have been used to distinguish between positive and negative hydration of ions [227, 231, 348]. Positively and negatively hydrated ions have been listed [178].

In making structural calculations of ion hydration, various models have been proposed [40, 47, 49, 50, 108, 130, 167, 344], and data from many experimental techniques applied. There are basically three different types of models [50]: (1) The coordination number is assumed to be 4, and there is no distinction between coordination and solvation numbers. (2) The coordination number is determined from x-ray measurements, and there is no distinction between coordination and solvation numbers. (3) The coordination number is taken from x-ray measurements, and a distinction is made between coordination and solvation numbers. For each of the three models, three models of the structure-broken region outside the first solvent layer around the ion are calculated. Models in which the coordination number is distinguished from the solvation number are numerically more consistent than are those that make no such distinction [50].

Primary solvation number has been defined as the number of solvent molecules that give up their own translational freedom and move with the ion relative to the surrounding solvent, or as the number of solvent molecules aligned in the force field of the ion [49]. Thus the primary solvation number is measured by those methods that detect the number of solvent molecules accompanying an ion in its movements through the solution.

Sometimes this detected number of molecules is less than the geometry of the solvated ions suggests. This lesser number is the coordination number of the ion, and is the number of solvent molecules *actually in contact with the ion.*

Figure 3.8 is a schematic representation of a univalent positive ion in solvent water. The ion is shown with a primary shell of solvent molecules completely aligned, and a secondary solvation shell of solvent molecules partly aligned. There can be several secondary shells of solvent molecules with degrees of alignment of solvent molecules decreasing with distance from the central ion. The ion and its retinue of solvating solvent molecules are immersed in solvent water with its free molecules and flickering clusters of hydrogen-bonded molecules. There is continuous exchange of molecules of solvent between adjacent regions of the system, although the rate of exchange is probably less rapid in regions near the ion, and especially in the primary solvation sphere, where the solvent molecules are strongly attracted by the ion.

In calculations involving the enthalpies and entropies of hydration and solvation, and the coordination numbers of ions in water solutions, three models were used [50]. (1) In the first coordination shell a coordination number of 4 was assumed, and no distinction was made between coordination number and

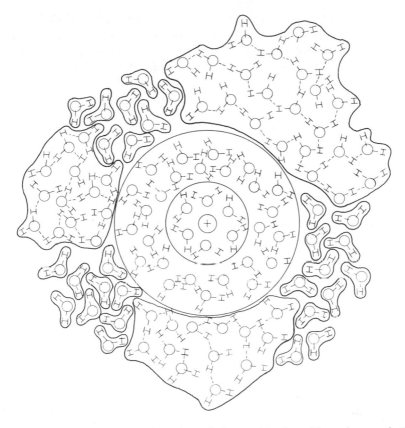

Fig. 3.8. A schematic representation of a univalent positive ion with a primary solvation shell of solvent molecules completely aligned, and a secondary shell of solvent molecules partially aligned. There can be several secondary shells of solvent molecules with different degrees of alignment of the solvent molecules. The ion and its retinue of solvating solvent molecules are immersed in solvent water with its free molecules and flickering clusters of hydrogen-bonded molecules.

solvation number for monovalent ions. In the second layer model 1 was divided into three subclasses. In model 1A the second or structure-broken region had molecules hydrogen-bonded to those in the first shell; and outside this structure-broken region, the rest of the solvent behaved as bulk solvent. In model 1B the structure-broken region consisted of dimers formed from monomers drawn from bulk water. In model 1C the structure-broken region consisted of monomers or of molecules hydrogen-bonded to molecules in the first shell. (2) In the first coordination shell, the coordination number ranged from 6 to 8, and there was no distinction between coordination number and solvation number for monovalent ions. In the second or structure-broken region, there were three submodels, as in model 1. (3) In the first coordination shell the coordination number ranged from 6 to 8, and a distinction was made between solvation number and coordination number. The second or structure-broken region was divided into submodels 3A, 3B, and 3C, as in model 1.

For calculations of enthalpies of solvation, models 3A and 3C were about equally consistent and, for calculations of entropies of solvation, model 3C was most consistent for anions and was satisfactorily consistent for cations. The number of water molecules in the primary solvation shell of monovalent ions was found to be six to eight, and the number of water molecules in the structure-broken region n_{SB} was calculated from the equation

$$n_{SB} = \frac{4\pi(r_i + 2r_w)^2}{\pi r_w^2} \tag{3.90}$$

where r_i is the radius of the ion, r_w is the radius of the water molecule, and πr_w^2 is the cross-sectional area of a water molecule. The numbers of water molecules in the structure-broken region of several monovalent ions are listed in Table 3.3.

Table 3.3 Number of Water Molecules n_{SB} in the Structure-Broken Region of Several Monovalent Ions

Ion	Li^+	Na^+	K^+	Rb^+	Cs^+	F^-	Cl^-	Br^-	I^-
n_{SB}	23.7	28.7	34.8	37.4	41.1	35.3	43.4	46.0	50.2

Polarography

Polarographic data on some complexed ions of copper, zinc, cadium, and rubidium were used to calculate [160] the approximate molecular weights of the ions from the Riecke formula [238], and from these the degree of hydration of the Cu^{2+}, Zn^{2+}, and Cd^{2+} ions. Lead did not hydrate.

In 17 F H_2SO_4 it was found [407] from half-wave potentials $F_{1/2}$ that Tl^+ ions bind 2.3 times as many water molecules as Cd^{2+} ions.

The hydration of certain bivalent metal ions in methanol-water solvents was studied [278] using polarography. With increasing water concentration each ion showed a regular decrease in half-wave potential $E_{1/2}$. Stability constants of the Cd^{2+}, Pb^{2+}, and Zn^{2+} ions in NH_4ClO_4 solutions were given for steps of the type: $M^{2+}A_n(H_2O)_m + H_2O \rightarrow M^{2+}A_{n-1}(H_2O)_{m+1} + A$. Tl^+ was not hydrated in these solutions. From formation constants determined polarographically for the oxalate complexes of copper(II) and cadmium(II) in light and heavy water, it was concluded [260] that light water was more strongly solvated than heavy water. The method of calculation used was that of De Ford and Hume [102], who defined certain $F_j(X)$ terms which made possible calculation of the stability constants for different complexes of a metal ion with different proportions of ligand. For the system the $F_j(X)$ terms are:

$$F_0(X) = \text{antilog}\left\{n/0.05916\ [(E_{1/2})_s - (E_{1/2})_c]\right.$$
$$\left. + \log\ (i_d)_s/(i_d)_c\right\} \tag{3.91}$$

$$F_0(X) = K_0 + K_1[\text{ox}] + K_2[\text{ox}]^2 + K_3[\text{ox}]^3 \tag{3.92}$$

$$F_1(X) = (F_0(X) - K_0)/[\text{ox}] = K_1 + K_2[\text{ox}] + K_3[\text{ox}]^2 \tag{3.93}$$

$$F_2(X) = (F_1(X) - K_1)/[\text{ox}] = K_2 + K_3[\text{ox}] \tag{3.94}$$

$$F_3(X) = (F_2(X) - K_2)/[\text{ox}] = K_3 \tag{3.95}$$

where K_0 is the formation constant for the zero complex and is zero, K_j is the formation constant of the complex $MX_j^{t(n-jb)}$ (where X^{b-} is the complex-forming substance of valence $-b$), $(E_{1/2})_s$ and $(i_d)_s$ are, respectively, the half-wave potential and the diffusion current of the simple metal ion M^{n+}, and $(E_{1/2})_c$ and $(i_d)_c$ are, respectively the half-wave potential and diffusion current of the complex ion. The original equations were in terms of activities $a_{\text{ox}} = C_{\text{ox}}f_{\text{ox}}$, but since the values of the several activity coefficients that occurred in the fundamental equations were not known, McMasters and co-workers [260] used the formation constants in terms of molarities [ox] of complexing oxalate.

$F_0(X)$ can be calculated from the measured half-wave potentials and diffusion currents of the simple and of the complexed ion by using (3.91). Knowing $F_0(X)$, $F_1(X)$ at various concentrations of [ox] can be calculated from (3.92). These values of $F_1(X)$ plotted against [ox] and extrapolated to [ox] = 0, give K_1 according to the second equality in (3.93). In a similar manner K_2 and K_3 can be calculated from (3.94) and (3.95).

Values of $1.87 \pm 0.12 \times 10^9$ and $3.27 \pm 0.13 \times 10^9$ were found for the overall formation constant K_2 for the dioxalato cuprate(II) complex in light and heavy water, respectively. It was concluded from these results that light water is more strongly solvating than heavy water. A similar interpretation was given to the data for the oxalato cadmium(II) complex.

For acetone-methanol, acetone-ethanol, and acetone-water systems, the general type of complex was found [279, 280] to be $[M(ROH)_i(MeCO)_{n-i}]^{2+}$, where n is the coordination number, M = Cd or Pb, and R = H, Me, or Et.

Table 3.4 presents the solvation numbers of Cd^{2+}, Pb^{2+}, and Zn^{2+} ions in formamide-methanol, formamide-ethanol, and formamide-water solvents [279, 280].

Table 3.4 Solvation Numbers of Cd^{2+}, Pb^{2+}, and Zn^{2+} in Formamide-Methanol, Formamide-Ethanol, and Formamide-Water Solvents

Ion[a]	*Solvent*	*Methanol*	*Ethanol*	*Formamide*	*Water*
Zn^{2+}	$HCONH_2$-MeOH	2	–	2	–
Pb^{2+}	$HCONH_2$-MeOH	2	–	2	–
Zn^{2+}	$HCONH_2$-EtOH	–	4	2	–
Pb^{2+}	$HCONH_2$-EtOH	–	4	2	–
Zn^{2+}	$HCONH_2$-H_2O	–	–	0	0
Pb^{2+}	$HCONH_2$-H_2O	–	–	0	0
Cd^{2+}	$HCONH_2$-H_2O	–	–	0	0

[a] For Cd^{2+} the coordination numbers in methanol were 1 and 2, and for ethanol 1, 2, and 3.

In polarography a constantly increasing potential is applied to a cell consisting of a small polarizable electrode, usually a dropping mercury electrode, and a large nonpolarizable electrode, usually a calomel electrode. A solution containing the ion to be reduced is introduced, along with a surface-active agent to suppress the maximum in the volt-microampere current, and an indifferent supporting electrolyte which eliminates the migration current due to migration of charged particles in the electric field. In this manner a current-voltage curve is obtained, in the shape of an elongated S, from which the diffusion current, which is a measure of the concentration of the ion being studied and of the half-wave potential characteristic of the ion, is obtained. Figure 3.9 illustrates how the pertinent quantities, diffusion current and half-wave potential, are obtained from a polarographic wave for 1.8×10^{-3} M Tl_2SO_4 in 0.1 N KCl as the supporting electrolyte, and containing a small concentration of acid fuchsin as a maximum suppressor. Ilkovic [195] and MacGillavry and Rideal [254] have derived theoretical expressions for the limiting diffusion current to the dropping mercury electrode in the presence of an indifferent electrolyte.

Further Thermodynamic Studies

Several general thermodynamic studies of ion hydration have been made.

It was observed [300] that singly charged d^{10} ions, Cu^+, Ag^+, and Au^+ showed more marked solvation effects than any other ions, including the isoelectronic species Zn^{2+}, Cd^{2+}, and Hg^{2+}; that the thermodynamic properties change more for hydration of anions than for cations of the same size; and that available data for anions do not show the monotonic variation with ionic size as do cations.

It was found [319] that to understand the associational behavior of KI in a variety of solvents and of bibivalent sulfates in water, account must be taken of the molecular nature of the solvent. Something radically different from primitive theories is needed [129] for the interpretation of ionization and solvation phenomena.

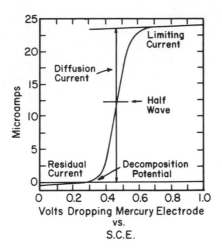

Fig. 3.9. Current-voltage curve for 1.8×10^{-3} M Tl_2SO_4 in 0.1 N. KCl Acid fuchsin used as a maximum suppressor.

Solvation energies of calcium, strontium, and barium chlorates, bromates, and amides were reported [123] and compared with those using a Born model. If hydrogen bonding in the solvent predominates over formation of ion-solvent interactions, solvents with least structure will show the greatest ion-solvating tendencies.

For alkali and halide ions in a formamide solution, experimental and absolute enthalpies were evaluated. The absolute enthalpies of solvation were explained on the basis of the interaction of the ion with six coordinated formamide molecules. The polarization energy of the solvent beyond the first coordination shell was taken into consideration.

The experimental real free energy a_i^s of species i in solvent s is the free-energy change in the process when an ion in field-free space is introduced into a large volume of solution s with no net electrical charge [80, 365]. From the Debye-Hückel theory the region of activity a_i is given by

$$a_i^s = a_i^{s,\phi} + RT \ln a_i \tag{3.96}$$

where $a_i^{s,\phi}$ is the standard real free-energy obtained by extrapolation, and is separated into bulk and surface contributions. The bulk contribution was identified with the "chemical" solvation energy which was the quantity required to separate the free energy of solvation of a salt into its ionic components.

Ionic solvation in mixed solvents and the changes in free energy accompanying the transfer of electrolytes from water to mixed aqueous solvents were discussed [116].

The thermodynamic treatment of a mixed fluid in an electrostatic field was applied [308] to preferential solvation and explained on the basis of partial molal free energies at infinite dilution.

Simultaneous consideration of the electrostatic interaction of ions and of their solvation was given [217] in calculating the individual and average activity coefficients of ions in solution. The changes in activity coefficients were caused by changes in enthalpy of solvation with concentration.

In the gas phase, equilibrium constants and thermodynamic functions for the reactions $H^+(H_2O)_{n-1} + H_2O \rightarrow H^+(H_2O)_n$ were determined [214]. The structures of the hydrates $H^+(H_2O)_n$ changed continuously, and no single structure showed dominant stability. In the lower hydrates ($n = 2$ to 6), all water molecules appeared to be equivalent, hence the notation $H_3O^+(H_2O)_n$ is inappropriate. A new shell is started beyond $n = 4$, or crowding of the first shell occurs. Na^+ ion solvation was also discussed.

The conditions for positive and for negative hydration of ions have been theoretically discussed [347], and the thermodynamic functions for the solvation of the ions calculated [281].

Near-hydration may be negative as well as positive and shows a greater temperature effect than for hydration, since most of the entropy characteristics of the variation in the structure of water are close-up effects. Near-hydration occurs from the freezing point up to some limiting temperature [229].

The qualitative observation that small, densely charged anions are more strongly solvated by protic than by dipolar aprotic solvents was corroborated by solvent activity coefficients [9]. The opposite is true for large polarizable anions. Methanol solvates very large and polarizable anions more than water. Dimethylformamide, dimethyl sulfoxide, and hexamethylphosphoramide solvate cations more than water. An equation has been derived (166) from the statistical thermodynamics of solvation relating the change in volume in solution, the free energy of solvation, and the isothermal compressibility of the solvent.

Thermal analysis was used [106] to determine the solvates of alkali metal methylates with methanol. Coordination numbers of 4, 6, 6, 8, and 8 were found, respectively, for Li^+, Na^+, K^+, Rb^+, and Cs^+.

A change in the hydration number of the lanthanide ions somewhere in the middle of the lanthanide series from neodymium to terbium is shown by the variation of the entropies of hydration [43]. Variations in the thermodynamic functions in complexation reactions are frequently caused by dehydration of the ions rather than the ligation. Values of the hydration numbers of La^{3+} and Lu^{3+} ions were estimated [155] to be 7.5 and 8.7, respectively. The ratio is 1:2, which is that found for the ratios of the entropies of hydration of the La^{3+} and Lu^{3+} ions. This might suggest that the hydration sphere is 20% greater in the heavier lanthanides.

A method has been proposed [282] for the separation of the overall enthalpy change into the components due to the inner and outer sphere complexes without using the stability constants of the complexes.

It has been shown [304] that for weakly hydrated salted-out cations, the relation between the separation and hydration of the salting-out cations passes through a minimum for all pairs investigated. For intensely hydrated cations the curve passes through a maximum [8].

An empirical relation was found [322] between the equivalent conductance of an aqueous electrolyte solution and the sum of the reciprocals of the ion hydration energies divided by the electrovalence of the ions.

The stability constants and the enthalpies and entropies of formation of the anisotropic and outer-sphere complexes of $Cu(en)_2 \cdot (H_2O)_2^{2+}$ were calculated [46] from electronic spectroscopic, calorimetric, and electron spin resonance (ESR) data.

The solvation of sodium salts in glycerol acetate binary mixtures with water was studied [20] by ^{23}Na NMR spectroscopy and thermodynamics. Heats of solution were given for NaCl, sodium tetraphenylboron, and tetraphenylarsonium chloride in a series of water-diacetin (glycerol diacetate) mixtures.

The hydration of ClO_4^-, $B(C_6H_5)_4^-$, and NO_3^- in various organic solvents was studied [216] by the extraction of the tetralkylammonium salts. One out of every three to five ClO_4^- ions was found to carry a water molecule in dichloroethane, 20% nitrobenzene, 80% benzene, and benzene solvent; 1.4 molecules of water were involved per nitrate ion.

The values of the selectivity coefficients $K_{B/A}$ for heterovalent exchange with a strong-base resin and radiotracers ReO_4^-, CrO_4^{2-}, and WO_4^{2-} versus macro Cl^-, and radiotracers $Cr(CN)_6^{3-}$, $Co(CN)_6^{3-}$, and $Fe(CN)_6^{3-}$ versus macro Cl^- show that in such systems, contrary to early ideas on the nature of resin selectivity, the direction of the exchange is determined by the superior hydration of the ions in the dilute external phase over that in the resin phase, and not by the ion pairing in the latter phase.

Calorimetrically, the enthalpies of formation of $CoCl_2$ and $CoBr_2$ complexes with pyridine (Py) and quinoline (Qu) were determined [249] in n- and t-butyl alcohols and in ethylene glycol. The complexes found are listed in Table 3.5, where the symbol S represents the solvent.

The magnitudes of the instability constants show that the complexes were very little dissociated. The solvation was of the $CoCl_2$ and $CoBr_2$ molecules rather than of the ions from the salts.

Diffusion

The Onsager-Fuoss equation is generally used [174, 302, 326] for calculating the diffusion coefficient D_{calc}. For a binary electrolyte the equation is

$$D_{calc} = 2000 \, RT \, \frac{\overline{M}}{C} \left(1 + C \frac{\partial \ln \gamma_\pm}{\partial C}\right) \tag{3.97}$$

Table 3.5 Complexes of CoCl₂ and CoBr₂ with Pyridine and Quinoline in *n*-and *t*-Butyl Alcohols

Solvent	Complex	Instability Constant K_{inst}
n-Butyl alcohol	$CoCl_2Py_2$	4.4×10^{-4}
	$CoBr_2Py_2$	7.0×10^{-4}
	$CoCl_2Qu_2$	2.5×10^{-4}
	$CoBr_2Qu_2$	3.5×10^{-4}
t-Butyl alcohol	$CoCl_2Py_2$	6.6×10^{-5}
	$CoBr_2Py_2$	2.8×10^{-5}
Ethylene glycol	$[CoPy_2S_2]Cl_2$	3.7×10^{-3}
	$[CoPy_2S_2]Br_2$	3.0×10^{-3}

where R is the gas constant in ergs per degree per mole, T is the absolute temperature, $\overline{M}$ is a function in the diffusion theory, C is the molar concentration of the solution, and $\gamma_\pm$ is the mean activity coefficient of the electrolyte.

The solvation number n of an electrolyte can be obtained from the calculated and observed values of the diffusion coefficient using the equation

$$D_{obs} = D_{calc} (1 - 0.018n) \tag{3.98}$$

Diffusion coefficients can be obtained from conductivity measurements as a function of time [171, 172], or from the Einstein relationship [49]. If K_B and K_T are the reciprocal of the resistances measured at the bottom and the top of the cell, respectively,

$$\ln (K_B - K_T) = \frac{t}{\tau} + \text{constant} \tag{3.99}$$

where t is the time, and τ is given by the equation

$$\frac{1}{\tau} = \frac{\pi^2 D}{a^2} \tag{3.100}$$

where a is the height of the cell. The slope of the line when $\ln(K_B - K_T)$ is plotted versus t is $1/\tau$, and using this slope D can be calculated from the equation [171, 172]

$$D = \frac{a^2}{\pi^2} \frac{1}{\tau} \tag{3.101}$$

This is D_{obs} of (3.98). The Einstein relation is

$$D = \overline{u}_{abs}kT \qquad (3.102)$$

where $\overline{u}_{abs}$ is the absolute mobility of the ion in centimeters per second per dyne at the absolute temperature T, and k is the Boltzmann gas constant. Other methods for obtaining D [D_{obs} of (3.98)] include the porous diaphragm approach [299, 375].

Applying (3.98) to the data for $CaCl_2$ in water, the hydration number of $CaCl_2$ was found [377] to be 24 which did not agree with 11.9 found from activity coefficients. From diffusion coefficient measurements degrees of hydration of metallic ions were found [145] that agreed with those found by other acceptable methods. For water at $25°C$ the hydration numbers of Li^+, Na^+, K^+, Rb^+, Cs^+, Tl^+, Ag^+, Cl^-, Br^-, I^-, Mg^{2+}, Ca^{2+}, Sr^{2+}, Ba^{2+}, Cu^{2+}, Zn^{2+}, Cd^{2+}, Co^{2+}, Fe^{2+}, and Mn^{2+} were found to be, respectively, 5, 3, 1, 1, 1, 1, 2, 1, 1, 1, 9, 9, 9, 8, 11, 11, 11, 13, 12, and 12. From diffusion and mobility studies, it was ascertained [146] that the number of water molecules m in the inner coordination sphere of an ion composed of n atoms is given by the expression

$$m = (n-1)/3 \qquad (3.103)$$

Diffusion measurements on cells both with [27] and without [170] parchment membranes have been used to measure hydration numbers of ions. In the former type of cell, large solvation numbers were found. Thus for Li^+, Na^+, K^+, H^+, Cl^-, Br^-, and I^-, the hydration numbers found were, respectively, 62, 44.5, 29.3, 5, 26.6, 29.6, and 21.4. Measurements on the cell without parchment membrane yielded hydration numbers for Cu^{2+} and Ni^{2+} of 7 to 8, and 10, respectively.

Of dialysis and diffusion methods, only diffusion methods gave a reliable degree of hydration [366]. Tl^+ as a reference ion was given a hydration number of 0 instead of 2 as assigned by Ulich [401].

In diffusion methods, reference substances such as allyl alcohol and pyridine were found [192, 288, 316] to be carried by the electrolytes and ions. Pyridine combined in large proportions with Ag^+ in the electrolysis of $AgNO_3$ solution. Other treatments of the hydration of ions by diffusion methods are plentiful [1, 16, 110, 252, 274, 334, 337, 341, 343].

Depending on the conditions diffusion methods yield widely divergent results, for example, Li^+ may have a solvation number of 5, 22, or 62. The latter value from a parchment paper diaphragm cell measures both primary and secondary solvation sheaths of the ion. The value of 22 was obtained by imposing a high potential across the diffusion cell, and includes perhaps as many as three solvation shells of the Li^+ ion.

Diffusion methods do not yield a consistent order of solvation of ions within a family. Ordinarily, the order in the alkali metal ions is $Li^+ > Na^+$

$> K^+$, and in the halogen family it is $I^- > Br^- > Cl^-$; however, in some cases $K^+ = Rb^+ = Cs^+$, $I^- = Br^- = Cl^-$, and $Mg^{2+} = Sr^{2+} = Ca^{2+} \cong Ba^{2+}$. In some cases, for example the halogens, the orders of solvation are opposite those based on other methods.

It appears that diffusion is one of the least consistent and least dependable methods of measuring solvation, and the data from this method are difficult to rationalize and interpret.

Isotopic Equilibrium

Characteristic internal vibrations, detected by spectral or other methods [119], indicated the existence of ionic hydrates $A(H_2O)_n$. The purpose of the investigation was to demonstrate the connection between the fractionation by ions [191] of the oxygen isotopes in solvent water and ionic hydration.

Ionic hydrates cause group vibrations of bound isotopic water molecules which differ in their zero-point vibrational levels because of differences in their masses. That the activity of the heavier relative to the lighter species is decreased because of such differences is shown by the equation

$$a \times 10^3 \equiv 1 - \frac{R}{R_0} = \sum_A m_A n_A (K_A - 1)/55.51 \qquad (3.104)$$

where a is the enrichment factor, R and R_0 are the ratio of the heavy to the light species in the solution and in the solvent, respectively, m_A is the molality of ion A, n_A is the hydration number of ion A, K_A is the intensity factor K_1/n_A, and K_1 is the equilibrium constant for the reaction

$$A(H_2O)_n + H_2O^* \rightleftharpoons A(H_2O)_{n-1}(H_2O^*) + H_2O \qquad (3.105)$$

where the heavier species is indicated by the asterisk.

In deriving (3.104) the following assumptions were made: (1) a solution of heavy and light water is ideal; (2) the gross isotopic composition of water is not altered by solute; (3) successive steps in the water replacement reactions differ only by a statistical factor. That is, the equilibrium constant K_1 for (3.105) is

$$K_1 = n_A K_A \qquad (3.106)$$

and for

$$A(H_2O)_{n-1}(H_2O^*) + H_2O \rightleftharpoons A(H_2O)_{n-2}(H_2O^*)_2 + H_2O \qquad (3.107)$$

$$K_2 = (n_A - 1)K_A/2 \qquad (3.108)$$

and so on.

In one procedure for determining the enrichment and intensity factors and thus the hydration number n_A of ion A, the solutes were dried to the anhydrous state and dissolved with minimum exposure to air in distilled water stored in Pyrex vessels. The solutions were degassed by repeated cycles of freezing, pumping, and thawing. CO_2 was then admitted to the equilibration vessel. After 5 to 30 days equilibration at a controlled temperature, the gas was sampled, dried, and measured for isotopic ratio using a mass spectrometer. The mass 46/mass 44 isotopic ratio reflects the relative activity of the water species via the equilibrium

$$H_2^{18}O(\text{soln}) + C^{16}O_2(g) \rightleftharpoons H_2^{16}O(\text{soln}) + C^{18}O_2(g) \tag{3.109}$$

Pure water put through the same procedure as above was the standard used for comparison. A single measurement of a showed a standard deviation of ±0.02.

In the enrichment factor determination, a reasonable and consistent distribution of the separate effects due to the cation could be made [119] only if the ions ClO_4^-, Cl^-, I^-, Na^+, and $Co(en)_3^{3+}$ were not hydrated in the detection of characteristic internal vibrations. At 25 and 4°C the following ions were found to be solvated to the respective values listed: Li^+, 1 and 2; Ag^+, 0.7 and 0.7; H^+, 4.6 and 2; and Mg^{2+}, 6.2 and 7.1. At 25°C the respective solvation numbers 19 ± 10 and 6.0 ± 0.5 were found for Cr^{3+} and Al^{3+}.

Complexes such as $CrCl_2(H_2O)_4^+$ and $Cr(H_2O)_6^{3+}$ were discussed [378] in relation to the effect of ion hydration and chloride complexing on the rate of exchange of chromium between chromium(II) and chromium(III), using radioactive techniques.

In mixed electrolyte solutions evidence was found [44] for the competitive hydration of cations.

A study of the hydration of ions using oxygen isotopes has been discussed [393].

Isotopic exchange equilibrium data on the $H_2^{18}O$ and $H_2^{16}O$ concentrations in the body of the liquid (K_1) and in the vicinity of the ions (K_2) and the energy (ΔE) were used to formulate criteria for negative hydration of ions [340]. The criteria were $K = (K_1/K_2) < 1$ and $\Delta E < 0$.

In acidic water-methanol solvents at 60°C, the composition of the Cr^{3+} ion varied from $Cr(H_2O)_{5.831}(MeOH)_{0.169}^{3+}$ to $Cr(H_2O)_{2.40}(MeOH)_{3.60}^{3+}$ when the water-methanol solvent ranged in composition from 0.154 to 0.982 mole fraction of methanol, respectively [200]. Separations were made [215] by ion-exchange methods in water-ethanol solvents of individual differently solvated species, $Cr(H_2O)_{6-n}(C_2H_5OH)_n$ (n = 0, 1, 2, and 3) present in equilibrated species. At 50 and 75°C the average number of ethanol molecules $\bar{n}$ per Cr^{3+}

is the same, thus giving a small enthalpy change $(0 \pm 0.5$ kcal mol$^{-1})$ for the solvent replacement reaction.

Solvent isotope effects and ionic hydration equilibria for H_2O-D_2O mixtures can be calculated from the structure difference between D_2O and H_2O and that between HDO and H_2O, and the relative amounts of H_2O, D_2O, and HDO [382].

Other isotopic studies of ion hydration have been made [15, 22, 314]. For example, in a study of the exchange of water between ^{18}O-labeled solvent and hexaaquorhodium(III) ion, the hydrated ions $Rh(H_2O)_6^{3+}$, $Rh(H_2O)_5(OH)^{2+}$, $Rh(H_2O)_5^{3+}$, and $Rh(H_2O)_4(OH)^{2+}$ were used [314] to explain the data.

By isotopic equilibrium methods no general trend of hydration with temperature is noticeable, although cations other than hydrogen are in general more hydrated at lower than at higher temperatures. This would be expected, since at lower temperatures decreased thermal agitation permits greater ion-molecule attraction. In the case of hydrogen, lower temperatures might permit stronger hydrogen bonding among water molecules, which would prevent the breaking down of the water structure by protons and thus hinder the formation of hydronium and other hydrated hydrogen ions. The hydration number exhibits a change with ionic charge, radius, and type, which is qualitatively reasonable [40].

ClO_4^-, Cl^-, I^-, and Na^+ were found to be unhydrated by this approach, despite evidence of hydration by other methods. The cation $Co(en)_3^{2+}$ with the completed inner sphere shows zero hydration, probably because the hydration number is the same as the number of water molecules that can be held in the inner sphere.

This procedure along with several others shows a general increase in hydration with increasing valence, but shows Li^+ with a relatively low hydration number, in contrast to several other methods which show Li^+ to be fairly highly hydrated and to NMR which shows breaking of water structure by Li^+.

Spectroscopic and Other Optical Methods

Change in intensity and displacement of absorption bands can be used [423] to obtain solvation numbers of ions and complexes. Hydration of ions can produce distribution of intensity along the water band for acids in water [321]. The degree of, and the change in, solvation can be determined from the change in shape of hypersensitive and normal absorption bands with concentration of electrolyte and temperature. Ion solvation can be correlated with the appearance of new bands, and their change in intensity with temperature, electrolyte concentration, and solvent composition.

Rotation of polarized light [79, 96], fluorescence [321], and specific refraction [14] are used in the determination of solvation.

Optical methods are based on the change in absorption bands of valence vibrations, and the change in rotational bands of solvent and complex molecules with the nature and concentration of electrolyte, the nature and composition of solvent, and the temperature.

The solvation of ions in methanol and in acetophenone was studied using optical rotation [79]. In methanol the solvation of Li^+, Na^+, K^+, Cl^-, and I^- was found to be 4, 2.2, 1.2, 1 and 0, respectively; in acetophenone these same ions were found to be solvated to the extent 1, 0.5, 0.5, 0, and 0.

Absorption spectroscopy using water solvent showed [18, 284] Cu^{2+}, Nd^{3+}, Co^{2+}, and VO^{2+} to have four, three, six, and five water molecules per ion, respectively. Spectrophotometric measurements using water solvent gave [143] 6 as the solvation number of Cu^{2+}, while spectrophotometric studies of the same ion in water-acetone and in water-ethanol yielded [143] solvation numbers of $4H_2O$-$2(CH_3)_2CO$ and $2H_2O$-$4MeOH$, respectively. Absorption spectra studies in water gave [19, 226] solvation numbers of 6, 9, and 8 for Nd^{3+}. Absorption spectra of Nd^{3+} in water-methanol and in water gave [226] solvation numbers of $4H_2O$-$2MeOH$, $2H_2O$-$4MeOH$ and $6H_2O$.

Infrared studies of water in crystalline $BaCl_2 \cdot 2H_2O$ have been made [166]. The infrared results indicate that the strongest hydrogen bond formed in this hydrate has the same enthalpy as the hydrogen bonds present in ice.

From refractive index and density measurements on solutions of strong electrolytes in solution, the primary and secondary hydration numbers of electrolytes were obtained [351]. These data are recorded in Table 3.6. By selecting the primary and secondary hydration numbers of one of the ions as a standard reference, the primary and secondary hydration numbers of the other ions were determined.

Table 3.6 Primary and Secondary Overall Hydration Numbers for Some Uni-Univalent Electrolytes[a]

Electrolyte	Overall Hydration Number	
	Primary	*Secondary*
NaCl	6	28
NaBr	6	26
KCl	5	34
KBr	5	22
KI	5	19

[a]From Takeski Sato and Koreaki Hayashi, *J. Phys. Soc. Jap.*, **15**, 1658 (1960).

A complete correlation was found [329] between the change in solvation number of ions, and also the change in the total interaction between the ions and molecules of the solvent, and the characteristics of the rotational Brownian movement of the molecules of the solvent, namely, the time of relaxation of reorientation τ and the coefficient of rotational diffusion D_r. Ions as small as

Li^+ had the greatest solvation numbers and exerted the greatest effect on methanol molecules. Of several alcohols Li^+ and Cl^- ions exerted the greatest inhibiting effect on methanol molecules and the least on amyl alcohol molecules.

Optical methods have been applied to solvation phenomena too extensively to mention the various investigations. A rather complex list of literature articles and a summary of the contents thereof through 1962 is given in Ref.178.

Optical methods do not require alteration of the sample, and are rapid and comparatively accurate. Since the bond and rotational frequencies are characteristic parameters for each component, spectroscopy allows observations on individual components of the sample rather than on the composite sample. Spectroscopic and other light-measuring methods permit determination of the structural features of the sample. In all cases spectral studies are not definitive. The spectroscopic approach has shown that the effect of electrolytes and temperature on water, assumed to be the same, are quite different.

Sound Velocity.   Compressibility

Liquid adiabatic compressibility β_a is related to the velocity v of sound (in centimeters per second) for a liquid of density ρ (in grams per milliliter) by the expression [11, 12]

$$\beta_a = 10^6 v^{-2} \rho^{-1} \quad bar^{-1} \tag{3.110}$$

The isothermal compressibility β is related to β_a by the equation

$$\beta = \beta_a \frac{C_p}{C_v} = \beta_a + \frac{a^2 T}{J C_p \rho} \tag{3.111}$$

where C_p and C_v are heat capacities at constant pressure and constant volume, respectively, and a is the coefficient of volume expansion. Replacing a in this equation by $-(1/\rho)(d\rho/dT)$ yields

$$\beta = \beta_a + [(d\rho/dT)_p]^2 T/\rho^3 J C_p \tag{3.112}$$

Using the approximation $1/\delta T(\rho T + \delta T - \rho T)$ for $d\rho/dT$, at $25°C$,

$$\beta = \beta_a + 0.07125(\rho_{30°} - \rho_{20°})/(\rho_{25°})^3 C_p \tag{3.113}$$

From measurements of ultrasonic velocities in water, methanol, ethanol, and electrolytes in these solvents, it was found that for all solvents and solutes examined the results could be reproduced by equations of the general type:

$$\beta_a = \beta_{0,a}[1 + n_2(S_0 + An_2 + Bn_2^2/n_1)] - 1 \tag{3.114}$$

where A and B are constants, n_1 and n_2 are the numbers of moles of solvent and solute present, respectively, and S_0 is the limiting solvation number.

Passynski [309] related the adiabatic compressibility of the solvent $\beta_{0,a}$ to the solution β_a by the equation

$$\beta_a = \beta_{0,a}(1 - Sn_2/n_1) \qquad (3.115)$$

where S is the primary solvation number of the electrolyte. A graph of S against n_2 extrapolated to $n_2 = 0$ gives the "true" solvation number S_0.

Expanding (3.114) binomially, the first two terms correspond to (3.115). The remaining terms, assuming that the solvated species retain their solvent compressibility, may be considered contributions to the compressibility of the solvated ion-solvent and solvated ion-solvated ion interactions.

The limiting solvation numbers of electrolytes were obtained by extrapolating (3.115) to $n_2 = 0$ and taking the initial slope of the plot which is proportional to S_0 according to the following equation obtained from (3.114) and (3.115):

$$Lt(\partial\beta_0/\partial n_2)_{n_2 \to 0} = -\beta_{0,a}S_0/(n_1)_0 \qquad (3.116)$$

Ultrasonic velocities in solvents and solutions have been calculated from the wavelengths of ultrasonic waves in the media [11, 12, 14, 18, 19, 66, 79, 143, 226, 284, 309, 329, 349-351, 392, 410, 426, 427] and, from these velocities and measured densities, the adiabatic and isothermal compressibilities were obtained for the solvents and solutions. From these compressibilities the solvation numbers of electrolytes were calculated [11, 12, 309, 349, 350, 392, 410, 426, 427]. Then by assuming the solvation numbers of certain reference ions, the solvation numbers of various ions were calculated.

Two types of ultrasonic interferometers used are those of Hubbard and Loomis [185] and of Debye and Sears [101]. In the latter instrument (Fig. 3.10), standing waves in the solution caused periodic density variations, which are made visible by the Schlieren effect. The interferometer body is of stainless steel, with optical-glass windows. A rectangular optical-glass cell resting on the upper surface of the crystal contained the solution. The diameter of the crystal was 2.5 cm, and its resonance frequency after mounting was 973.2 kc sec⁻¹. The crystal was driven via a power amplifier, by a Colpitt's oscillator of carefully matched frequency which was kept constant by a second crystal. Connections were made to the gold-sputtered lower face to the crystal through a screened 75-Ω coaxial cable and via the cable screening to the steel body. The upper surface of the solution served as the reflector. The half-wavelength was measured by illuminating the solution with a collimated beam of light and counting the number of nodes within a known distance using a cathetometer. The temperature was 25.0 ± 0.1°C controlled by a thermostated enclosure. This

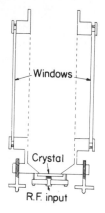

Fig. 3.10. Debye and Sears interferometer.

ultrasonic interferometer has the advantages that the liquid surface reflector may readily be made parallel to the upper crystal face, and that variations in mains-supply voltage merely cause small changes in the intensity of the Schlieren bands rather than produce spurious changes in voltmeter reading as in some models.

In water two sets of ion hydration numbers were found at 25°C based on the assumptions: (*a*) the limiting solvation numbers S_0 of K^+ = S_0 of Cl^- = 3.2, and (*b*) S_0 of NO_3^- = 2. Table 3.7 gives data for water at 25°C [12].

Table 3.7 Limiting Ion Solvation Numbers at 25°C

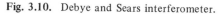

	Na^+	K^+	Li^+	NH_4^+	Ag^+	Mg^{2+}	Cl^-	NO_3^-	Br^-	I^-	OH^-	SO_4^{2-}
(*a*) $S_0(K^+) = S_0(Cl^-)$ in water:												
S_0	3.9	3.2	2.7	1.1	3.1	7.0	3.2	2.9	2.6	2.1	6.1	8.8
(*b₁*) $S_0(NO_3^-) = 2$ in water:												
S_0	4.8	4.1	3.6	2.0	4.0	7.9	2.3	2.0	1.7	1.2	5.2	7.9
(*b₂*) $S_0(NO_3^-) = 2$ in methanol:												
S_0	3.9	3.6	3.3	3.3	–	–	0.9	2.0	1.7	2.4	–	–
(*b₃*) $S_0(NO_3^-) = 2$ in ethanol:												
S_0	9.0	–	1.4	0.9	–	–	1.3	2.0	2.0	2.3	–	–

Many data have been accumulated on the solvation of ions using compressibility methods [10, 21, 91, 124, 198, 294, 305, 312].

As in other methods the solvation numbers of ions depend on the reference ion chosen and on the solvent. In general the solvation numbers of cations for similar temperatures and concentrations correspond more closely in water and in methanol than in ethanol. Under similar conditions the solvation numbers of anions agree more closely in methanol and in ethanol than in water.

Sometimes the solvation number of an ion differs with different investigators. Thus the solvation number of NH_4^+ in water from Ref. 12 is 1.1 or 3, depending on whether the reference standard for solvation is selected as $K^+ = Cl^- = 3.2$ or as $NO_3^- = 2$. From Ref. 21 the solvation number of the ion is 0.

By adiabatic compression, in opposition to other methods, all investigators find a solvation number for Li^+ ion that is low compared to other alkali metal ions. NMR indicated that structure breaking dominated structure making by this ion. If the concept of dynamic solvation [179] is accepted, and if this loosely constructed solvation permits some compressibility of the solvated ion, the incompressibility assumed in the adiabatic compressibility method would account for this reverse in relative solvation number of Li^+ compared to other alkali metal ions.

Only the primary solvation shells of ions are detected by the adiabatic compressibility approach.

Effective Volume

Darmois [94] calculated the primary hydration number n of an ion using the equation

$$(n_1 - n_2 n)v_1 + n_2 V_2 = V \tag{3.117}$$

where V is the volume of the solution containing a total number of water molecules n_1 of which n are attached to each of the n_2 salt molecules present in volume V as primary hydration water, v_1 is the volume of one free water molecule, and V_2 is the Stokes volume of one hydrated ion. The Stokes volume is often too small [48] and yields hydration numbers that may be too small.

It has been suggested [48, 90] that a better method would be to express the Stokes volume in terms of the crystallographic volumes of the ions v_i and of the compressed water molecules nv_1^h, using the equation

$$V = (nv_1^h + v_i)n_2 + (n_1 - n_2 n)v_i \tag{3.118}$$

Introducing the apparent molal volume ϕ and solving for n gives

$$n = \frac{\phi - v_i}{v_1^h - v_1} \tag{3.119}$$

From data for the pressure created by the coulombic ionic field in a surrounding dielectric at a mean distance $r_i + r_w$ from the center of the ion [421], and from data for compressibility of water at these pressures [65], the $\nu_1{}^h - \nu_1$ term was evaluated. By using literature values [40, 257, 371], the solvation numbers n for Li^+, Na^+, K^+, F^-, and Cl^- were found to be, respectively, 2.5, 4.8, 1.0, 4.3, and 0.

A method was outlined [163] for obtaining V^e_{max}, the maximum volume of water decreased by electrostriction, and ν^e, the magnitude of the electrostriction; using these values the hydration numbers n of electrolytes were calculated at 20°C employing the equation

$$n = \frac{(100\nu^e/V^e_{max})(\text{molecular weight of solute})}{18} \tag{3.120}$$

The hydration numbers found are listed in Table 3.8.

Table 3.8 Hydration Numbers n Calculated by (3.120)

Electrolyte	LiCl	LiBr	LiI	NaCl	NaBr	NaI	KCl	KBr
n	30	51	66	22	37	42	34	46
Electrolyte	KI	RbCl	RbBr	RbI	CsCl	CsBr	CsI	
n	47	49	38	28	45	87	96	

The large hydration numbers found were discussed in terms of clathrates around the ions.

The empirical Tait equation [390] has been used to develop methods [296, 391, 392, 428] for interpreting the nature of the solute and the hydration number of the solution molecules.

Based on the incompressible-shell model, an equation was obtained [309, 410] for calculating the hydration number h of salts. The equation was

$$h = V_h/V_1{}^{p0} \tag{3.121}$$

where V_h is the volume of water bound to 1 mol of solute in solution, and $V_1{}^{p0}$ is the molar volume of water at the standard pressure of 1 bar. Many hydration numbers were listed at different temperatures. For example, at 25°C the solvation numbers of LiCl, NaCl, KCl, NH_4Cl, and $MgCl_2$ were, respectively, 4.24, 4.98, 4.44, 1.50, and 10.40.

If each ion is originally surrounded by water molecules in such a way that the density of water prevails everywhere, then the first layer will have its center at $r_1 = r_0 + r_w = \overline{r}$ and, for each subsequent layer, it is assumed that the distance of each layer increases by the diameter of a water molecule, namely, 2.76 Å [158]. This distance holds strictly only for ions equal in size to the water molecule. At the relatively large distances from the ion involved here, this difference is not important. The number of water molecules X_n in the nth layer surrounding an ion was calculated as follows. For each sphere of radius r_n there are approximately $\frac{1}{2}X_n$ water molecules with centers at r_n, plus all the molecules at $r < r_n$, plus the intrinsic volume of the ion V^0. Hence

$$X_1 = \frac{8}{3}\frac{\pi}{v_w}r_1^{\,3} - 2\frac{V^0}{v_w} \tag{3.122}$$

where v_w is the molar volume of water at a given temperature [157], and subscript 1 refers to the first layer. Therefore

$$X_n = \frac{8}{3}\frac{\pi}{v_w}r_n^{\,3} - 2\frac{V^0}{v} - 2\sum_{a=1}^{a=n-1} X_a \tag{3.123}$$

Data calculated are shown in Table 3.9.

Table 3.9 Number of Water Molecules in the nth Layer in the Uncompressed State at 25°C

Layers, n	Ion			
	I^-	Cs^+	K^+	Li^+
1	7.3	5.1	3.8	2.0
2	22	18	16	11
3	35	30	27	22
4	60	53	48	41

It has been shown [39] that the correlation [292] of partial molal volumes of ions with the continuum model of Born can be explained by an isomorphic replacement of water molecules in a simple cubic lattice by ions whose sizes range from smaller to only slightly larger than water molecules. The arguments resulted in optimum values for coordination numbers of 6 to 8 for ions of radius between 1 and 2 Å [39].

Solvation numbers of lanthanides in aqueous solution were obtained [305, 307] using the equation

$$\overline{V}_h = \overline{V}_2^{\,0} - n^0 \overline{V}^0 \tag{3.124}$$

where $\overline{V}_h$ is the molar volume of the hydrated ion, $\overline{V}_2^{\,0}$ is the partial molal volume at infinite dilution of the lanthanide salts in water, $\overline{V}^0$ is the partial molal volume of water at infinite dilution, and n^0 is the solvation number. The values of $\overline{V}_h$ were calculated from the relation between the B coefficient of the Jones-Dole [204] equation and $\overline{V}_h$, namely $\overline{V}_h = B/2.5$ [376].

The values of B and $\overline{V}_2^0$ were found in the literature [368]. Solvation numbers of ions were then calculated using $n^0 = 1$ for chloride ion and $\overline{V}_2^0 = 18.60$ [306]. For the lanthanides the following solvation numbers were determined: $La^{3+}(8.5)$, $Pr^{3+}(9.5)$, $Nd^{3+}(9)$, $Sm^{3+}(9.5)$, $Tb^{3+}(10)$, $Dy^{3+}(11)$, $Ho^{3+}(11)$, $Er^{3+}(11)$.

From apparent molal volumes of electrolytes in water [210] and in H_2SO_4 [127, 151], solvation numbers n of ions have been determined using in water the solvation numbers of H^+ and OH^- as 1 and 0, respectively, and in H_2SO_4 the solvation number of Na^+ as 3 [128]. By using the relationship between osmotic coefficient and apparent molar volume, the solvation numbers of ions in H_2SO_4 were also obtained [127, 151]. If the solution is treated as containing a single electrolyte species of molality m, the variation in osmotic coefficient with electrolyte concentration may be expressed as

$$\phi = 1 + \phi_{el} + b\Sigma m \tag{3.125}$$

where ϕ_{el} is the electrostatic contribution to the osmotic coefficient, Σm is the total concentration of ionic species, and b is the osmotic coefficient parameter related to the solvation number s of the electrolyte by

$$b = (r + s)^2/40.8 - r/20.4 \tag{3.126}$$

where r is the ratio of the apparent molar volume of the electrolyte to the molar volume of the solute, 54 ml, and s is the solvation number of the electrolyte [127, 245]. Osmotic coefficients from freezing-point data [32] were used in determining the values of b. Solvation numbers obtained from cryoscopic and density measurements were compared [126], and are shown in Table 3.10.

Table 3.10 Solvation Numbers of 0.2 M Cations in H_2SO_4 from Cryoscopic and Density Measurements at $25°C$

Cation	NH_4^+	H_3O^+	Li^+	Na^+	K^+	Ca^{2+}	Sr^{2+}	Ba^{2+}	Ag^+
Solvation (cryoscopic)	1.2	1.8	2.3	3.0	2.1	—	—	6.5	2.1
Solvation (density)	1	0.3	2.0	3	2	8	8	5	1.5

Dielectric Properties

By assuming the first shell of water molecules to be dielectrically saturated with respect to positive ions but unsaturated with respect to negative ions, the molar dielectric depression in aqueous solutions of ions has been discussed

[175]. That the dielectric constant depends on the salt concentration in the manner

$$D_s = D_{H_2O} - \delta c \qquad (3.127)$$

was found from the results of the measurements made at centimeter wavelengths and extrapolated to zero frequency. In (3.127), D_s is the dielectric constant of the electrolyte solution, D_{H_2O} is the dielectric constant of pure water, c is the electrolyte concentration, and δ is given by

$$\delta = 1.5 \left[\frac{(D_{H_2O} - D_{\infty,ions})}{1000} V_2 - \frac{(D_{H_2O} - D_{\infty,H_2O})}{1000} n V_{H_2O} \right] \qquad (3.128)$$

where $D_{\infty,ions}$ and D_{∞,H_2O} are taken as 2 and 5.5, respectively, V_2 and V_{H_2O} are the molar volumes of solute and water, respectively, and n is the primary hydration number. The value for the hydration of electrolytes was calculated [168] using (3.127) and (3.128), and from these values the minimum hydration numbers of the individual ions were obtained [175] by taking the δ parameter for sodium chloride as $(\delta^+ + \delta^-)/2$, and by taking into account that a small amount of water outside the first hydration sphere is also dielectrically saturated. These minimum hydration numbers of ions are listed in Table 3.11.

Table 3.11   Minimum Hydration Numbers of Positive Ions

Ion	H^+	Li^+	Na^+	K^+	Rb^+	Mg^{2+}	Ba^{2+}	La^{3+}
n	10	6	4	4	4	14	14	22

The assumption that positive ions are surrounded by a dielectrically saturated first shell of water molecules, while negative ions have their first shell of water molecules completely unsaturated, does not give a quantitative description of the decrease in dielectric constant with added electrolyte. The fields in the neighborhood of all monovalent ions except H^+ are such that the first shell is far from saturated [156]. The dielectric constants of aqueous electrolyte solutions were obtained [156] by integrating over the spatial distribution of the local dielectric constants using the appropriate procedure for dispersed systems. By applying the dielectric constant change of the water molecules as a rough measure of their immobilization, a mean hydration number h was determined from the equation

$$h = \sum_P \overline{n}_P \left(1 - \frac{D_{wp} - n^2}{D_0 - n^2} \right) \qquad (3.129)$$

where $\overline{n}_P$ is the number of water molecules that can fit into the pth shell, D_{wp} is the mean dielectric constant of water molecules in the layer, and D_0 is the dielectric constant of water at zero field strength. From D_{w1}, the value of D for the first-layer water molecules, rough estimates of the mean hydration number h were made and are recorded in Table 3.12.

Table 3.12 The Hydration Number h Determined from D_{w1}

Ion	H^+	Li^+	Na^+	K^+	Cs^+	F^-	Cl^-	I^-
h	4	6	8	10	12	10	13	16

X-Ray

The variation in the structure of water in ionic solutions was investigated [372] using x-ray liquid diffraction curves of water in these solutions. X-ray diffraction patterns were associated with different structural elements in solutions of several salts [318]. Radial distribution functions of NaOH, HCl, and H_3PO_4 in solution have been studied, but the peak resolution in the distribution functions was insufficient to permit the realization of quantitative information [33-35, 122].

Distribution functions for two KOH solutions were found [63] to possess primary peaks with maxima at about 2.87 Å for a 18.8% solution and 2.92 Å for a 11.4% solution. A region of decreased electronic density then occurred, followed by a peak at 4.75 Å. The peaks were identified from information on the structure of water and the ionic radii of the species involved. The nearest-neighbor distance in water corresponded to 4.75 Å. The length of the tetrahedral edge is the second nearest-neighbor distance in tetrahedral water. Based on the nearest-neighbor distance of 2.92 Å, the calculated value of this length is 4.75 Å, in agreement with the observed value. The maximum of the nearest-neighbor peak for liquid water at 30°C is at 2.94 Å. The effective radius of water, 1.38 Å, and the radii of the K^+ and OH^- ions, 1.33 Å for each, are very similar, and one would expect only a slight change in peak position in KOH solution. Since each primary peak has only one maximum, it was thought that the primary peak in the KOH distribution function includes nearest-neighbor water molecules, K^+-H_2O neighbors, and OH^--H_2O neighbors. Nearest neighbors were found constantly to change position with other molecules in solution. The K^+ ion was found to have a hydration number of 4 from an analysis of peak areas in terms of the numbers of molecules around OH^- and around K^+ together with the number of nearest-neighbor water molecules around any other water molecule. Subsequent x-ray investigations gave hydration numbers for K^+, Li^+, OH^- and Cl^- as 4, 4, 6, and 8, respectively [62]. X-ray analyses have yielded [363] coordination numbers for Li^+, Na^+,

K^+, OH^-, and Cl^-. X-ray analysis of sulfate solutions [359] suggests that K^+ and Ca^{2+} have six octahedrally coordinated water molecules. Similar studies give 6 molecules of water in the first hydration shell of Mg^{2+}, and in the second shell in dilute solutions 10 to 11 molecules of water [336]. X-ray studies [233, 422] of aqueous solutions of $ZnCl_2$ and $ZnBr_2$ show the coordination number of Zn^{2+} to be 4. X-ray diffraction studies were used to derive electron radial distribution functions for six concentrated aqueous alkaline-earth chlorides [5]. Analysis of these functions indicated that Mg^{2+}, Ca^{2+}, Sr^{2+}, and Ba^{2+} ions were hydrated with 6.5 to 8.0 molecules of water. Mg^{2+} ions significantly alter the structure of water, as do Ca^{2+} ions to an even greater extent, as evidenced by second nearest-neighbor coordinations. The hydration of Mg^{2+} and Ca^{2+} ions appears to be independent of the concentrations of the ions in solution from N_{MgCl_2} = 0.071 and N_{CaCl_2} = 0.055 to saturation.

Intensity curves for x-ray scattering were obtained [358] at 20°C for water and aqueous solutions of $MnSO_4$, $ZnSO_4$, and $CuSO_4$ by photographic and scintillation methods, using molybdenum and silver radiations. A LiF crystal was used for monochromatization, and lead foil was used for fluorescent radiation. The radial distribution curves gave a coordination number of 4.8 for water molecules in pure water. Mn^{2+} and Zn^{2+} ions in sulfate solutions were hydrated octahedrally, and Cu^{2+} ions were situated in the center of a distorted octahedron. With increased concentration of electrolyte, an increase in the coordination number of the water molecules in solution was observed.

Nuclear Magnetic Resonance

Various NMR techniques have been applied to the determination of solvation numbers [178]. Solvent molecules can exist in several environments in solution. Arbitrarily, these environments may be classified into bulk regions in which solvent molecules are outside the range of ionic influence, secondary solvation regions, and primary solvation regions [75].

For a very slow exchange of solvent molecules among these regions, several peaks would be expected in the NMR spectrum of the solvent nuclei, corresponding to the different interactions. Ordinarily, there is a very rapid exchange of solvent molecules among the various regions, so that the separate resonance signals expected for each environment are time-averaged to a single peak whose shift from the pure solvent resonance peak represents the mean effect of the different environments.

Various NMR techniques have been used to determine the solvation of several ions; the most direct of these techniques is determination of the areas of resonance peaks of the bulk solvent and the bound solvent which is coordinated with the ions. Initially $H_2^{17}O$ NMR spectra of several aqueous solutions gave peaks attributed [199] to bulk and bound solvent. The ^{17}O showed [420]

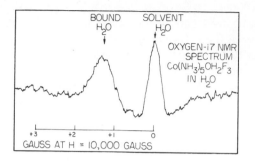

Fig. 3.11. $1.2 \, M \, \text{Co(NH}_3)_5\text{OH}_2\text{F}_3$ enriched in oxygen 17, dissolved in ordinary water.

large chemical shifts, hence was used as a probe. The ions studied were Al^{3+}, Be^{2+}, Ga^{3+}, and $(\text{NH}_3)_5\text{Co(H}_2\text{O})^{3+}$, the last ion being selected since the number of water molecules attached to the Co^{3+} is known [333], and because of the long exchange time of bulk water with bound water. For aqueous solutions of these ions, the ^{17}O NMR spectra showed that only a small amount of Co^{2+} had to be added to the solutions before separate peaks appeared, which could be attributed to bulk and bound solvent molecules [199]. It had been shown [361, 362] that paramagnetic ions cause large shifts in the resonance adsorption of other nuclei. A great many more bulk molecules come in contact with Co^{2+} ions than the water molecules bound to Al^{3+}, Be^{2+}, Ga^{3+}, or $(\text{NH}_3)_5\text{Co(H}_2\text{O})^{3+}$, hence the bulk water resonance signal undergoes a large shift while no significant effect on the shift of the bound molecules is noted. Figure 3.11 [199] shows the separation of the bound water and bulk water ^{17}O NMR signals. The number of water molecules in the primary sphere can be calculated either by comparing the areas under the two ^{17}O resonance peaks, or by measuring the shift of the resonance peak produced by the paramagnetic ion in the presence and absence of the diamagnetic ion.

Because of unfavorable signal-to-noise ratio, the precision of measurement was not sufficiently high to determine [199] the number of water molecules associated with Al^{3+}, Be^{2+}, and Ga^{3+}. However, the coordination of Al^{3+} and Be^{2+} was determined using the peak-area comparison method by increasing the signal-to-noise ratio using water of greater ^{17}O enrichment and the sideband detection technique. Solvation numbers of 5.9 and 4.2 for Al^{3+} and Be^{2+} were found in solutions of Al^{3+} and Be^{2+} containing Co^{2+} ions.

By using the chemical shift method [199], the solvation numbers of Al^{3+}, Be^{2+}, and Cr^{3+} ions were determined [7]. A single H_2^{17}O NMR signal is observed in an aqueous solution of a paramagnetic ion in which there is a rapid exchange of water molecules between the first coordination sphere of the ion and the bulk solvent. The ratio of the concentration of the paramagnetic ion to the total labile water present determines the shift of the resonance signal from its position in pure water. A decrease in the amount of water available to interact with the paramagnetic ion and, consequently, a change in the labile $H_2^{17}O$ resonance position is produced by the addition of a second ion for which the water in the first coordination sphere is nonlabile. This change permits the calculation of the number of moles of labile water in the solution containing the diamagnetic and paramagnetic ions. The known total of moles of water minus the number of moles of labile water divided by the moles of diamagnetic ion gives the solvation number of the diamagnetic ion. The solvation numbers of Al^{3+} and Be^{2+} were found to be 5.9 and 3.8, respectively, using dysprosium(III) as the paramagnetic ion whose effect on water was known [246]. The equation used in calculating the hydration numbers was

$$\delta m'_{H_2O}/m_{Dy(III)} = \delta^0 m^0_{H_2O}/m^0_{Dy(III)} \qquad (3.130)$$

where m_x represents the millimoles of species x in solution, and the superscript 0 indicates the quantity measured in the reference solution with only dysprosium(III) present. The only unknown quantity in (3.130) is m'_{H_2O}, the number of millimoles of labile water in the solution containing the diamagnetic ion since the different shifts δ were measured and dysprosium(III) was added in known molalities m. After the millimoles of labile water were determined, the solvation of the diamagnetic ions were calculated as described above.

The chemical shift method of determining the solvation numbers may be applied to paramagnetic ions if the magnetic influence of these ions does not extend beyond the first solvation sphere [7]. This method, with modification, was also applied to the Cr^{3+} ion where the magnetic influence does extend beyond the first solvation sphere producing a shift in the labile $H_2^{17}O$ resonance. A solvation number of 6.8 was obtained [6]. This value is greater than 6, the solvation number found by the isotopic dilution method [190], and the discrepancy and validity of the assumption made in obtaining the value of 6.8 was discussed [6].

A solvation number of 3 was found for $(CH_3)_3Pt^+$ using the chemical shift method, and a solvation number of 2 for $(NH_3)_2Pt^+$ from bulk and bound resonance signals [153, 154].

By using the chemical shift method, cation hydration numbers in hydrate melts have been determined [275] for melts of calcium nitrate tetrahydrate with anhydrous potassium nitrate, for tetramethylammonium nitrate, and for magnesium nitrate. It was found in these melts that Ca^{2+} cation was selectively hydrated over monovalent ions, but that Mg^{2+} ions were selectively solvated over Ca^{2+} ions.

V^{4+} and Ga^{3+} ions have been assigned the respective solvation numbers 3.8 ± 0.2 [324] and 6.8 ± 0.26 [121] from chemical shifts.

The temperature coefficients of the proton shifts of water have been used to determine hydration numbers [93, 218, 261-263, 409]. The proton shift of water is strongly temperature-dependent [353], and is an average of the bonded and nonbonded environments of the proton. Water vapor shows a high field shift, since it has few hydrogen bonds. Over a temperature range of 5 to 95°C, the chemical shift of pure water measured against gaseous ethane as a reference was found to be linearly dependent on the temperature t°C, as given by the equation

$$\delta_{H_2O} = 0.00956t - 4.38 \quad \text{ppm} \tag{3.131}$$

The ethane proton shift is independent of temperature, hence it is used as a reference. The single resonance signal observed for water protons in an aqueous electrolyte solution is the weighted average of those in the different environments experienced by the protons. A proton residing in the normal water structure has a shift δ_N identical with the shift δ_{H_2O} of the proton in pure water at the same temperature. Water molecules solvating an ion have a proton resonance signal shifted relative to that of pure water, and this shift δ_S depends on both the cation and the anion. The average time a proton spends in any environment is proportional to its instantaneous mole fraction, hence the proton shift δ_{H_2O} observed for an aqueous solution is represented by

$$\delta_{H_2O} = X_N\delta_N + X_S\delta_S \tag{3.132}$$

where X_N and X_S are mole fractions of protons in normal and hydrated forms of water, assuming water molecules beyond the first solvation sphere are unaffected by the ions. If h represents the effective solvation number, that is, the moles of solvated water per stoichiometric mole of salt, and m is the stoichiometric molality, (3.132) becomes

$$h = (55.5/m)[(\delta_{H_2O} - \delta_N)/(\delta_S - \delta_N)] \tag{3.133}$$

The above equation also predicts a linear relation between shift and molality if h and δ_N remain independent of concentration, and indicates that at some temperature $\delta_{H_2O} = \delta_N$, at which temperature $\delta_S = \delta_N$ also. If the shift of the proton in the solvated environment does not vary with temperature, the solvation number of the salt can be found from (3.133) by substituting into it the experimentally determined values of m and the various δ's. This method was used to determine the solvation numbers of several electrolytes [93, 261, 409]. Thus LiCl, NaCl, KCl, RbCl, CsCl, $MgCl_2$, $CaCl_2$, and $Al(NO_3)_3$ were found to have the respective solvation numbers: 3.2, 4.5, 4.6, 4.0, 3.9, 8.2, 9.5, and 13.4.

The large hydration numbers of Mg^{2+}, Ca^{2+}, and Al^{3+} salts were attributed either to the hydration of the anions or to a contribution by a second hydration layer.

The ^{7}Li NMR chemical shifts in water were downfield, temperature-independent, and linear with mole fraction of salt up to 0.3 mole fraction [4]. Progressive polarization of Li^+ ion by an increasing proportion of anions was used to explain the observed shifts. The data are incompatible with the concept of a tightly bound hydration shell or a tightly united ion pair, and suggest a solution structure of a relatively random mixture of ions and molecules with no permanent interaction between them. The average number of competitive sites around the lithium was calculated using

$$\delta = \delta^0 Xm \qquad (3.134)$$

where δ is the observed and δ^0 is the limiting chemical shift, X is the number of sites, and m is the mole fraction of salt. δ^0 applied to the pure salt. For values of $m < 0.3$, a plot of δ/δ^0 versus m had a slope $X = 2$. This value of X does not so much represent [4] the classic hydration number as the average number of sites accessible to halide substitution or, possibly, the effectiveness of the water molecules in excluding anions from the cation.

Studies of the solvation of Mg^{2+} ion when $Mg(ClO_4)_2$ was dissolved in methanol-water [388] and in methanol [295] have shown Mg^{2+} to be solvated by 5.7 ± 0.02 methanol molecules at $-75°C$ in the former solvent, and by 6 methanol molecules in the latter solvent. The former measurements were made from the resonance peak area associated with the bulk and bound hydroxyl protons with methanol and with the water proton. Association of the Mg^{2+} and ClO_4^- ions was thought to cause deviation of the solvation number from 6. At $-85°C$ the amount of solvation water relative to solvation methanol was observed and was attributed to a reequilibration of water between solvated cation and solvent. The solvation number of 6 was obtained from the molar ratio of $Mg(ClO_4)_2$ and methanol, and the ratio at low temperatures of the well-separated peaks of the OH protons of methanol bound to Mg^{2+} and those in bulk methanol. The solvation number remained constant at high concentrations of $Mg(ClO_4)_2$ and at very low temperatures, indicating little "close" ion association. The symmetry of the OH quadruplet of methanol molecules in the solvation shell suggested a regular octahedral configuration in the first solvation shell under the experimental conditions used. Had the ClO_4^- ion penetrated the first solvation shell of the Mg^{2+} ion there would have been a decrease in the calculated solvation number.

An excellent technique [199] for determining whether the exchange of protons of the solvation shell arises from protons or from whole molecules was found to be the addition of a small amount of $Cu(ClO_4)_2$ to the $Mg(ClO_4)_2$-methanol system. The bulk OH and CH_3 proton resonances were broadened by added $Cu(ClO_4)_2$, while the proton resonances of solvation shell methanol molecules were only slightly affected. This and the broadening and disappearance of CH_3 and OH proton signals of the solvation shell molecules with increase in temperature indicated that the whole molecule exchanges in solution.

The NMR bulk-bound signal area method indicates that the solvation number of Mg^{2+} is 6 in aqueous acetone [138, 273, 415] and in methanolic acetone [273] solutions. Direct evidence has been obtained [74] of anion shifts in the hydroxyl proton of methanol. In liquid ammonia the solvation number of Mg^{2+} ion was found [384] to be 5.

By the bulk-bound signal area ratio method, the solvation number of Co^{2+} has been determined as 6 in anhydrous methanol [250], in water [269], in N,N-dimethylformamide [268], and in acetonitrile [271]. No consistent trend of solvation number with either temperature or concentration was observable in these solutions.

By using the bulk-bound signal area ratio method, Al^{3+} has been found to have a solvation number of 6 in dimethylformamide [290], in anhydrous dimethyl sulfoxide [395], in aqueous dimethyl sulfoxide [141], and in water [120, 133, 139, 142, 270, 355]. In the spectrum of aqueous acetonitrile solutions of $Al(ClO_4)_3$, separate resonance peaks corresponding to different hydrated Al^{3+} ions have been observed [380]. By using the above method, solvation numbers of 4 for Be^{2+} in dimethylformamide [272] and in aqueous and aqueous acetone solutions [139], 6 for Ga^{3+} [136, 272, 355], (a minimum of 4), 6 for In^{3+} [139], 3.9 for Sc^{3+} [137, 140], 2.4 for Y^{3+} [137, 140], 2.9 for Th^{4+} [137, 140], 2 for Sn^{2+} [132], and about 4 for UO_2^{2+} in aqueous mixed solvent systems [134, 135] have been determined.

Mixing a salt of the metal ion being investigated and a very soluble salt of a metal ion with which the coordinated water has been shown to be labile has been suggested [370] as a NMR technique for the determination of solvation numbers of metal ions in aqueous solutions. The method still involved analysis of bulk and bound water peaks. Ion pairing was evidenced by the solvation number data obtained for Al^{3+} and Ga^{3+}.

Utilizing integral measurements of resonance signals has proved to be an accurate method for the determination of hydration numbers from measurements of total hydrogen content of materials dissolved in D_2O [234]. In this method the material must contain some nonexchangable protons and no paramagnetic ions.

The selective broadening of the NMR peaks of ethyl alcohol in aqueous denaturated alcoholic mixtures was used to study the solvation of several paramagnetic ions [357]. Varying amounts of D_2O were added to ethanolic solutions of $Cu(ND_3)_2 \cdot 3H_2O^{2+}$, $CuCl_2$, $MnCl_2$, and $Cr(NO_3)_3 \cdot 9H_2O$. Selective solvation was predicated from the ratio of the amplitudes of the proton resonances of the CH_2 and CH_3 groups with varying amounts of added D_2O. For Cu^{2+} a solvation number of 120 was obtained.

Hydration numbers of several cations and anions in alkaline solutions have been determined from proton relaxation times as a function of salt concentration. Taking the hydration number of K^+ to be 6, the hydration numbers of the following ions were obtained [114]: Li^+, 1 ± 1; Na^+, 3.6 ± 1; Rb^+, 9 ± 2; Cs^+, 14.6 ± 2; F^-, 9.9 ± 2; Cl^-, 13.2 ± 2; Br^-, 16.2 ± 2; I^-, 21.8 ± 2.

Solvation numbers for the cations and anions of HCl, NaCl, KCl, $MgCl_2$, and $CaCl_2$ in water, and for $CoCl_2$ and $CuCl_2$ in methanol have been calculated [429, 430], using the correlation between relaxation time and salt concentration. In methanol the solvation numbers found for Co^{2+}, Cu^{2+}, and Cl^- ions were, respectively, 4, 6, and 8.

From proton relaxation times using a modified Stokes' equation [67], hydration numbers ($\pm 40\%$) were determined for Li^+ (5), Na^+ (3), K^+ (1), Mg^{2+} (6), Zn^{2+} (12), Al^{3+} (16), Cl^- (2), OH^- (4), SO_4^{2-} (5), and ice (−5). It is stated that the value of ice is to be subtracted per monovalent ion.

From proton magnetic resonance (PMR) chemical shifts, effective hydration numbers of cations and anions were obtained [176]. A solution model divided the chemical shift into four terms: bond-breaking, structural, polarization, and electrostatic. Effective hydration numbers in which weak interactions with a large number of water molecules are replaced by stronger interactions with a limited number of water molecules were found for the ions: Li^+ (4.0), Na^+ (3.1), K^+ (2.1), Rb^+ (1.6), Cs^+ (1.0), Ag^+ (2), F^- (1.6); Cl^-, Br^-, I^-, NO_3^-, ClO_4^- (all 0). It was concluded that the concept of a complete hydration sphere of tightly bound water was not compatible with the data and that, with increasing ionic radius, the effectiveness of these ions in coordinating with water decreases. The data for Li^+ suggest a structure-making effect. The larger halide ions tend to break down water structure, and only F^- forms a chemically bonded hydrate.

A technique based on the kinetically distinguishable water molecule exchange with a hydrated Mn^{2+} ion and the relationship between this exchange and the PMR relaxation time has been developed [385]. Kinetically, it might be able to distinguish between water molecules associated with the anion, those in the primary hydration sphere of a cation, and those associated only with other water molecules, by adding a probe ion to the solution capable of reacting with water molecules. For kinetically distinguishable water molecules, the reaction will proceed as the sum of parallel reactions characterized by different rate constants. For the method to be successful a probe species must have the following properties. It must react directly with water molecules; the rate constant for the reaction between the probe and the primary hydration sphere water must differ considerably from the rate of reaction of all other types of water molecules; the probe's lifetime must be related to a precisely measureable experimental quantity. Mn^{2+} ion proved to be a suitable probe. The procedure was to compare the linewidths of two solutions each of which contained the probe and a common anion, but one of which contained a cation of known hydration number and the other of which contained the cation of interest. The ratio of the proton linewidths to the primary hydration number of the line studied was related by the empirical equation

$$\frac{W_{AB} - W_A}{W_{AB'} - W_{A'}} = \frac{[H_2O]_{AB'} - n_{A'}[A']}{[H_2O]_{AB} - n_A[A]}$$
(3.135)

where subscript AB refers to the solution containing the cation of known hydration number, AB' refers to the solution containing the cation of interest, and A and A' refer to solutions of the same compositions as AB and AB' except for the absence of Mn^{2+} ion. W_{AB}, W_A, $W_{AB'}$, and $W_{A'}$, are measured linewidths in a single comparison, [A] and [A'] are cation concentrations, $[H_2O]_{AB}$ and $[H_2O]_{AB'}$ are water concentrations, and n_A and $n_{A'}$ are hydration numbers of the two cations. Equation (3.135) was tested for temperature, concentration, and anion using the ions Al^{3+}, Be^{2+}, NH_4^+, and H^+, the primary hydration numbers of which are known. The technique gave primary hydration numbers for the cations: Mg^{2+} (3.8), Ca^{2+} (4.3), Sr^{2+} (5.0), Ba^{2+} (5.7), Zn^{2+} (3.9), Cd^{2+} (5.6), Hg^{2+} (4.9), and Pb^{2+} (5.7). The doubly charged series shows a direct correlation between ionic radius and primary hydration number. The effect of water structure in determining the structures of ions was suggested as the reason for the relatively small hydration numbers obtained by this method. Other hydration numbers found [383] by this method are Ga^{3+} (6) and Th^{4+} (10 ± 0.2). The latter is of interest in relation to the recently prepared [78, 291] complexes of thorium with a coordination number of 10. The hydrations of Ga^{3+} and Th^{4+} were found to be independent of temperature. The temperature dependence of the linewidth of Ni^{2+} was used to determine [387] a solvation number of 6 for this ion. There has been some polemical discussion [276, 386] in the literature concerning the validity of the method used in the determination of the primary hydration of the above ions.

Since the protons of histidine utilize the first hydration sphere sites of Co^{2+}, these sites become inaccessible to water. Since the PMR of coordinated and free water at constant Co^{2+} concentration is a single resonance signal, the displacement of water by histidine causes a shift in water PMR towards the cobalt-free resonance line. This method has been used [253] to estimate the number of available sites.

The number of first hydration shell sites of Co^{2+} occupied by water has been determined from PMR shifts in aqueous solutions of paramagnetic metal ions [251]. Paramagnetic ions in solution cause a chemical shift Δw, related to the coordination number of the metal ion by the equation

$$\Delta w = pqw_1 \, S(S + 1) \, g|B|/3kT\gamma_1 \, A$$
(3.136)

where q is the coordination number, $p = [M]/[H_2O]$, S is the spin quantum number for the electron interactions between the paramagnetic ion and the protons, w_1 is the Lamour frequency for the electrons, g is the splitting factor, A is the hyperfine interaction constant, and

$$B = \frac{N_c \mu_{\text{eff}}^2}{r_e^6} \qquad (3.137)$$

where N_c is the number per unit volume of type i spins which can interact at one time with spin j, μ_{eff} is the effective moment of the ion, and r_e is the distance between i and j. The other terms in (3.136) have their usual meanings. A ligand introduced into metal ion solution causes displacement of water from the first hydration sphere, thus producing a difference in the chemical shift compared to the ligand-free solution. Linear plots of water proton shifts versus [Co^{2+}] were obtained for solutions with and without excess ligand (NaH_2PO_4). In the phosphate solution the slope was 0.83 as great as for the ligand-free solution, thereby indicating that one out of six water sites is occupied by the phosphate group. In RNA-Co^{2+} solutions, Co^{2+} binds to one phosphate of RNA.

Evidence for a four-coordinated complex was found when tetrahydrofuran was added to a solution of sodium tetrabutyl aluminate, while a 1:1 complex of sodium-diethyl ether and of sodium-triethylamine became apparent when the respective ligands diethyl ether and triethylamine were added to the solution of sodium tetrabutyl aluminate in cyclohexane. The proton chemical shifts of the ligands were used to make these determinations [352].

Analysis of the contribution of the N^{14} spin lattice relaxation to the proton line shape gave a range of 20 to 40 for the solvation number of the electron for a dilute solution of potassium in liquid ammonia [313].

Molal shifts Δm of small cations in methanol and water were explained [373] by a model which takes account of nearest-neighbor interactions only. The resulting equation is

$$\Delta m = nM/1000 \; (\Delta_c + \Delta_h) \qquad (3.138)$$

where n is the solvation number, M is the molecular weight of the solvent, Δ_c is the cation complex shift, and Δ_h is a desolvation characteristic of the solvent. For small and/or polyvalent cations, n appears to have a value of approximately 6. Rb^+ and Cs^+ showed deviations from predicted behavior. The deviations and the molal shifts of larger halide ions were attributed to disordered primary solvation shells, so that the effective solvation numbers of these ions are less than the number of nearest-neighbor solvent molecules. Δ_h was found to be 1.5 ppm for cations in methanol at $-69°C$ and in water at $25°C$, and was thought possibly to be due to structure making by small cations and structure breaking by large anions. Δ_c the direct ion-molecule interaction shift is assumed to be the complex shift measured in acetonitrile. It characterizes individual ions and may be estimated reasonably accurately by an electrostatic model. It is independent of temperature within experimental error, which is reasonable for an interaction energy in excess of 15 kcal mol^{-1} of solvate [295].

By using the influence of salts on the proton resonance chemical shifts of the solvents, the solvation of a variety of ions by the dipolar aprotic solvents acetonitrile, sulfolane, and dimethyl sulfoxide was investigated [85]. Supplemental measurements by infrared spectroscopy showed that in general anions affect only the C–H stretching frequencies, and cations both and C–C and particularly the C≡N stretching frequencies of acetonitrile. The order of specific solvation of the halide ions in acetonitrile was Cl$^-$ > Br$^-$ > I$^-$, which is the order observed for the same ions in protic solvents, contrary to what is generally accepted for the order of solvation of these ions in aprotic solvents.

It has been found [100], using NMR chemical shifts, that $AgNO_3$ complexes, solvated with two methyl cyanide solvent molecules, exchange one of the methyl cyanide molecules for a heterocyclic compound when the latter is added as a donor compound. If AS_2 represents two solvent molecules (S), the equilibrium for the reaction of AS_2 and the donor D is written:

$$AS_2 + D \rightleftharpoons DAS + S \tag{3.139}$$

The heterocyclic donors used were 1,4-dioxin, 1,4-dithiin, 1,4-benzodioxin, and 1,4-benzodithiin. All the compounds had equilibrium constants greater than 4.

Conclusions

The solvation number of any ion depends on the method of measurement. Some methods of measurement, for example, mobility measurements, denote the number of molecules of solvent moving with the ion, while other methods, for example, adiabatic compressibility measurements, give only the primary solvation. This difference in the results of various methods of investigation is a source of confusion that is more imaginary than factual. Another source of confusion is that the choice of the reference ion and its assumed extent of solvation determine the magnitude of solvation of all other ions. This artifice has to be resorted to, because most methods used in solvation measurements yield the total solvation of the electrolyte and not that of individual ions. These factors, combined with the fact that there are no accepted structures for even common solvents and for solvated ions, make the field of ion solvation difficult to really comprehend.

Because of the limitations of Stokes' law, measurements of ion solvation by mobility methods are erroneous for ions whose sizes in solution are small compared to the molecules of the medium. From x-ray studies of ion size in the crystalline state, there is no a priori way of knowing whether Stokes' law applies, since ion size in crystals as measured by x-rays may not necessarily coincide with the size of the ion in solution. It has been suggested [48] that ion solvation measured by mobility methods might be less than the actual case, because of the loss through Brownian motion of molecules of solvation by the ion. It has been pointed out [232] that, while the results of the study of transport coefficients of aqueous solutions may be interpreted in terms of solution structure effects, it is impossible to use such data to distinguish among the various proposed models to explain these same structural effects, because of the difficulties associated with the molecular theory of transport coefficients.

Spectroscopic and diffraction methods give structure data which average the contributions of various structures, and are therefore not definitive in the determination of absolute solution structure [232]. The chief difficulty arises from the short lifetimes, 10^{-12} sec, characteristic of hydrogen-bonded structures in hydrogen-bonded solvents and their solutions [232]. Such high concentrations are required for Raman and infrared methods, that these are further complicated by spectral bond components resulting from both ion-ion and ion-solvent interactions. A weakness of NMR methods is that the structural models used generally ignore the anion. One exception to this rule was discussed [85], and there are others. The limitation described above concerning transport coefficients [85] is also present in the interpretation of NMR relaxation data. Use of the area under the NMR absorption peak to obtain the molecules nearest an ion suffers from the fact that a separate peak is not in practice obtained for monovalent ions [50].

The approach to the model near an ion through the interpretation of physical properties of ionic solutions, such as density, dielectric constant, compressibility, heat capacity, mobility, ionic vibration potential, and so on, seldom gives information on individual ions [50].

References

1. A. W. Adamson, *J. Phys. Chem.*, **58**, 514 (1954).
2. A. I. Agaev and A. I. Dzhabarov, *Tr. Azerb. Goz. Univ. Ser. Khim.*, **1959**, 8.
3. L. H. Ahrens, *Geochim. Cosmochim. Acta*, **2**, 155 (1952).
4. J. W. Akitt and A. J. Downs, *Chem. Commun.*, **1966**, 222.
5. J. N. Albright, *J. Chem. Phys.*, **56**(8), 3783 (1972).
6. M. Alei, *Inorg. Chem.*, **3**, 44 (1964).
7. M. Alei and J. A. Jackson, *J. Chem. Phys.*, **41**, 3402 (1964).
8. Yu. P. Aleshko-Ozhevskii, *Zh. Phys. Khim.*, **45**(4) (1971); *Russ. J. Phys. Chem.*, **45**(4), 554(988) (1971).
9. R. Alexander and A. J. Parker, *J. Am. Chem. Soc.*, **89**, 5549 (1967).
10. D. S. Allam and W. H. Lee, *J. Inorg. Nucl. Chem.*, **13**, 218 (1960).

11. D. S. Allam and W. H. Lee, *J. Chem. Soc.*, **1964**, 6049.
12. D. S. Allam and W. H. Lee, *J. Chem. Soc., A*, **1966**, 5.
13. D. S. Allam and W. H. Lee, *J. Chem. Soc., A*, **1966**, 426.
14. R. Amiot, *C. R. Acad. Sci., Paris*, **243**, 1311 (1956).
15. L. B. Anderson, University Microfilms, Ann Arbor, Mich., no. 62-117; *Diss. Abstr.*, **22**, 3002 (1962).
16. G. A. Andreev, *Dokl. Akad. Nauk SSSR*, **145**, 358 (1962).
17. E. Angelescu and D. Motoo, *Commun. Acad. Repub. Pop. Rom.*, **3**, 267 (1953).
18. I. I. Antipova-Karataeva, Yu. I. Kutsenko, and G. I. Yatsun, *Zh. Neorg. Khim.*, **7**, 1913 (1962).
19. E. Antonescu, *Stud. Cercet. Chim.*, **15**, 645 (1967).
20. E. M. Arnett, H. C. Ko, and R. J. Minasz, *J. Phys. Chem.*, **76**(17) 2474 (1972).
21. H. Asai, *J. Phys. Soc. Jap.*, **16**, 761 (1951).
22. T. Asano, *Radioisotopes*, **14**, 363 (1965).
23. A. M. Azzam, *Z. Phys. Chem.* (Frankfurt am Main), **33**, 23 (1962); **32**, 309 (1962).
24. J. Baborovsky, *Z. Phys. Chem.*, **129**, 129 (1927).
25. J. Baborovsky, *Z. Phys. Chem.*, **A168**, 135 (1934).
26. J. Baborovsky and J. Velisek, *Chem. Listy*, **21**, 227 (1927).
27. J. Baborovsky, O. Viktorin, and A. Wägner, *Collect. Czech. Chem. Commun.*, **4**, 200 (1932).
28. H. W. Baldwin and H. Taube, *J. Chem. Phys.*, **33**, 206 (1960).
29. J. A. Bard, J. O. Wear, R. F. Griffin, and E. S. Amis, *Electroanal. Chem.*, **8**, 419 (1964).
30. S. Barnartt, *Quart. Rev. Chem. Soc.*, **7**, 84 (1953).
31. S. J. Bass and R. J. Gillespie, *J. Chem. Soc.*, **1960**, 814.
32. S. J. Bass, R. J. Gillespie, and J. V. Oubridge, *J. Chem. Soc.*, **1960**, 837.
33. O. Bastiansen and C. Finbak, *Tidsskr. Kjemi, Bergv. Met.*, **3**, 98 (1943).
34. O. Bastiansen and C. Finbak, *Tidsskr. Kjemi, Bergv. Met.*, **4**, 40 (1944).
35. O. Bastiansen and C. Finbak, *Tidsskr. Kjemi, Bergv. Met.*, **4**, 50 (1944).
36. T. W. Bauer, H. L. Johnson, and E. C. Kerr, *J. Am. Chem. Soc.*, **72**, 5174 (1950).
37. L. Benjamin and V. Gold, *Trans. Faraday Soc.*, **50**, 797 (1954).
38. K. Bennewitz, *Z. Elektrochem.*, **33**, 540 (1927).
39. S. W. Benson and C. S. Copeland, *J. Phys. Chem.*, **67**, 1194 (1963).
40. J. D. Bernal and R. H. Fowler, *J. Chem. Phys.*, **1**, 515 (1933).
41. J. D. Bernal and R. H. Fowler, *J. Chem. Phys.*, **1**, 538 (1933).
42. A. M. Beronius and P. Beronius, *Z. Phys. Chem.*, (Frankfurt am Main) **79**, 83 (1972).
43. S. L. Bertha and G. R. Choppin, *Inorg. Chem.*, **8**, 613 (1969).
44. R. H. Betts and A. N. McKenzie, *Can. J. Chem.*, **30**, 146 (1952).
45. N. Bjerrum and L. Zeichmeister, *Chem. Ber.*, **56B**, 894 (1923).
46. V. V. Blokhin, V. F. Anufrienko, Yu. A. Makashev, and V. E. Mironov, *Russ. J. Phys. Chem.*, **45**(7), 1062(1860) (1971).

47. J. O'M. Bockris, *Quart. Rev. Chem. Soc.,* **3**, 173 (1949).
48. J. O'M. Bockris and B. E. Conway, *Modern Aspects of Electrochemistry,* Butterworths, London, 1954, p. 67.
49. J. O'M. Bockris and A. K. N. Reddy, *Modern Electrochemistry,* Vol. 1, Plenum, New York, 1970.
50. J. O'M. Bockris and P. P. S. Saluja, *J. Phys. Chem.,* **76**, 2298 (1972).
51. P. S. Bogoyovlenski, *Dokl. Akad. Nauk. SSR,* **101**, 185 (1955).
52. P. S. Bogoyovlenski, *Zh. Fiz. Khim.,* **32**, 2035 (1958).
53. M. Boudart, *J. Am. Chem. Soc.,* **74**, 1531, 3556 (1952).
54. F. Bourion and O. Hun, *C. R. Acad. Sci., Paris,* **198**, 1921 (1934).
55. F. Bourion and O. Hun, *C. R. Acad. Sci., Paris,* **202**, 2149 (1936).
56. F. Bourion and E. Rouyer, *C. R. Acad. Sci., Paris,* **188**, 626 (1929).
57. F. Bourion and E. Rouyer, *C. R. Acad. Sci., Paris,* **196**, 1111 (1933).
58. F. Bourion and E. Rouyer, *C. R. Acad. Sci., Paris,* **197**, 52 (1933).
59. F. Bourion and E. Rouyer, *C. R. Acad. Sci., Paris,* **198**, 1944 (1934).
60. F. Bourion and E. Rouyer, *C. R. Acad. Sci., Paris,* **198**, 1490 (1934).
61. F. Bourion and E. Rouyer, *C. R. Acad. Sci., Paris,* **201**, 65 (1935).
62. G. W. Brady, *J. Chem. Phys.,* **28**, 464 (1958).
63. G. W. Brady and J. T. Krause, *J. Chem. Phys.,* **27**, 304 (1957).
64. L. Brewer, L. A. Bromley, P. W. Gillis, and N. L. Lafgren, *Chemistry and Metallurgy of Miscellaneous Materials: Thermodynamics,* L. L. Quill, Ed., McGraw-Hill, New York, 1950, p. 165.
65. P. W. Bridgman, *J. Chem. Phys.,* **3**, 597 (1935).
66. G. Brink, *Spectrochim. Acta,* **28A(6)**, 1151 (1972).
67. S. Broersma, *J. Chem. Phys.,* **28**, 1158 (1958).
68. H. C. Brown, L. P. Eddy, and R. Wong, *J. Am. Chem. Soc.,* **75**, 6275 (1953).
69. H. C. Brown and W. J. Wallis, *J. Am. Chem. Soc.,* **72**, 6265 (1953).
70. H. C. Brown and W. J. Wallis, *J. Am. Chem. Soc.,* **75**, 6279 (1953).
71. J. Bucher, R. M. Diamond, and B. Chu, *J. Phys. Chem.,* **76**, 2459 (1972).
72. G. Buckbock, *Z. Phys. Chem.,* **55**, 563 (1906).
73. A. D. Buckingham, *Discuss. Faraday Soc.,* **24**, 151 (1957).
74. R. N. Bulter, E. A. Phillpott, and M. C. R. Symons, *Chem. Commun.,* **1968**, 371.
75. J. Burgess and M. C. R. Symons, *Quart. Rev. Chem. Soc.,* **3**, 276 (1968).
76. B. P. Burylev, *Zh. Fiz. Khim.,* **41**, 676 (1967).
77. M. N. Buslaeva and O. Ya. Samoilov, *Zh. Strukt. Khim.,* **2**, 551 (1961); **4**, 682 (1963).
78. E. Butler, *Z. Chem.,* **5**, 199 (1967).
79. M. Cardier, *C. R. Acad. Sci., Paris,* **214**, 707 (1942).
80. B. Case and R. Parsons, *Trans. Faraday Soc.,* **63**, 1224 (1967).
81. J. Chanu, *J. Chim. Phys.,* **55**, 733, 743, (1958).
82. S. S. Chin, *Zh. Fiz. Khim.,* **26**, 960 (1952).
83. S. S. Chin, *Zh. Fiz. Khim.,* **26**, 1125 (1952).
84. G. R. Choppin and A. J. Graffeo, *Inorg. Chem.,* **4**, 1254 (1965).

85. J. R. Coetzee and W. R. Sharpe, *J. Soluton Chem.*, **1**(1), 77 (1972).
86. R. E. Connick and D. N. Fiat, *J. Chem. Phys.*, **39**, 1349 (1963).
87. B. E. Conway and J. O'M. Bockris, *Modern Aspects of Electrochemistry*, Vols. 1 and 2, Academic, New York, 1954.
88. B. E. Conway and J. O'M. Bockris, *Modern Aspects of Electrochemistry*, Vol. 3, Butterworths, London, 1964.
89. B. E. Conway, in *Modern Aspects of Electrochemistry*, J. O'M Bockris and B. E. Conway, Eds., Butterworths, London 1964, Chap. 2.
90. B. E. Conway, R. E. Verrall, and J. E. Desnoyers, *Z. Phys. Chem.*, (Leipzig), **230**, 157 (1965).
91. V. B. Cory, *Phys. Rev.*, **64**, 350 (1943).
92. A. M. Couture and K. J. Laidler, *Can. J. Chem.*, **35**, 202 (1957).
93. R. W. Creekmore and C. N. Reilley, *J. Phys. Chem.*, **73**, 1563 (1969).
94. E. Darmois, *J. Phys. Radium*, **2**, 2 (1941).
95. E. Darmois, *Chem. Zentr.*, **[II]**, 259 (1942).
96. E. Darmois *J. Phys. Radium*, **8**, 1 (1944).
97. E. Darmois, *J. Chim. Phys.*, **43**, 1 (1946).
98. P. K. Das, *J. Indian Chem. Soc.*, **36**, 613 (1959).
99. P. B. Das, P. K. Das, and D. Patrick, *J. Indian Chem. Soc.*, **36**, 761 (1959).
100. K. K. Deb, J. E. Bloor, and T. C. Cole, *Inorg. Chem.*, **11**(10), 2428 (1972).
101. P. Debye and F. W. Sears, *Proc. Nat. Acad. Sci. U.S.*, **18**, 409 (1932).
102. D. D. DeFord and D. N. Hume, *J. Am. Chem. Soc.*, **73**, 5321, 5323 (1951).
103. R. B. Dennison and B. D. Steele, *Phil. Trans.*, **205**, 449 (1906); *Z. Phys. Chem.*, **57**, 110 (1906).
104. M. Depas, J. J. Leventhal, and L. Friedman, *J. Chem. Phys.*, **49**, 5543 (1968).
105. P. H. Dike, *Rev. Sci. Instrum.*, **2**, 379 (1931).
106. S. I. Drakin, R. Kh. Kurmalieva, and M. Kh. Karapet'yants, *Reakt. Sposobnost Org. Soedin.*, **2**, 267 (1965).
107. A. K. Dzhabarov and A. I. Agaev, *Uch. Zap. Azerb. Goz. Univ. Ser. Fiz. Mat. Khim. Nauk.*, no. 2, 77 (1960).
108. D. D. Eley and M. G. Evans, *Trans. Faraday Soc.*, **34**, 1093 (1938).
109. H. J. Emeleus, and J. S. Anderson, *Modern Aspects of Inorganic Chemistry*, Van Nostrand, New York, 1943.
110. M. I. Emel'yanow and A. Sh. Agishev, *Zh. Strukt. Khim.*, **6**, 909 (1965).
111. A. Eucken, *Z. Elektrochem.*, **52**, 6 (1948).
112. A. Eucken and M. Eigen, *Z. Elektrochem.*, **55**, 343 (1951).
113. M. B. Everhard and P. M. Gross, Jr., *U.S. Dep. Commer. Off. Tech. Ser., P. B. Rep.*, **1145**, 174 (1959).
114. B. P. Fabricand, S. S. Goldberg, R. Leiferaard, and S. G. Unger, *Mol. Phys.*, **7**, 425 (1964).
115. K. Fajans, *Ber. Deut. Phys. Ges.*, **21**, 549, 709 (1919).
116. D. Feakins, *Phys. Chem. Processes Mixed Aqueous Solvents*, **1967**, 71.

117. D. Feakins and C. M. French, *J. Chem. Soc.,* **1957**, 2581.

118. D. Feakins and R. P. T. Tompkins, *J. Chem. Soc., A,* **1967**, 1458.

119. H. M. Feder and H. Taube, *J. Chem. Phys.,* **20**, 1335 (1952).

120. D. Fiat and R. E. Connick, *J. Am. Chem. Soc.,* **90**, 608 (1968).

121. D. Fiat and R. E. Connick, *J. Am. Chem. Soc.,* **88**, 4754 (1966).

122. C. Finbak, *Tidsskr. Kjemi, Bergv. Met.,* **4**, 77 (1945).

123. A. Finch, P. J. Gardner, and C. J. Steadman, *J. Phys. Chem.,* **71**, 2996 (1967).

124. F. Fittipaldi, *Nature,* **210**, 835 (1966).

125. R. Flatt and A. Jordon, *Helv. Chim. Acta,* **16**, 388 (1927).

126. R. H. Flowers, R. J. Gillespie, and E. A. Robinson, *J. Chem. Soc.,* **1960**, 4327.

127. R. H. Flowers, R. J. Gillespie, and E. A. Robinson, *J. Chem. Soc.,* **1960**, 845.

128. R. H. Flowers, R. J. Gillespie, and E. A. Robinson, *J. Chem. Soc.,* **1960**, 848.

129. H. S. Frank, Ref. 319, p. 53.

130. H. S. Frank and W. Y. Wen, *Discuss. Faraday Soc.,* **24**, 133 (1957).

131. E. G. Franklin and H. P. Cady, *J. Am. Chem. Soc.,* **26**, 499 (1904).

132. A. Fratiello, D. D. Davis, S. Peak, and R. E. Schuster, *J. Phys. Chem.,* **74**, 3730 (1970).

133. A. Fratiello, V. Kubo, R. E. Lee, S. Peak, and R. E. Schuster, *J. Inorg. Nucl. Chem.,* **32**, 3114 (1970).

134. A. Fratiello, V. Kubo, R. E. Lee, and R. E. Schuster, *J. Phys. Chem.,* **74**, 3726 (1970).

135. A. Fratiello, V. Kubo, and R. E. Schuster, *Inorg. Chem.,* **10**, 744 (1971).

136. A. Fratiello, R. E. Lee, and R. E. Schuster, *Inorg. Chem.,* **9**, 82 (1970).

137. A. Fratiello, R. E. Lee, and R. E. Schuster, *Inorg. Chem.,* **9**, 391 (1970).

138. A. Fratiello, R. E. Lee, V. M. Nishida, and R. E. Schuster, *Chem. Commun.,* **1968**, 173.

139. A. Fratiello, R. E. Lee, V. M. Nishida, and R. E. Schuster, *J. Chem. Phys.,* **48**, 3705 (1968).

140. A. Fratiello, R. E. Lee, V. M. Nishida, and R. E. Schuster, *J. Chem. Phys.,* **50**, 3624 (1969).

141. A. Fratiello and R. E. Schuster, *Tetrahedron Lett.,* **1967**, 4641.

142. A. Fratiello and R. E. Schuster, *J. Chem. Educ.,* **45**, 91 (1968).

143. N. J. Friedman and R. A. Plane, *Inorg. Chem.,* **2**, 11 (1963).

144. W. S. Fyfe, *J. Chem. Soc.,* **1952**, 2023.

145. E. N. Gapon, *Z. Anorg. Allg. Chem.,* **168**, 125 (1927).

146. E. N. Gapon, *J. Russ. Phys. Chem. Soc.,* **60**, 237 (1928).

147. F. J. Garrick, *Phil. Mag.,* **9**, 131 (1930); **10**, 76 (1930).

148. L. Genow, *Bulg. Akad. Nauk Izv. Khim. Inst.,* **4**, 251 (1951).

149. L. Genow, *C. R. Akad. Bulg. Sci.,* **9**(2), 23 (1956).

150. R. J. Gillespie, J. V. Oubridge, and C. Solomons, *J. Chem. Soc.,* **1957**, 1804.

151. R. J. Gillespie and S. Wasif, *J. Chem. Soc.*, **1953**, 215.
152. R. J. Gillespie and R. F. M. White, *Can. J. Chem.*, **38**, 1371 (1960).
153. G. E. Glass, W. B. Schwabacher, and R. S. Tobias, *Inorg. Chem.*, **7**, 2471 (1968).
154. G. E. Glass and R. S. Tobias, *J. Am. Chem. Soc.*, **89**, 6371 (1967).
155. E. Glueckauf, *Trans. Faraday Soc.*, **51**, 1235 (1955).
156. E. Glueckauf, *Trans. Faraday Soc.*, **60**, 1639 (1964).
157. E. Glueckauf, *Trans. Faraday Soc.*, **61**, 914 (1965).
158. E. Glueckauf, *Trans. Faraday Soc.*, **64**, 2423 (1968).
159. E. Glueckauf and G. P. Kitt, *Proc. Roy. Soc.*, **A228**, 322 (1955).
160. Ta. P. Gokhshtein, *Tr. Kom. Anal. Khim. Akad. Nauk SSSR, Otd. Khim. Nauk*, **2**(5), 5 (1949).
161. R. Gopel and O. N. Bhatnagar, *J. Phys. Chem.*, **69**, 2382 (1965).
162. A. R. Gordon, *Ann. Rev. Phys. Chem.*, **1**, 59 (1950).
163. S. Goto, *Bull. Chem. Soc. Jap.*, **37**, 1685 (1964).
164. R. G. Griffin, E. S. Amis, and J. O. Wear, *J. Inorg. Nucl. Chem.*, **28**, 543 (1966).
165. H. G. Grimm, *Hand-Jahrb. Chem. Phys.*, **27**, 518 (1927).
166. Yu. V. Gurikov, *Teor. Eksp. Khim.*, **4**, 61 (1968).
167. R. W. Gurney, *Ionic Processes in Solution*, Dover, New York, 1962.
168. G. H. Haggis, J. B. Hasted, and T. J. Buchanan, *J. Chem. Phys.*, **20**, 1452 (1952).
169. H. F. Halliwell and S. C. Nyburg, *Trans. Faraday Soc.*, **59**, 1126 (1963).
170. G. Hander and H. Mörk, *J. Phys. Chem.* (Leipzig), **A190**, 81 (1942); *Chem. Zentr.*, **1**, 2098 (1942).
171. H. S. Harned and D. M. French, *Ann. N.Y. Acad. Sci.*, **44**, 267 (1945).
172. H. S. Harned, A. S. Keston, and J. G. Donelson, *J. Am. Chem. Soc.*, **58**, 989 (1936).
173. H. S. Harned and R. L. Nuttall, *J. Am. Chem. Soc.*, **69**, 736 (1947).
174. H. S. Harned and B. B. Owen, *The Physical Chemistry of Electrolyte Solutions*, 2nd ed., Reinhold, New York, 1956, pp. 86-90.
175. J. B. Hasted, D. M. Ritson, and C. H. Collie, *J. Chem. Phys.*, **16**, 1 (1948).
176. J. C. Hindman, *J. Chem. Phys.*, **36**, 1000 (1962).
177. J. F. Hinton, L. S. McDowell, and E. S. Amis, *Chem. Commun.*, **1966**, 776.
178. J. F. Hinton and E. S. Amis, *Chem. Rev.;* **71**(6), 627, 1971.
179. J. F. Hinton, E. S. Amis, and W. Mettetal, *Spectrochim. Acta*, **A25**, 119 (1969).
180. W. Hittorf, *Z. Phys. Chem.*, **39**, 612 (1901).
181. W. Hittorf, *Z. Phys. Chem.*, **43**, 239 (1903).
182. A. M. Hogg, R. M. Haynes, and P. Kebarle, *J. Am. Chem. Soc.*, **88**, 28 (1966).
183. A. M. Hogg and P. Kebarle, *J. Chem. Phys.*, **43**, 449 (1965).
184. J. Horvichi and G. Okamoto, *Sci. Papers Inst. Phys. Chim. Res., Tokyo*, **28**, 231 (1936).
185. J. C. Hubbard and A. L. Loomis, *Phil. Mag.*, **5**, 1177 (1928).

186. R. F. Hudson and B. Saville, *J. Chem. Soc.,* **1955,** 4114.
187. O. Hun, *C. R. Acad. Sci., Paris,* **198,** 740 (1934).
188. O. Hun, *C. R. Acad. Sci., Paris,* **201,** 547 (1935).
189. O. Hun, *C, R. Acad. Sci., Paris,* **202,** 1779 (1936).
190. J. P. Hunt and H. Taube, *J. Chem. Phys.,* **18,** 757 (1950).
191. J. P. Hunt and H. Taube, *J. Chem. Phys.,* **19,** 602 (1951).
192. A. Hunyar, *J. Am. Chem. Soc.,* **71,** 3552 (1949).
193. A. Indelli and E. S. Amis, *J. Am. Chem. Soc.,* **82,** 332 (1960).
194. V. P. Il'insku and A. T. Uverskaya, *Sb. Tr. Inst. Prikl. Khim.,* **112,** 41 (1958).
195. D. Ilkovic, *Collect. Czech. Chem. Commun.,* **6,** 498 (1934).
196. D. J. Ives and G. L. Jans, *Reference Electrodes,* Academic, New York, 1961.
197. K. Jablezynski and A. Balczewski, *Rocz. Chem.,* **12,** 880 (1932).
198. L. G. Jackopin and E. Yeager, *J. Phys. Chem.,* **70,** 313 (1966).
199. J. A. Jackson, J. F. Lemons, and H. Taube, *J. Chem. Phys.,* **32,** 553 (1960).
200. J. C. Jayne and E. L. King, *J. Am. Chem. Soc.,* **86,** 3989 (1964).
201. D. Jelenkow and L. Genow, *Bulg. Akad. Wiss., Nachr. Chem. Inst.,* **3,** 67 (1955).
202. D. Jelenkow and L. Genow, *Ber. Bulg. Akad. Wiss.,* **7,** 37 (1954).
203. G. Jones and G. M. Bollinger, *J. Am. Chem. Soc.,* **53,** 411 (1931).
204. G. Jones and M. Dole, *J. Am. Chem. Soc.,* **51,** 2950 (1929).
205. G. Jones and R. C. Josephs, *J. Am. Chem. Soc.,* **50,** 1049 (1928).
206. A. F. Kapustinski, I. I. Lipilina, and O. Ya. Samoilov, *Zh. Fiz. Khim.,* **30,** 896 (1956).
207. A. F. Kapustinski and O. Ya. Samoilov, *Zh. Fiz. Khim.,* **26,** 918 (1952).
208. A. F. Kapustinski and O. Ya. Samoilov, *Izv. Akad. Nauk SSSR,* **1950,** 337.
209. M. V. Kaulgud, *Z. Phys. Chem.* (Frankfurt am Main), **47,** 24 (1965).
210. A. L. Kaz'min, *J. Phys. Chem. USSR,* **11,** 585 (1938).
211. P. Kebarle, *Advan. Chem. Ser.,* **72,** 24 (1968).
212. P. Kebarle, R. M. Haynes, and S. K. Searles, *Advan. Chem. Ser.,* **58,** 210 (1966).
213. P. Kebarle and A. M. Hogg, *J. Chem. Phys.,* **42,** 798 (1965).
214. P. Kebarle, S. K. Searles, A. Zolla, J. Scarborough, and M. Arshadi, *J. Am. Chem. Soc.,* **89,** 6393 (1967).
215. D. W. Kemp and E. L. King, *J. Am. Chem. Soc.,* **89,** 3433 (1967).
216. T. Kenjo and R. M. Diamond, *J. Phys. Chem.,* **76,** 2454 (1972).
217. N. E. Khomutov, *Zh. Fiz. Khim.,* **41,** 2258 (1967).
218. P. S. Knapp, R. O. Waite, and E. R. Malinowski, *J. Chem. Phys.,* **49,** 5459 (1968).
219. F. Kohlrausch and L. Holbron, *Das Leitvermongen der Elektroyte,* Teubner, Leipzig and Berlin, 1916, p. 241.
220. E. Koizumi and H. Miyamoto, *Bull. Chem. Soc. Jap.,* **29,** 250 (1956).
221. Y. Kondo, T. Goto, I. Suo, and N. Tokura, *Bull. Chem. Soc. Jap.,* **39,** 1230 (1966).

222. B. P. Konstantinov and V. P. Troshim, *Zh. Prikl. Khim.,* **36**, 449 (1963).
223. G. Kortum and J. O'M. Bockris, *Textbook of Electrochemistry,* Vol. 1, Elsevier, New York, 1951, p. 136.
224. G. Kortüm and J. O'M. Bockris, *Textbook of Electrochemistry,* Vol. 2, Elsevier, New York, 1951, pp. 672-675.
225. P. P. Kosakevich, *Z. Phys. Chem.,* **143**, 216 (1929).
226. N. A. Kostrumina and T. V. Ternovaya, *Zh. Neorg. Chem.,* **12**, 700 (1967).
227. G. A. Krestov, *Zh. Strukt. Khim.,* **3**, 137 (1962).
228. G. A. Krestov, *Zh. Struk. Khim.,* **3**, 402 (1962).
229. G. A. Krestov and V. K. Abrosimov, *Zh. Strukt. Khim.,* **8**, 522 (1967).
230. G. A. Krestov and V. I. Klophov, *Zh. Strukt. Khim.,* **4**, 502 (1963).
231. G. A. Krestov, *Zh. Strukt. Khim.,* **3**, 402 (1962); G. A. Krestov and V. I. Klophov, *Zh. Strukt. Khim.,* **4**, 502 (1963).
232. C. V. Krishnan and H. L. Friedman, *J. Phys. Chem.,* **74**, 2356 (1970).
233. R. F. Kruh and C. L. Standley, *Inorg. Chem.,* **1**, 941 (1962).
234. R. J. Kula, D. L. Rabenstein, and G. H. Reed, *Anal. Chem.,* **37**, 1783 (1965).
235. K. J. Laidler and C. Pegis, *Proc. Roy. Soc.,* **A241**, 80 (1957).
236. P. C. Lammers and K. Braunstein, *J. Phys. Chem.,* **71**, 2626 (1967).
237. F. W. Lampe, F. H. Field, and J. L. Franklin, *J. Am. Chem. Soc.,* **79**, 6132 (1957).
238. E. Lang and A. Eichler, *Z. Phys. Chem.,* **129**, 285 (1937).
239. W. M. Latimer, *Chem. Rev.,* **18**, 349 (1936).
240. W. M. Latimer, *J. Am. Chem. Soc.,* **73**, 1480 (1951).
241. W. M. Latimer and R. M. Buffington, *J. Am. Chem. Soc.,* **48**, 2297 (1926).
242. W. M. Latimer, P. W. Schutz, and J. F. G. Hicks, Jr., *J. Chem. Phys.,* **2**, 82 (1934).
243. V. I. Lebed and V. V. Alksandrov, *Elektrochimiya,* **1**, 1359 (1965).
244. W. H. Lee, in *The Chemistry of Nonaqueous Solvents,* J. J. Lagowski, Ed., Academic, New York, 1967.
245. W. H. Lee, in *The Chemistry of Nonaqueous Solvents,* Vol. II, J. J. Lagowski, Ed., Academic, New York, 1967, p. 166.
246. W. B. Lewis, J. A. Jackson, J. F. Lemons, and H. Taube, *J. Chem. Phys.,* **36**, 694 (1962).
247. I. I. Lipilina, *Dokl. Akad. Nauk SSSR,* **102**, 525 (1955).
248. I. I. Lipilina and O. Ya. Samoilov, *Dokl. Akad. Nauk SSSR,* **98**, 99 (1954).
249. W. A. Logachev and V. I. Dulova, *Russ. J. Phys. Chem.,* **45**(9), 1311(2325) (1971).
250. Z. Luz and S. Meiboom, *J. Chem. Phys.,* **40**, 1058 (1964).
251. Z. Luz and R. G. Shulman, *J. Chem. Phys.,* **43**, 3750 (1965).
252. D. W. McCall, D. C. Douglass and E. W. Anderson, *Ber. Bunsenges, Phys. Chem.,* **67**, 336 (1963).
253. C. C. McDonald and W. D. Phillips, *J. Am. Chem. Soc.,* **85**, 3736 (1963).
254. D. MacGillavry and E. K. Rideal, *Rec. Trav. Chim. Pays-Bas.,* **56**, 1013 (1937).

255. D. A. McInnes, *The Principles of Electrochemistry*, Reinhold New York, 1939, p. 93.

256. D. A. McInnes and T. B. Brighton, *J. Am. Chem. Soc.*, **47**, 994 (1925).

257. D. A. McInnes and M. O. Dayhoff, *J. Am. Chem. Soc.*, **74**, 1017 (1952).

258. D. A. McInnes and E. R. Smith, *J. Am. Chem. Soc.*, **45**, 2246 (1923).

259. J. M. McIntyne and E. S. Amis, *J. Chem. Eng. Data*, **13**(3), 371 (1968).

260. D. L. McMasters, J. C. DiRaimondo, L. M. Jones, R. P. Lindley, and E. W. Zeltmann, *J. Phys. Chem.*, **66**, 249 (1962).

261. E. R. Malinowski and P. S. Knapp, *J. Chem. Phys.*, **48**, 4989 (1968).

262. E. R. Malinowski, P. S. Knapp, and B. Feuer, *J. Chem. Phys.*, **45**, 4274 (1966).

263. E. R. Malinowski, P. S. Knapp, and B. Feuer, *J. Chem. Phys.*, **47**, 347 (1967).

264. M. M. Mann, A. Hustralid, and J. T. Late, *Phys. Rev.*, **58**, 340 (1940).

265. M. Marezio, H. A. Pettinger, and W. H. Zachariasen, *Acta Crystallogr.*, **14**, 234 (1961).

266. B. F. Markov, *Ukr. Khim. Zh.*, **23**, 706 (1957).

267. D. M. Matthews, J. O. Wear, and E. S. Amis, *J. Inorg. Nucl. Chem.*, **13**, 298 (1960).

268. N. A. Matwiyoff, *Inorg. Chem.*, **5**, 788 (1966).

269. N. A. Matwiyoff and P. E. Darley, *J. Phys. Chem.*, **72**, 2659 (1968).

270. N. A. Matwiyoff, P. E. Darley, and W. G. Movius, *Inorg. Chem.*, **7**, 2173 (1968).

271. N. A. Matwiyoff, and S. V. Hooker, *Inorg. Chem.*, **6**, 1127 (1967).

272. N. A. Matwiyoff and W. G. Movius, *J. Am. Chem. Soc.*, **89**, 6077 (1967).

273. N. A. Matwiyoff and H. Taube, *J. Am. Chem. Soc.*, **90**, 2796 (1968).

274. I. V. Matyash, A. I. Toryanik, and V. I. Vashkichev, *Zh. Strukt. Khim.*, **5**, 777 (1964).

275. C. T. Maynihan and A. Fratiello, *J. Am. Chem. Soc.*, **89**, 5546 (1967).

276. S. Meiboom, *J. Chem. Phys.*, **46**, 410 (1967).

277. S. L. Melton and E. S. Amis, *J. Chem. Eng. Data*, **13**(3), 429 (1968).

278. P. K. Migal and N. Kh. Grinberg, *Zh. Neorgan. Khim.*, **6**, 727 (1961).

279. P. K. Migal and N. Kh. Grinberg, *Zh. Neorgan. Khim.*, **7**, 527 (1962).

280. P. K. Migal and N. Kh. Grinberg, *Zh. Neorgan. Khim.*, **7**, 531 (1962).

281. W. A. Millen and D. W. Watts, *J. Am. Chem. Soc.*, **89**, 6051 (1967).

282. V. E. Mironov, Yu. A. Makashev, and V. V. Blokhim, *Russ. J. Phys. Chem.*, **45**(4), 431(772) (1971).

283. K. P. Mischenko, *Zh. Fiz. Khim.*, **26**, 1736 (1952).

284. K. P. Mischenko and K. S. Pominov, *Zh. Fiz. Khim.*, **31**, 2026 (1957).

285. K. P. Mischenko and V. V. Sokolov, *Zh. Strukt. Khim.*, **4**, 184 (1963).

286. H. Miyamoto, *Nippon Kagaku Zasshi*, **80**, 4, 825 (1959); **81**, 54, 1376 (1960).

287. M. D. Monica and U. Lamanna, *J. Phys. Chem.*, **72**, 4329 (1968); M. D. Monica, U. Lamanna, and L. Senatore, *J. Phys. Chem.*, **72**, 2124 (1968).

288. J. L. R. Morgan and C. W. Kanolt, *J. Am. Chem. Soc.*, **28**, 572 (1906).

289. J. Morgan and B. E. Warren, *J. Chem. Phys.*, **6**, 666 (1938).

290. W. G. Movius and N. A. Matwiyoff, *Inorg. Chem.*, **6**, 847 (1967).

291. E. L. Muetterties, *J. Am. Chem. Soc.*, **88**, 305 (1966).

292. P. Mukerjee, *J. Phys. Chem.*, **65**, 740, 744 (1961).

293. N. Muller, and R. C. Reiter, *J. Chem. Phys.*, **42**, 3265 (1965).

294. M. S. Murty, *Indian J. Pure Appl. Phys.*, **3**(5), 156 (1965).

295. S. Nakamura and S. Meiboom, *J. Am. Chem. Soc.*, **89**, 1765 (1967).

296. W. Nernst and P. Drude, *Z. Phys. Chem.*, **15**, 79 (1894).

297. W. Nernst, C. C. Gerard, and E. Oppermann, *Nachr. Akad. Kgl. Ges. Wiss. Goettingen,* **56**, 86 (1900).

298. E. R. Nightingale, Jr., *J. Phys. Chem.*, **63**, 1381 (1929).

299. J. H. Northrop and M. L. Anson, *J. Gen. Physiol.*, **12**, 543 (1929).

300. R. M. Noyes, *J. Am. Chem. Soc.*, **84**, 513 (1962); **86**, 971 (1964).

301. R. M. Noyes, *J. Phys. Chem.*, **69**, 3183 (1965).

302. L. Onsager and R. M. Fuoss, *J. Phys. Chem.*, **36**, 2689 (1932).

303. T. G. Owe Berg, *Ann. N.Y. Acad. Sci.*, **125**, 298 (1965).

304. P. Oyari, *Russ. J. Phys. Chem.*, **45**(4), 546(976) (1971).

305. J. Padova, *Bull. Res. Counc. Isr. Sec. A,* **10**, 63 (1961).

306. J. Padova, *J. Chem. Phys.*, **39**, 1552 (1963).

307. J. Padova, *J. Phys. Chem.*, **71**, 2347 (1967).

308. J. Padova, *J. Phys. Chem.*, **72**, 796 (1968).

309. A. G. Passynski, *Acta Physicochim. USSR,* **8**, 385 (1938).

310. L. Pauling, *J. Am. Chem. Soc.*, **49**, 765 (1927).

311. L. Pauling, *The Nature of the Chemical Bond,* 3rd ed., Cornell University Press, Ithaca, N.Y., 1960.

312. L. A. Petrenko and Yu. A. Petrenko, *Zh. Strukt. Khim.*, **8**, 212 (1967).

313. R. A. Pinkowitz and T. J. Swift, *J. Chem. Phys.*, **54**, 2858 (1971).

314. W. Plumb and G. M. Harris, *Inorg. Chem.*, **3**, 542 (1964).

315. S. Poczopko, *Rocz. Chem.*, **34**, 1245, 1255 (1960); **36**, 103, 111, 295, 303 (1962).

316. E. Pogany, *Magy. Kem. Foly.*, **48**, 85 (1942).

317. I. S. Povinov, *Zh. Fiz. Chem.*, **31**, 1926 (1957).

318. J. A. Prins, *J. Chem. Phys.*, **3**, 72 (1935).

319. J. E. Prue, *Discuss. Int. Sympo. Electrochem. Soc., Toronto, 1964,* **1966**, 163.

320. M. Randall and F. D. Rossini, *J. Am. Chem. Soc.*, **51**, 323 (1929).

321. C. S. Rao, *Curr. Sci.,* **4**, 649 (1936); *Phil. Mag.,* **24**, 87 (1973).

322. A. K. Reikhardt, *Russ. J. Phys. Chem.*, **45**(7), 1052(1845) (1971).

323. H. Remy, *Z. Phys. Chem.*, **118**, 161 (1925); **124**, 41 394 (1926).

324. J. Reuben and D. Fiat, *Inorg. Chem.*, **6**, 579 (1967).

325. E. Riecke, *Z. Phys. Chem.*, **6**, 564 (1890).

326. R. A. Robinson and C. L. Chia, *J. Am. Chem. Soc.*, **74**, 2776 (1952).

327. R. A. Robinson and R. H. Stokes, *Trans. Faraday Soc.*, **45**, 612 (1949).

328. R. A. Robinson and R. H. Stokes, *Electrolyte Solutions,* Butterworths, London, 1955.

329. G. P. Roshchina, A. S. Kaurova, and I. D. Kosheleva, *Zh. Strukt. Khim.*, **9**, 3 (1968).

330. E. Rouyer, *C. R. Acad. Sci., Paris*, **198**, 742 (1934).
331. E. Rouyer, *C. R. Acad. Sci., Paris*, **198**, 1156 (1934).
332. E. Rouyer and O. Hun, *C. R. Acad. Sci., Paris*, **196**, 1015 (1933).
333. A. C. Rutenberg and H. Taube, *J. Chem. Phys.*, **20**, 825 (1952).
334. A. J. Rutgers, W. Rigole, and V. Hendrix, *Chem. Weekbl.*, **59**, 33 (1963).
335. B. H. van Ruyven, *Chem. Weekbl.*, **54**, 636 (1958).
336. A. I. Ryss and I. V. Radchenko, *Zh. Strukt. Khim.*, **6**, 449 (1965).
337. J. Salvinien and B. Brun, *C. R. Acad. Sci., Paris*, **259**, 565 (1964).
338. O. Ya. Samoilov, *Izv. Akad. Nauk. SSSR, Otd. Khim. Nauk*, **1952**, 398.
339. O. Ya. Samoilov, *Izv. Akad. Nauk. SSSR*, **1952**, 627.
340. O. Ya. Samoilov, *Dokl. Akad. Nauk. SSSR*, **102**, 1173 (1956).
341. O. Ya. Samoilov, *Zh. Fiz. Khim.*, **29**, 1582 (1955).
342. O. Ya. Samoilov, *Izv. Akad. Nauk. SSSR, Otd. Khim. Nauk.*, **1956**, 1415. *Bull. Acad. Sci. USSR, Div. Chem. Sci.*, **1956**, 1449.
343. O. Ya. Samoilov, *Discuss. Faraday Soc.*, **24**, 141, (1957).
344. O. Ya. Samoilov, *Structure of Aqueous Electrolyte Solutions and Hydration of Ions*, Consultants Bureau, New York, 1965.
345. O. Ya. Samoilov, K. Y. Hu, and T. A. Nosova, *Zh. Strukt. Khim.*, **1**, 131 (1960).
346. O. Ya. Samoilov, K. Y. Hu, and T. A. Nosova, *Zh. Strukt. Khim.*, **1**, 404 (1960).
347. O. Ya. Samoilov and G. G. Malenkov, *Zh. Strukt. Khim.*, **8**, 618 (1967).
348. O. Ya. Samoilov and V. A. Yashkichev, *Zh. Strukt. Khim.*, **3**, 334 (1962).
349. T. Sasaki and T. Yasunaga, *Bull. Chem. Soc. Jap.*, **28**, 269 (1955).
350. T. Sasaki, T. Yasunaga, and H. Fujihara, *J. Chem. Soc. Jap., Pure Chem. Sect.*, **73**, 181 (1952).
351. T. Sato and K. Hayashi, *J. Phys. Soc. Jap.*, **15**, 1658 (1960).
352. E. Schaschel and M. C. Day, *J. Am. Chem. Soc.*, **90**, 503 (1968).
353. W. G. Schneider, H. J. Bernstein, and J. A. Pople, *J. Chem. Phys.*, **28**, 601 (1958).
354. H. Schneider and H. Strethlow, *Z. Elektrochem.*, **66**, 309 (1962).
355. R. E. Schuster and A. Fratiello, *J. Chem. Phys.*, **47**, 1554 (1967).
356. V. F. Serveeva, *Izv. Uyssh. Ucheb. Zaved., Khim. Khim. Tekhnol*, **5**, 905, 908 (1962).
357. Yu. Shamoyan and S. A. Yan, *Dokl. Akad. Nauk SSSR*, **152**, 677 (1963).
358. I. M. Shapovalov and I. V. Radchenko, *Zh. Strukt. Khim.*, **12**(5), 769 (1973).
359. I. M. Shapovalov, I. V. Radchenko, and K. M. Lesovitskaya, *Zh. Strukt. Khim.*, **4**, 10 (1963).
360. J. Sherman, *Chem. Rev.*, **11**, 93 (1932).
361. R. G. Shulman, *J. Chem. Phys.*, **29**, 945 (1958).
362. R. G. Shulman and B. J. Wyluda, *J. Chem. Phys.*, **30**, 335 (1959).
363. A. F. Skryshevskii, *Str. Fiz. Svoistva Veshchestv. Zhidk. Sostoyanii*, **1954**, 27; Ref. *Zh. Khim., Abstr.*, **1956**, 32076; *Chem. Abstr.*, **52**, 17910, (1958).
364. G. B. Smith, C. A. Hollingsworth, and D. H. McDaniel, *J. Chem. Educ.*, **38**, 489 (1961).

365. G. Somsen, *Rec. Trav. Chim. Pays-Bas*, **85**, 317, 526 (1966).
366. H. Spandau and C. Spandau, *Z. Phys. Chem.*, **192**, 211 (1942).
367. F. H. Spedding and K. C. Jones, *J. Phys. Chem.*, **70**, 2450 (1966).
368. F. H. Spedding and M. J. Pikal, *J. Phys. Chem.*, **70**, 2430, 2440, (1966).
369. G. L. Starobinets, A. B. Chizhevskaya and T. L. Dubovik, *Vestsi. Akad. Navuk Belarus., SSR, Ser. Khim. Navuk.*, **1965**, 110.
370. J. F. Stephens and G. K. Schweitzer, *Spectrosc. Lett.*, **1**, 373 (1968); **3**, 11 (1970).
371. G. W. Stewart, *J. Chem. Phys.*, **7**, 381 (1939).
372. G. W. Stewart, *J. Chem. Phys.*, **7**, 869 (1939).
373. G. W. Stockton and J. S. Martin, *J. Am. Chem. Soc.*, **94**, 6921 (1972).
374. R. H. Stokes, *Trans. Faraday Soc.*, **44**, 295 (1948).
375. R. H. Stokes, *J. Am. Chem. Soc.*, **72**, 763 (1950).
376. R. H. Stokes and R. Mills, *Viscosities of Electrolytes and Related Properties*, 1st ed, Pergamon, London, 1965.
377. R. H. Stokes and R. A. Robinson, *J. Am. Chem. Soc.*, **70**, 1870 (1948).
378. H. van der Stratten and A. H. W. Aten, Jr., *Rec. Trav. Chim. Pays-Bas*, **73**, 89 (1954).
379. H. Strethlow and H. M. Koepf, *Z. Electrochem.*, **62**, 372 (1958).
380. L. D. Supran and N. Sheppard, *Chem. Commun.*, **1967**, 832.
381. J. Sutton, *Nature*, **160**, 235 (1952).
382. C. G. Swain, R. F. W. Bader, and E. R. Thornton, *Tetrahedron*, **10**, 200 (1960).
383. T. J. Swift, O. G. Fritz, and T. A. Stephenson, *J. Chem. Phys.*, **46**, 406 (1967).
384. T. J. Swift and H. H. Lo, *J. Am. Chem. Soc.*, **89**, 3987 (1967).
385. T. J. Swift and W. G. Sayre, *J. Chem. Phys.*, **44**, 3567 (1966).
386. T. J. Swift and W. G. Sayre, *J. Chem. Phys.*, **46**, 410 (1967).
387. T. J. Swift and G. P. Weinberger, *J. Am. Chem. Soc.*, **90**, 2023 (1968).
388. J. H. Swinehart and H. Taube, *J. Chem. Phys.*, **37**, 1579 (1962).
389. R. W. Taft, Jr., *J. Am. Chem. Soc.*, **82**, 2965 (1960).
390. P. G. Tait, *The Physics and Chemistry of the Voyage of H.M.S. Challenger*, Part IV, 1888.
391. G. Tammann, *Uber die Beziehungen Zwischen den inneren Kräften und Eigenshaften Lösungen*, Voss, Leipzig, 1907, p. 36.
392. K. Tamura and T. Sasaki, *Bull. Chem. Soc. Jap.*, **36**, 975 (1963).
393. H. Taube, *J. Phys. Chem.*, **58**, 523 (1954).
394. E. G. Taylor and C. A. Kraus, *J. Am. Chem. Soc.*, **69**, 1731 (1947).
395. S. Thomas and W. L. Reynolds, *J. Chem. Phys.*, **44**, 3148 (1966).
396. V. P. Troshin, *Electrokhimiya*, **2**, 232 (1966).
397. Ya. I. Tur'yan, *Zh. Fiz. Khim.*, **38**, 1853 (1964).
398. Ya. I. Tur'yan, *Zh. Fiz. Khim.*, **38**, 1690 (1964).
399. H. Ulich, *Trans. Faraday Soc.*, **23**, 388 (1927).
400. H. Ulich, *Z. Elektrochem.*, **36**, 497 (1930).
401. H. Ulich, *Z. Elektrochem.*, **36**, 505 (1930).
402. H. Ulich, *Z. Phys. Chem.*, **168**, 141 (1934).

403. C. L. van P. Van Eck, H. Mendel, and W. Boog, *Discuss. Faraday Soc.*, **24**, 200 (1957).

404. V. P. Vasilev, E. K. Zolotarev, A. F. Kapustykoku, K. P. Mischenko, P. A. Podgornaya, and K. B. Yakimirshii, *Russ. J. Phys. Chem.*, **34**, 1763 (1960).

405. E. J. W. Verwey, *Rec. Trav. Chim. Pays-Bas*, **61**, 127 (1942).

406. W. Ves Childs and E. S. Amis, *J. Inorg. Nucl. Chem.*, **16**, 114 (1960).

407. A. A. Vleck, *Collect. Czech. Chem. Commun.*, **16**, 230 (1951).

408. V. O. Vodovenko, I. G. Slglobova, and D. N. Suglobov, *Radiokhimya*, **1**, 637 (1959).

409. F. J. Vogrin, P. S. Knapp, W. L. Flint, A. Anton, G. Highberger, and E. R. Malinowski, *J. Chem. Phys.*, **54**, 178 (1971).

410. Y. Wada, S. Shimbo, M. Oda, and J. Nagumo, *Oyo Buturi*, **17**, 257 (1948).

411. P. Walden and E. J. Birr, *Z. Phys. Chem.*, **A153**, 1 (1931).

412. E. W. Washburn, *J. Am. Chem. Soc.*, **31**, 322 (1909); *Z. Phys. Chem.*, **68**, 513 (1909).

413. E. W. Washburn, *Principles of Physical Chemistry*, McGraw-Hill, New York, 1921, p. 276.

414. E. W. Washburn and E. B. Millard, *J. Am. Chem. Soc.*, **37**, 694 (1915).

415. R. G. Wawro and T. J. Swift, *J. Am. Chem. Soc.*, **90**, 2792 (1968).

416. J. O. Wear and E. S. Amis, *J. Inorg. Nucl. Chem.*, **24**, 903 (1962).

417. J. O. Wear, J. T. Curtis, Jr., and E. S. Amis, *J. Inorg. Nucl. Chem.*, **24**, 93 (1962).

418. J. O. Wear, C. V. McNully, and E. S. Amis, *J. Inorg. Nucl. Chem.*, **18**, 48 (1961).

419. J. O. Wear, C. V. McNully, and E. S. Amis, *J. Inorg. Nucl. Chem.*, **19**, 278 (1961).

420. H. E. Weaver, B. M. Tolbert, and R. C. LaForce, *J. Chem. Phys.*, **23**, 1956 (1955).

421. T. J. Webb, *J. Am. Chem. Soc.*, **48**, 2589 (1926).

422. D. Wertz, R. M. Lawrence, and R. F. Kruh, *J. Chem. Phys.*, **43**, 2163 (1965).

423. W. Weyle, *Beih. Z. Ver. Deut. Chem.*, No. **18** (1935); *Angew. Chem.*, **48**, 573 (1935).

424. E. Whalley, *J. Chem. Phys.*, **38**, 1400 (1963).

425. E. Wicke, M. Eigen, and Th. Ackermann, *Z. Phys. Chem.* (Frankfurt am Main), **1**, 340 (1954).

426. T. Yasunaga, *J. Chem. Soc. Jap., Pure Chem. Sect.*, **72**, 87 (1951).

427. T. Yasunaga and T. Sasaki, *J. Chem. Soc. Jap., Pure Chem. Sect.*, **72**, 89 (1951).

428. T. Yasunaga, *J. Chem. Soc. Jap., Pure Chem. Sect.*, **72**, 87 (1951).

429. P. A. Zagorets, V. I. Ermakov, and A. P. Granow, *Zh. Fiz. Khim.*, **37**, 1167 (1963).

430. P. A. Zagorets, V. I. Ermakov, and A. P. Granow, *Zh. Fiz. Khim.*, **39**, 4 (1965).

431. Ya. Zeldovick, *Acta Physicochim. USSR*, **1**, 961 (1935).

432. E. K. Zolotarev and V. E. Kalinini, *Zh. Neorg. Khim.*, **7**, 1225 (1962).

Chapter **IV**

INFLUENCE OF SOLVENTS ON SPECTROSCOPY

Michel Jauquet and Pierre Laszlo

1 SCOPE OF THIS REVIEW

In this chapter we concern ourselves with the spectral modifications induced by a change in solvent. Line positions, intensities, as well as band shapes, can and do vary. Solvent shifts are often useful, for instance, as empirical measurements of solvent "polarity," which can be correlated with kinetic data.

In order to provide the reader with an introduction to the subject keyed to the practical problems usually encountered in the laboratory, we have restricted ourselves to consideration of liquid solvents, to the exclusion of solid and gaseous solvents. Furthermore, we have chosen to concentrate on general solvation phenomena rather than cases of specific complexation, such as hydrogen bonds or charge-transfer complexes, on which excellent books are available [73, 151, 163, 185, 226]. Moreover, there is increasing evidence that there is no lack of continuity between the magnitude of a spectral shift in going from an isolated molecule to a weakly interacting medium, or to a strongly associating solvent. For instance, plotting $\Delta \nu / \nu$ for the O–H stretching vibration of phenol, C_6H_5OH, against the corresponding quantity for the $\nu(N–H)$ vibration of pyrrole, C_4H_4NH, yields a straight line going from nonpolar solvents such as carbon tetrachloride to strongly associating solvents such as acetone [27, 52].

Yet more convincing are the observations in binary solvent mixtures: the $\nu(C–I)$ vibration of equatorial iodocyclohexane undergoes a *continuous* change when the solvent is changed from cyclohexane to acetone, or from carbon tetrachloride to acetonitrile [190]. This type of behavior is fairly general, and a good indication that similar factors operate whether one is dealing with weakly interacting or strongly interacting solvents. For instance, it has been shown for the *presumed* iodine-benzene charge-transfer complex in n-hexane-benzene mixtures that a *single* Raman line is obtained, whose frequency gradually shifts from 210.0 cm^{-1} in pure n-hexane to 204.6 cm^{-1} in pure benzene [201]. The resolution of the experiment being of the order of 0.1 cm^{-1}, it would assuredly have detected two iodine species in equilibrium, free and complexed iodine. Likewise, dielectric relaxation times show that the benzene-iodine species has a lifetime shorter than the duration of a molecular rotation [168].

Systematic use of solvent variation to effect desired changes in the spectrum of a solute is not covered. These highly specialized techniques, such as obtaining band polarizations and the position of chromophores relative to a molecular axis by the stretched-film technique [84, 157, 210, 243, 244, 245] or by recording the electronic spectrum in a liquid crystalline solvent [136], determining internuclear distances from nuclear magnetic resonance (NMR) spectra in liquid crystalline solvents [59, 139], and obtaining the relative signs of NMR coupling constants by line splitting or coalescence as a result of solvent variation [74], are not within the province of this article.

We have tried to emphasize the physical mechanisms responsible for a given effect, to indicate the limits of current knowledge, and to describe the practical uses to which these solvent effects have been and can be put. Our viewpoint is

Table 4.1[a] Some of the More Common Solvents Used in Spectroscopy

Solvent	E	η	ϵ	$\dfrac{\epsilon-1}{2\epsilon+1}$	$\dfrac{\epsilon-1}{\epsilon+2}$	$\dfrac{n^2-1}{n^2+2}$	μ (D)	η at 20°C (cp)
Hexane	30.9	1.375	1.9	0.188	0.231	0.229	0	0.33
Cyclohexane	31.2	1.426	2.0	0.200	0.250	0.256	0	1.02 (17°C)
Carbon tetrachloride	32.5	1.460	2.2	0.222	0.286	0.274	0	0.969
Carbon disulfide	32.6	1.626	2.6	0.258	0.348	0.354	0	0.36
Toluene	33.9	1.497	2.4	0.241	0.318	0.293	0.36	0.58
Benzene	34.5	1.501	2.3	0.232	0.302	0.295	0	0.652
Diethyl ether	34.6	1.353	4.2	0.340	0.516	0.217	1.2	0.24
1,4-Dioxane	36.0	1.422	2.2	0.223	0.286	0.254	0	1.3
Tetrahydrofuran	37.4	1.405	7.4	0.405	0.681	0.245	1.7	0.53
Chlorobenzene	37.5	1.525	5.6	0.377	0.605	0.306	1.69	0.80
Bromobenzene	37.5	1.660	5.4	0.373	0.595	0.369	1.70	105
Ethyl acetate	38.1	1.372	6.0	0.385	0.625	0.228	1.78	0.455
Pyridine	40.2	1.509	12.3	0.441	0.790	0.299	2.19	0.974
Hexamethylphosphoramide	40.9	1.456	29.6	0.475	0.905	0.272	4.31	3.47
Ethyl methyl ketone	41.3	1.379	18.5	0.461	0.854	0.231	2.7	0.43
Nitrobenzene	42	1.553	34.8	0.479	0.918	0.320	4.22	2.0
Acetone	42.2	1.359	20.7	0.465	0.868	0.220	2.88	0.32
Dimethylformamide	43.8	1.427	36.7	0.480	0.922	0.257	3.8	0.84
Dimethyl sulfoxide	45.0	1.478	48.9	0.485	0.941	0.283	4.0	1.98
Chloroform	39.1	1.443	4.7	0.356	0.552	0.265	1.01	0.58
Methylene dichloride	41.1	1.425	8.9	0.420	0.725	0.255	1.60	0.43
t-Butyl alcohol	43.9	1.384	12.2	0.441	0.789	0.234	1.66	3.3 (30°C)

Table 4.1[a] Some of the More Common Solvents Used in Spectroscopy (continued)

Solvent	E	n	ϵ	$\dfrac{\epsilon - 1}{2\epsilon + 1}$	$\dfrac{\epsilon - 1}{\epsilon + 2}$	$\dfrac{n^2 - 1}{n^2 + 2}$	$\mu(D)$	η at 20°C (cp)
Acetonitrile	46	1.344	37.5	0.480	0.924	0.212	3.92	0.36
Nitromethane	46.3	1.394	38.6	0.481	0.926	0.239	3.46	0.647
Isopropyl alcohol	48.6	1.377	18.3	0.460	0.852	0.230	1.66	2.5
Acetic acid	51.1	1.372	6.2	0.388	0.634	0.227	1.74	1.29
Ethyl alcohol	51.9	1.361	24.3	0.470	0.886	0.221	1.69	1.20
Diethylene glycol	53.8	1.447	29.4	0.475	0.904	0.267	2.28	35.7
Methyl alcohol	55.5	1.329	32.6	0.477	0.913	0.203	1.70	0.597
Formamide	56.6	1.448	109.5	0.493	0.973	0.268	3.37	3.764
Water	63.1	1.333	78.5	0.491	0.963	0.206	1.85	1.002

[a]E, Solvent polarity parameter of Reichardt and Dimroth [C. Reichardt and K. Dimroth *Fortschr. Chem. Forsch.*, **11**, 1 (1968)]; n, dynamic viscosity; n, refractive index at 5,892.6 Å; ϵ, dielectric constant; μ, dipole moment.

Table 4.2. Ultraviolet Solvent Absorptions (mμ)

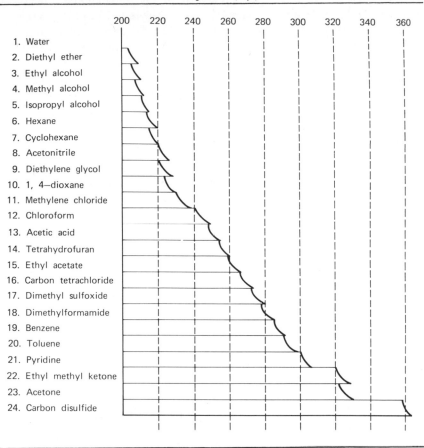

	200	220	240	260	280	300	320	340	360
1. Water									
2. Diethyl ether									
3. Ethyl alcohol									
4. Methyl alcohol									
5. Isopropyl alcohol									
6. Hexane									
7. Cyclohexane									
8. Acetonitrile									
9. Diethylene glycol									
10. 1, 4–dioxane									
11. Methylene chloride									
12. Chloroform									
13. Acetic acid									
14. Tetrahydrofuran									
15. Ethyl acetate									
16. Carbon tetrachloride									
17. Dimethyl sulfoxide									
18. Dimethylformamide									
19. Benzene									
20. Toluene									
21. Pyridine									
22. Ethyl methyl ketone									
23. Acetone									
24. Carbon disulfide									

illustrative rather than exhaustive: We have preferred to concentrate on electronic spectroscopy, vibrational spectroscopy, and NMR, rather than offer a super-ficial survey of scores of examples drawn from the various spectroscopic techniques. Finally, we have eliminated from consideration theories or explana-tions that are not consistent with the *time scale* of a particular spectroscopy:

"A particular method observes accurately only those dynamical phenomena which are on the time-scale of the measurements. The importance of this seemingly obvious point cannot be underlined too strongly. For instance, it has recently been shown that it is futile to extrapolate chemical shift data with the aim of predicting the momentary relative orientations between weakly interacting molecules: the time-scale of the measurements is of the order of fractions of seconds, whereas the time-scale of the orientational processes is about 10^{-8}-10^{-12}sec. As further example, consider the results of nuclear magnetic relaxation studies and infrared or Raman bandwidth determinations, respectively. In these measurements, certain correlation times or life times are obtained which represent time integrals over the rotational-translational or rotational-vibrational motions of the molecules. These correlation times, of the order of 10^{-8}-10^{-12}sec, are comparable to the time-scales of the molecular motion, but since they represent integrals, they have the obvious disadvantage of not being very sensitive to the detailed characteristics of the molecular motion. The same holds for integrated absorption coefficients" [204].

Choice of a spectral solvent may be facilitated by consulting Tables 4.1 to 4.4 which list some of the more common spectroscopic solvents and their properties.

Table 4.3 Infrared Solvent Absorptions[a] (Å)

Cyclohexane	3000-2850	1480-1430	910-850	—	
Carbon tetrachloride	—	—	820-720	—	
Carbon disulfide	2200-2140	1595-1460	—	—	
Benzene	3100-3000	1820-1800	1490-1450	1050-1020	680
Diethyl ether	3000-2650	1500-1010	850-830	—	
1,4-Dioxane	3700-2600	1750-1700	1480-1030	910-830	
Tetrahydrofuran	3050-2630	1500-742	—	—	
Pyridine	3500-3000	1620-1420	1230-980	780-650	
Dimethylformamide	3000-2700	1780-1020	870-860	680-650	
Chloroform	3020-3000	1240-1200	805-650	—	
Acetonitrile	3700-3500	2350-2250	1500-1350	1060-1030	930-910
Nitromethane	3100-2800	1770-1070	925-910	690-650	
Isopropyl alcohol	3600-3200	1540-1090	990-960	830-650	
Methyl alcohol	4000-2800	1500-1370	1150-970	700-650	
Water	3650-2930	1750-1580	930-650	—	
Heavy water	2780-2200	1280-1160	—	—	

[a]These solvents cannot be used in the indicated regions, with a cell thickness of 0.1 mm.

Table 4.4 Molar Diamagnetic Susceptibilities of Solvents

Solvent	$-\chi \times 10^{-6}$
Hexane	74.3
Cyclohexane	66.1
Carbon tetrachloride	66.8
Carbon disulfide	42.2
Toluene	66.1
Benzene	54.85
Diethyl ether	55.1
1,4-Dioxane	51.1
Chlorobenzene	70.0
Bromobenzene	78.1
Ethyl acetate	54.2
Pyridine	49.3
Ethyl methyl ketone	45.6
Nitrobenzene	61.8
Acetone	33.8
Chloroform	58.8
Methyl dichloride	46.6
t-Butyl alcohol	57.3
Acetonitrile	27.6
Nitromethane	21.1
Isopropyl alcohol	47.6
Acetic acid	31.8
Ethyl alcohol	33.72
Diethylene glycol	38.8
Methyl alcohol	21.40
Formamide	21.9
Water	12.9
Dimethyl sulfoxide	43.5

2 SOLVENT SHIFTS OF ELECTRONIC SPECTRA

It is a common observation to find a shift in the position of the absorption maximum on going from the gas phase to a solution, or simply by changing the nature of the solvent. Can this shift be related to the electronic and vibrational energy levels of the solute? How does it depend on the ordering of solvent molecules in a cage surrounding the solute? What else can be learned from this displacement about the polarity of the solvent or the intervening relaxation times?

In order to answer these questions, we start by examining the empirical evidence, that is, the magnitude of these shifts. We go on to consider semi-empirical models which have proved of at least limited applicability. Finally, we see how the observations can be reconciled with the best theoretical models available.

Our vocabulary includes the traditional phrases *blue shift* (or hypsochromic) and *red shift* (bathochromic) as no more than a convenient shorthand. We express all shifts in reciprocal centimeters (cm^{-1}): $1\ cm^{-1} = 2.8593\ cal\ mol^{-1}$.

Example 1

A classical case is that of the $n \rightarrow \pi^*$ transition of ketones. For instance, the results [60] listed in Table 4.5 were obtained for the shift in the absorption maximum for acetone and for di-*t*-butyl ketone

Table 4.5   Absorption Maximum for Acetone and for Di-*t*-butyl ketone in Various Solvents

Solvent	$\begin{matrix} O \\ \parallel \\ C \\ H_3C \diagup \diagdown CH_3 \end{matrix}$	$\begin{matrix} O \\ \parallel \\ C \\ (CH_3)_3C \diagup \diagdown C(CH_3)_3 \end{matrix}$
CCl_4	35,870	33,545
$(CH_3)_2SO$	36,380	33,660
$(CH_3)_2CHOH$	36,670	33,780
$HCONH_2$	36,820	34,000
$HOCH_2CH_2OH$	37,080	33,920
CH_3COOH	37,285	34,010
CF_3CH_2OH	37,735	34,270
$HCOOH$	38,270	34,410

The transition energy *increases* as the solvent becomes more polar, going from carbon tetrachloride to formic acid. Note that the effect on di-*t*-butylketone is less pronounced than for acetone, the respective ranges being 865 and 2,400 cm^{-1}. One is tempted to implicate steric hindrance to solvation [82] in the case of the former ketone. To be more precise, a protic solvent forms a hydrogen bond to the oxygen lone pairs, thus lowering the energy of the n state. To a first approximation, the energy of the π^* state is not modified.

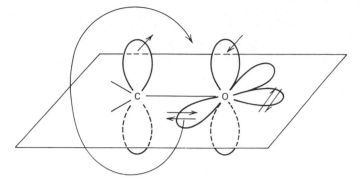

Fig. 4.1. Diagram illustrating an $n \rightarrow \pi^*$ transition for a carbonyl group.

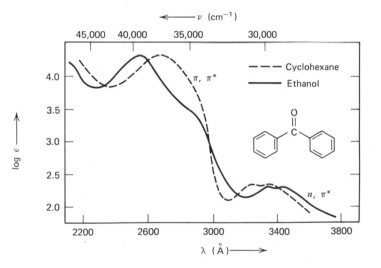

Fig. 4.2. Absorption spectrum of benzophenone in ethanol and cyclohexane at 25°C. [From N. J. Turro, in *Molecular Photochemistry,* Benjamin (1965) p. 45.]

One witnesses a blue shift as the solvent becomes a better hydrogen-bond donor.

An alternate, albeit equivalent, description focuses on the dipole moment of the ketone in its ground state, which we label μ_0. In accord with the familiar resonance picture:

μ_0 is directed from the carbon (positive) to the oxygen. The π^* excited state corresponds to promotion of *one* n electron into the π^* antibonding molecular orbital (Fig. 4.1). Removal of one electron from the oxygen atom can be seen from molecular orbital arguments to imply a considerable contribution of the $>\!\!C^- - O^+$ resonance form, with a reversal in direction of the dipole moment μ^1 characteristic of the π^* excited state.

Conversely, in the $\pi \to \pi^*$ absorption of ketones, μ^1 is not only colinear with μ_0 but also has increased magnitude with respect to μ_0. The configuration of solvent molecules around the solute is the same in the ground and in the excited state (Franck-Condon principle), thus stabilizing the $\pi \to \pi^*$ transition, whereas the $n \to \pi^*$ transition is destabilized. In going from a nonpolar solvent such as cyclohexane to a polar solvent such as ethanol, the $n \to \pi^*$ absorption undergoes a blue shift, while the $\pi \to \pi^*$ absorption undergoes a red shift (Fig. 4.2).

Example 2

One should not assume that solvent shifts of electronic transitions are observed only with polar molecules. Part of the spectrum of coronene:

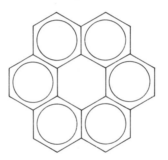

is shown in Fig. 4.3, both in the gas phase and in a benzene solution. Shifts of ca. 1000 cm^{-1} in the positions of the absorption maxima are seen.

The molecule is nonpolar in the ground state (μ_0 = 0), but acquires polarity in the excited states ($\mu^i \neq 0$).

3 SPECTRAL SHIFTS DUE TO FRANCK-CONDON STRAIN

These have been described in beautiful language by Bayliss and McRae [23]:

"Consider a solute molecule in its ground state which is in equilibrium (modified by thermal motion) with the surrounding solvent molecules that form its cage. The solvation energy of this equilibrium state involves (a) a packing factor depending on the geometry of the solute and solvent molecules, and (b) a factor which depends on the degree of mutual orientation interaction if solute and solvent are polar or if there is hydrogen bonding between them.

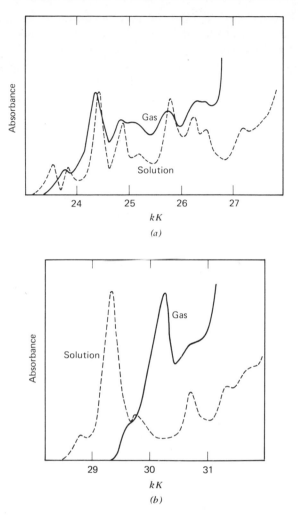

Fig. 4.3. (*a*) Absorption spectrum of the *a* band of coronene. (*b*) Absorption spectrum of the *p* band of coronene. [From J. Aihara, K. Ohno and H. Inokuchi, *Bull. Chem. Soc. Jap.*, **43**, 2437 (1970).]

To the extent that the size, charge distribution, and dipole moment of the solute molecule are different in the excited state, the configuration of the solvent cage in equilibrium with the excited state is different from that of the ground state cage. Now it is the essence of the Franck-Condon principle that an optical transition occurs within a time that is short compared with the period of nuclear motions. At the instant of its formation, *i.e.* when it is in what might be called the *Franck-Condon state,* the excited solute molecule is momentarily surrounded by a solvent cage whose size and orientation are those that are appropriate to the ground state. The equilibrium excited configuration is only reached subsequently by a process of relaxation, which requires at least several molecular vibration periods ($\sim10^{-13}$ sec) as far as readjustment for size is concerned, and a time of the order of 10^{-11} sec [234] if a readjustment of the solvent orientation is required. The lifetime of the excited state is known to be of the order of 10^{-8} sec, which is ample for equilibrium with the solvent to be established before deactivation eventually occurs. (This time factor may be of importance in comparing solvent effects in absorption and in fluorescence). The Franck-Condon excited molecule and its solvent cage are thus in a state of strain whose energy is necessarily greater than that of the equilibrium state, and the excited state solvation energy s' to be used in finding $\Delta\nu$ is less than the equilibrium value (and in certain cases it may even be < 0). In discussing this strain, it is convenient to use the general term *Franck-Condon strain,* while its two components can be termed *packing strain,* and *orientation strain."*

Let us consider an isolated molecule A. It can be in one of two electronic energy states: the ground state A_0 of energy E_0, or the first excited state A^1 of energy E^1 (the same argument applies to an excited triplet state).

Even though A_0 and A^1 are two electronic states of the same molecule, it is necessary to consider these two states A_0 and A^1 as two different molecular species whose electronic distributions, dipole moments, geometries, and so on, are in principle different.

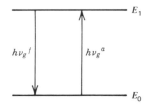

In the absorption or emission of a quantum of frequency

$$\nu = \frac{(E^1 - E_0)}{h} \quad ,$$

a redistribution of electron density takes place, triggered by the promotion of an electron from the lower to the upper state. The permanent dipole moment changes from μ_0 to μ^1; in the general case, the moments μ_0 and μ^1 are not parallel.

Let us now consider how this description must be altered in the presence of solvent molecules. The solute molecule interacts with the entire surrounding set of polar or polarizable molecules. These interactions affect the position of its electronic levels, which are lowered by an energy stabilization W_0 or W^1.

Furthermore, for each electronic state of the solute, one has to consider two distinct levels: one corresponding to the equilibrium geometry (relaxed) and one corresponding to the nonequilibrium geometry resulting from application of the Franck-Condon principle (strained). It is recalled that this principle states that, to a first approximation, nuclei are motionless during the time of an electronic excitation; the local arrangement of solvent molecules around the solute does not change. Immediately after excitation, the excited singlet state A^1 is characterized by a configuration of solvent molecules identical to that for the ground state A_0. Likewise, the ground state resulting from emission of a photon has the same configuration of solvent molecules as that in the initial excited singlet state.

With time, these strained Franck-Condon states relax by adopting their own solvent configuration, that of minimum potential energy. The rotational relaxation time for the solvent molecules is in the range of 10^{-12} to 10^{-10} sec at room temperature. The lifetime of an excited singlet state is of the order of 10^{-8} sec.

From this very general analysis, the accompanying schematic picture can be drawn:

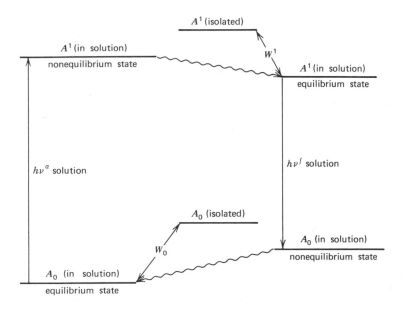

Consider absorption of a quantum by the A_0 equilibrium state in solution; the Franck-Condon A^1 state is reached (left part of diagram). A relaxation must necessarily follow the excitation, and a relaxed excited state is obtained in which another solvation equilibrium has been established:

If the relaxation rate is of the same order of magnitude as the fluorescence decay time, the fluorescence spectrum should display an observable time dependence. The correlation time τ_c characteristic of the reorientation of the solvent molecules can be adapted to the lifetime of the excited singlet simply by changing the nature of the solvent until the right viscosity is found.

Ware and co-workers [227] studied the phthalimide chromophore in various alcoholic solvents. The eloquent results of their experiment are shown in the Figs. 4.4 and 4.5.

A nanosecond flash lamp is used to excite fluorescence of 4-aminophthalimide. The fluorescence spectrum is detected after a variable delay time. In more viscous solvents such as glycerol, the same experiment can be done at higher

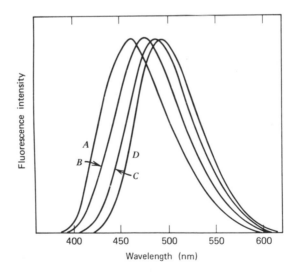

Fig. 4.4. Time-resolved fluorescence spectrum of 4- aminophthalimide at -70°C in n-propyl alcohol. A, 4 nsec; B, 8 nsec; C, 15 nsec; D, 23 nsec. [From W. R. Ware, S. K. Lee, G. J. Brant, and P. P. Chow, *J. Chem. Phys.*, **54**, 4729 (1971).]

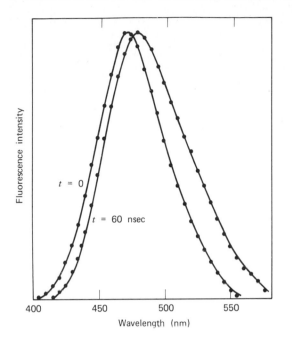

Fig. 4.5. Time-resolved fluorescence spectrum of 4-aminophthalimide in glycerol at 0°C. Data taken with the single-photon method. $\Delta t = 60$ nsec. [From W. R. Ware, S. K. Lee, G. J. Brant, and P. P. Chow, *J. Chem. Phys.*, **54**, 4729 (1971).]

temperatures (ca. 0°C). The parameter $[\bar{\nu}(t) - \bar{\nu}_\infty]$, [where $\bar{\nu}(t)$ is the frequency of the spectral maximum at a time t after excitation, and $\bar{\nu}_\infty$ is that with full relaxation, at high temperature] decreases exponentially with time. The corresponding relaxation times τ_R

$$\bar{\nu}(t) - \bar{\nu}_\infty \propto \exp\,(-t/\tau_R)$$

are plotted against reciprocal temperature in Fig. 4.6 and compared with the dielectric relaxation times for the alcoholic solvent. In the words of Ware et al.,

"the qualitative picture that emerges from this work is that the strong temperature dependence of the emission maximum in the steady-state spectra is in fact the result of a time-dependent variation of the energy gap between the first excited singlet state and the ground state. It appears that after vibrational relaxation has thermally equilibrated the excited singlet to what might be called the Franck-Condon state (with respect to solvation), the interactions of the solvent shell with the excited molecules are such as to lower the energy gap as reorientation takes place to accommodate the new dipole moment

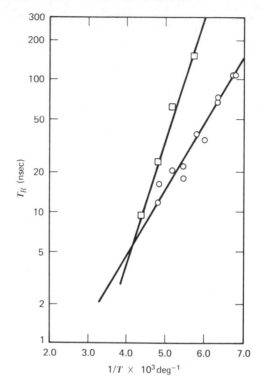

Fig. 4.6. Relaxation times as a function of $1/T$ for 4-aminophthalimide in n-propyl alcohol. ○, $[\bar{\nu}(t) - \bar{\nu}_\infty]$; □, Dielectric relaxation time of n-propyl alcohol. (From W. R. Ware, S. K. Lee, G. J. Brant, and P. P. Chow, *J. Chem. Phys.*, **54**, 4729 (1971).]

and moment direction. This appears to be in addition to a fast subnanosecond relaxation process that energywise is just as important."

These authors suggest that this subnanosecond relaxation process is associated with an exciplex interaction; an exciplex is a complex formed between the A^1 excited state and solvent molecules.

Calculation of the Solute-Solvent Interaction

During the electronic excitation of the solute, dipoles are induced in the polarizable solvent molecules. These dipole-induced dipole interactions have been related to macroscopic properties of the solvent, treated as a continuous polarizable medium. McRae [144] has proposed the following function of the refractive index n:

$$\frac{n^2 - 1}{2n^2 + 1}$$

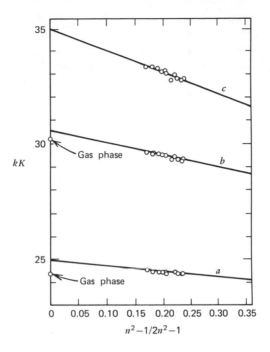

Fig. 4.7. ´Solvent effect for coronene. a, a band; b, p band; c, β band. [From J. Aihara, K. Ohno, and H. Inokuchi, *Bull. Chem. Soc. Jap.*, **43**, 2438 (1970).]

to account for this general red shift, in proportion to the oscillator strength of the transition $f = 4.32 \times 10^{-9} \int \epsilon \, dv$. If the transition energies for various bands of coronene are plotted against this term $(n^2 - 1)/(2n^2 + 1)$ in various solvents, good correlations are obtained (Fig. 4.7).

The slopes of these lines are also proportional to the difference in polarizability $a_0 - a^i$ between ground and excited states. Note, however, that the points representative of the gas-phase spectrum ($n = 1$) are off, which serves to indicate the limits of this extremely simplified treatment.

Yet it is instructive to describe briefly its derivation.

Theory of Band Shifts Based on the Onsager Model

In Onsager's theory [174], the solvent consists of molecules with a dipole moment μ_s and an isotropic polarizability a_s. In order for this model to be applicable, the following relation must be verified:

$$\phi(\epsilon) = \left(\frac{\epsilon - 1}{\epsilon + 2}\right) - \frac{d}{M}\left(\frac{4\pi N}{M}\right)\left(a_s + \frac{\mu_s^2}{3kT}\right)$$

in which ϵ is the static dielectric constant for the solvent, d is its density, M is its molecular weight, N is the number of molecules per unit volume (in cubic centimeters), and T is the temperature.

Let us now dissolve a polar molecule, characterized by a permanent dipole moment μ_0. It is stabilized by solvation; the permanent dipole from the solute induces a reaction field in the solvent. This electrostatic field $\mathbf{R}_0$ is given by the expression

$$\mathbf{R}_0 = \frac{\mu_0}{a_0^3} \frac{2(\epsilon - 1)}{2\epsilon + 1}$$

where a_0 is the radius of the spherical cavity occupied by the solute. The stabilization energy resulting from the interaction of μ_0 with the reaction field $\mathbf{R}_0$ is

$$E_0 = \mu_0 \cdot \mathbf{R}_0 = \frac{\mu_0^2}{a_0^3} \frac{2(\epsilon-1)}{2\epsilon+1}$$

The solute molecule now absorbs a quantum of electromagnetic energy, and is pushed into an excited state characterized by a dipole moment μ^i. During this electronic transition the solvent molecules could not reorient themselves, and the reaction field remains at its initial $\mathbf{R}_0$ values. The excited state thus benefits from a stabilization energy:

$$E^i = \mu^i \cdot \mathbf{R}_0 = \frac{\mu^i \mu_0}{a_0^3} \frac{2(\epsilon-1)}{2\epsilon+1}$$

assuming nonexpansion or contraction of the solute cavity after the electronic excitation (in fact, excited states are generally expanded in volume, because of promotion of electrons into antibonding states; the resulting "cage strain," as a result of the internal pressure of solvent molecules, is believed not to vary markedly from solvent to solvent). We also temporarily assume that μ_0 and μ^i are point dipoles located at the center of the spherical cavity.

It follows that the spectral shift in going from a solvent 1 to a solvent 2 is given by

$$-\Delta E_{1\to2} = (\mu^i - \mu_0)\mu_0 \frac{\Delta f(\epsilon)}{a_0^3} \quad 1\to2$$

where

$$f(\epsilon) = \frac{2(\epsilon-1)}{2\epsilon+1} \ .$$

Extension of the Onsager Model to Nonpolar Solutes and Solvents

Bayliss and McRae [21, 144] showed that the dispersion interaction between solute and solvent molecules—even when they are devoid of permanent dipole moments—is proportional to the $(n^2 - 1)/(2n^2 + 1)$ term, where n is the refractive index of the medium. This term differs from the reaction field term in that n^2 equal ϵ_∞, the high-frequency dielectric constant.

For a nonpolar solute in a polar solvent, in addition to London forces, another component originates in the polarization fluctuations of the dipolar continuum around the cavity. This solvent Stark effect [20] is proportional to

$$\frac{(2\epsilon + 1)\ (\epsilon - 1)}{\epsilon} .$$

Thus the various cases can be summarized as follows:

	Solute	
Solvent	*Nonpolar*	*Polar*
Nonpolar	Dispersion $(n^2 - 1)/(2n^2 + 1)$	Dispersion plus Stark plus reaction field $(\epsilon - 1)/(2\epsilon + 1)$
Polar	Dispersion plus Stark $\dfrac{(2\epsilon + 1)(\epsilon - 1)}{\epsilon}$	Dispersion plus Stark plus reaction field $(\epsilon - 1)/(2\epsilon + 1)$

For a polar solute in a polar solvent, McRae [144] has proposed the following terms:

$$\Delta\nu = A\frac{n^2 - 1}{2n^2 + 1} \qquad \text{dispersion}$$

$$\Delta\nu = B\frac{n^2 - 1}{2n^2 + 1} \qquad \text{induction by the solute dipole}$$

$$\Delta\nu = C\left(\frac{\epsilon - 1}{\epsilon + 2} - \frac{n^2 - 1}{n^2 + 2}\right) \qquad \text{dipole-dipole}$$

$$\Delta\nu = D\left(\frac{\epsilon - 1}{\epsilon + 2} - \frac{n^2 - 1}{n^2 + 2}\right) \qquad \text{induction by the solvent dipole}$$

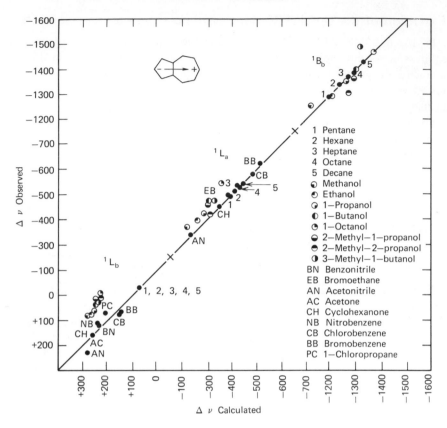

Fig. 4.8. Comparison of calculated and observed shifts for azulene. [From W. W. Robertson, A. D. King, Jr., and O. E. Weigang, Jr., *J. Chem. Phys.*, **35**, 464 (1961).]

An example of their application is azulene [194]; the values of the parameters are listed in Table 4.6.

Table 4.6   McRae Parameters for Azulene (cm^{-1})

Transition	^1L_b	^1L_a	^1B_b
$A + B$	−131	−2734	−7154
C	+286	+209	0
D	0	0	0

It is possible to correlate extremely well the experimental band positions in a series of protic and aprotic solvents (Fig. 4.8).

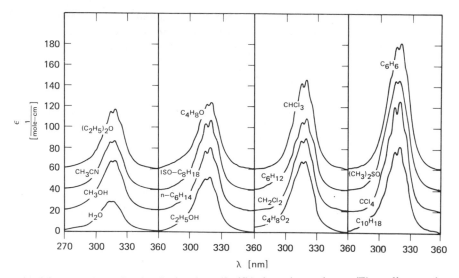

Fig. 4.9. The absorption band of carbon disulfide in various solvents. (The ordinate scales are displaced by 20 units of ϵ.) [From P. Baraldi, P. Mirone, and E. S. Guidetti, *Z. Naturforsch.*, **26a**, 1852 (1971).]

4   ELECTRONIC SPECTRAL INTENSITIES

Pronounced changes in the intensity of an absorption band are sometimes found to occur on a change in solvent. For instance, the absorption band of carbon disulfide is markedly solvent-dependent; the integrated intensity measured in benzene is twice that in ethanol and three times that in water solution. There is a relatively smooth variation between the different solvents used (Fig. 4.9) [15].

Another type of intensity change can be observed due to specific relationships between the absorption band of the solute and electronic transitions in the solvent molecules; intensity borrowing can be observed when there is partial or total overlap. For instance, the intensity of the 0-0 band of the $^1B_{2u}$-$^1A_{1g}$ transition of benzene increases in carbon tetrachloride in which the benzene transition borrows intensity from the nearby intense solvent transition [195].

What is the explanation for such variations [22]? Let us consider the supermolecule (A, B), where A is a solute molecule and B is a solvent molecule interacting weakly. Let us focus our attention on a transition of the type

$$a_1 \beta_0 \leftarrow a_0 \beta_0$$

in which an electron is promoted from the a_0 to the a_1 state in the solute without any electronic excitation occurring simultaneously in the solvent molecule. The intensity of this transition is proportional to the square of the transition moment M given by

$$M = \langle a_1 \beta_0 |M_s| a_0 \beta_0 \rangle$$

where $M_s = M_A + M_B$, and M_A and M_B are the transition moment operators for A and B. The interaction potential V represents the (A, B) interaction

$$H_{\text{sum}} = H_A + H_B + V$$

This perturbation alters the wave functions according to: $(a_j \beta_i)_{\text{sum}} = a_j \beta_i + \sum_{i,j \neq p,q} C_{jipq} a_p \beta_q$ to the first order where the coefficients C_{jipq} are defined by

$$C_{jipq} = \frac{\langle a_p \beta_q |V| a_j \beta_i \rangle}{(E_p^A + E_q^B) - (E_j^A + E_i^B)} \tag{4.1}$$

It then follows that the transition moment M_{1000} for the above $a_1 \beta_0 \leftarrow a_0 \beta_0$ transition is given by

$$(M_{1000})_{\text{sum}} = \langle a_1 \beta_0 + \Sigma C_{pq10} a_p \beta_q |M_s| a_0 \beta_0 + \Sigma C_{pq00} a_p \beta_q \rangle$$

which can be developed as

$$(M_{1000})_{\text{sum}} = \langle a_1 |M_A| a_0 \rangle \quad \text{that for the isolated molecule}$$

$+ C_{1000}[\langle a_1 |M_A| a_1 \rangle - \langle a_0 |M_A| a_0 \rangle]$ a term proportional to the difference in dipole moment between the ground and the excited state of the solute

$+ C_{j010} \langle a_j |M_A| a_0 \rangle$

$+ C_{j000} \langle a_j |M_A| a_1 \rangle$ contributions from higher excited states of the molecule

$+ (C_{0i10} + C_{1i00}) \langle \beta_i |M_B| \beta_0 \rangle$ contributions from electronic transitions of the solvent molecule (4.2)

Intensity borrowing is thus possible either from other electronic transitions in the solute ($a_j \leftarrow a_0; a_j \leftarrow a_1$) or from electronic transitions in the solvent ($\beta_i \leftarrow \beta_0$). The relative weights of these various contributions [see (4.2)] depend on the values of C_{jipq} coefficients, and therefore on the magnitudes of the

ΔE terms occurring in their denominators [see (4.1)]. Typically, for ultraviolet-visible transitions in the range 10,000 to 50,000 cm^{-1} (λ_{max} = 2,000 – 10,000 Å), and a transition moment M_{pq} of 0 to 14 D, the following values are characteristic:

1. $E_1{}^A - E_0{}^A \cong 10^4$ cm^{-1} → C_{1000} = 0 to 10^{-1}

2. $E_j{}^A - E_1{}^A$ is sometimes as small as 10^3 cm^{-1}, and C_{j010} can be very large.

3. $E_j{}^A - E_0{}^A > 4.10^4$ cm^{-1}; the corresponding C_{j000} coefficient can usually be neglected.

4. $(E_i{}^B - E_0{}^B) + (E_1{}^A - E_0{}^A)$ is so large that the C_{1i00} coefficient is negligible.

5. The $(E_i{}^B - E_0{}^B) - (E_1{}^A - E_0{}^A)$ term can be extremely small, when the solvent possesses a transition close in energy to that of the solute; the C_{0i10} coefficient is then very large.

In summary, there are two main origins of solvent-dependent intensities. The phenomenon occurs when the dipole moment of the excited state of the solute differs significantly from that of the ground state, or when intensity borrowing is feasible from an adjoining solvent transition. In the former case dispersion interactions alter the transition moment according to [137]:

$$M_{pq}^{sol} = M_{pq}^{gas} + \frac{n^2 - 1}{n^2 + \frac{1}{2}} a^3 W_{pq}$$

where W_{pq} is a vector describing the effect of the dispersion interactions on the transition moment, n is the refractive index, and a is the radius of the spherical cavity in which the solute is immersed.

The ratio

$$\frac{A^{sol} - A^{gas}}{(n^2 - 1)/(2n^2 + 1)}$$

where $A = \int \epsilon \, d\bar{v}$ is the integrated intensity of the band, can be shown to be a linear function of $(n^2 - 1)/(2n^2 + 1)$, in fair agreement with observation (Fig. 4.10).

5 ULTRAVIOLET BAND SHAPES

The ultraviolet band shape is the envelope of all the transitional frequencies available to the solute. There is a distribution of frequencies due to fluctuations

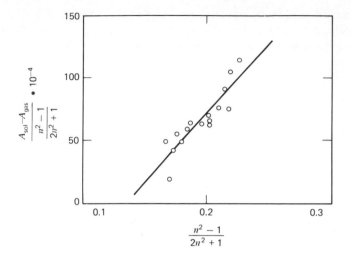

Fig. 4.10. Dependence of the integrated intensity on the refractive index of the solvent. [From P. Baraldi, P. Mirone, and E. S. Guidetti, *Z. Naturforsch.*, **26a**, 1852 (1971).]

in the local environment. For instance, consider the gas-to-liquid frequency shift for the absorption *maximum* of a polar molecule dissolved in a liquid [154, 155]:

$$\nu_e - \nu_g = \pm \frac{1}{h} \left[\overline{R}(\mu^i - \mu_0) + \frac{(\mu^i - \mu_0)^2}{2} K(n^2) \right] + \Delta\nu_d$$

where the plus sign refers to absorption and the minus sign to emission, $\Delta\nu_d$ is the dispersion contribution to the shift, and $\overline{R}$ is the mean value of the reaction field R. The classical theory of fluctuations gives the probability of finding a reaction field of value R in volume element dV as

$$\phi(R) \, dV \sim \exp\left\{ \frac{(R - \overline{R})^2}{2\,[K(\epsilon_0) - K(n^2)]\,kT} \right\} dV$$

where the dielectric (or refractive index) function K is defined by:

$$K(x) = \frac{f(x)}{(1-a)[f(x)]}$$

where a is the solute polarizability, and

$$f(x) = \frac{2(x-1)}{2x+1} \frac{1}{a^3}$$

By combining the above expressions, the line shape $\phi(\nu)$ is given by

$$\phi(\nu) \sim \exp\left\{ - \frac{h^2 (\nu - \bar{\nu})^2}{2[K(\epsilon) - K(n^2)]} \frac{1}{(\mu_0 - \mu^i)^2 \, kT} \right\}$$

or yet more simply:

$$\phi(\nu) \sim \exp\left[\frac{h(\nu - \bar{\nu})^2}{2\nu_{ST} \, kT} \right]$$

where ν_{ST} is an experimentally accessible quantity:

$$\nu_{ST} = (\nu_1{}^a - \nu_g{}^a) + (\nu_g{}^f - \nu_1{}^f)$$

where the subscripts a and f refer to absorption and fluorescence, respectively.

The mean-square broadening $\overline{(\nu - \bar{\nu})^2}$ is then equal to $\nu_{ST}kT/h$. The order of magnitude of ν_{ST}, for normal systems at ambient temperatures, is 2000 cm^{-1}, leading to $\overline{\nu - \bar{\nu}} = 600$ cm^{-1}, or $\Delta\nu_{1/2} \cong 1.500$ cm^{-1}.

6   INFRARED GROUP FREQUENCIES

Solvent effects on infrared absorption maxima have been well studied. Yet knowledge of them is still imperfect, even though changes of 5 to 50 cm^{-1} are currently observed, amounting to relative changes $\Delta\nu/\nu$ normally in the range 0 to 2%. For instance, the C=O stretch in formaldehyde absorbs at 1745 cm^{-1} in the gas phase, is displaced to 1736 cm^{-1} in n-hexane, to 1732 cm^{-1} in carbon tetrachloride, and to 1728 cm^{-1} in chloroform [116]. Another example is iodocyclohexane; the equatorial iodine conformer is characterized by a C–I stretch at 661 cm^{-1} in cyclohexane solution versus 654.5 cm^{-1} in acetone [190].

It is recalled that the duration of a vibration is short with respect to motion of the solvent molecules. Their presence influences the frequency of a particular vibrator, because both the direction and the magnitude of the transition moment can be affected. The simplest case is that of an ensemble of n polar solute molecules, with a dipole moment μ. In the gas phase, at zero pressure, all the molecules have the same frequency ω_0, n-fold degenerate and therefore infinitely sharp.

In a polar or polarizable medium, solute-solvent forces tend to orient the dipoles in order to minimize the potential:

$$V = - \, \mathbf{E} \cdot \mu = - \mathbf{E} \cdot \mu \, \cos \, \theta$$

where $\mathbf{E}$ is the instantaneous local electric field acting on the solute from the surrounding molecules and making an angle θ with the direction of μ.

Local order in the solvent molecules around the solute competes with thermal fluctuations. The degree of local order is measured by $\mu \cdot E/kT$; when $\mu \cdot E/kT \gg 1$, the effective potential $\overline{V}$ reduces to

$$\overline{V} = - s \frac{2\mu^2}{R^3}$$

where s is a coefficient of mutual molecular orientation, and R is the average solute-solvent distance [77].

As a result of solvation interactions, the vibrational degeneracy is lifted, and the frequency ω_0 splits into n slightly shifted values $\omega_1, \omega_2, \cdots, \omega_n$, the average value of which $\omega = \overline{\omega}$ gives the position of the band, and the distribution of which yields the band shape.

What happens to an arbitrary normal vibration of a molecule, whose frequency $\omega_0 = (K/M)^{1/2}$? M and K are the intramolecular inertial and force constant coefficients respectively; their dimensions are homogeneous with a reduced mass, and with a force divided by a distance, respectively.

In addition to the intramolecular springlike force characteristic of the vibrator, one has to consider the k_i force originating from the V_i potential [212]:

$$k_i = \frac{\delta^2 V_i}{\delta q_i^2}$$

where q_i is the normal internal coordinate corresponding to ω_i. Since $V_i = E_i \cdot \mu \cos \theta_i$,

$$k_i = - E_i \frac{\delta^2}{\delta q_i^2} (\mu \cos \theta_i) \equiv - a V_i$$

where

$$a = (\frac{\delta \theta}{\delta q_i})^2 - \frac{1}{\mu} \frac{\delta^2 \mu}{\delta q_i^2}$$

E_i can be treated as a constant, because the solvent cage is essentially motionless for the duration of the vibration. Hence the frequency shift is

$$\omega_0 = (K/M)^{1/2} \rightarrow \omega_i = [(K + k_i)/M]^{1/2}$$

If the perturbation $k_i \ll K$,

$$\omega_i \cong \omega_0 - \tfrac{1}{2} k_i/M = \omega_0 - a V_i/2M\omega_0$$

and the shift $\Delta\omega$ is given by

$$\Delta\omega = \omega_i - \omega_0 = -aV_i/2M\omega_0$$

Its direction depends on the sign of the coefficient a. If

$$\overline{V} = -s\,\frac{2\mu^2}{R^3}$$

is taken as a negative quantity (solvation lowers the potential energy), $\Delta\omega$ will be positive provided a is also positive. Depending on the relative magnitudes of

$$(\frac{\delta\theta}{\delta q_i})^2 \quad \text{and} \quad \frac{1}{\mu}\,\frac{\delta^2\mu}{\delta q_i^2}\ ,$$

a can be either positive or negative. Correspondingly, a blue shift $(a > 0)$ or a red shift $(a < 0)$ results.

The following cases are interesting.

1. A normal mode without alteration of the group dipole magnitude, only of its orientation:

$$\frac{1}{\mu}\,\frac{\delta^2\mu}{\delta q_i^2} \cong 0 \quad \text{and} \quad (\frac{\delta\theta}{\delta q_i})^2 \cong a > 0$$

A blue shift will be observed. This frequency increase is normal; the solute experiences a strain from the partial loss of the maximum stabilization afforded by the solvent cage in its optimum, ground-state configuration. Bending modes, scissoring modes, rocking modes, and so on, are typical. An example is the out-of-plane deformation of halogenobenzenes [195] (Table 4.7).

Table 4.7 Solvent Effects on the Out-of-Plane Deformation Frequency of Halogenobenzenes

Compound		γ_{C-X} (cm^{-1})
Fluorobenzene	Gas	220
	Neat	245
Chlorobenzene	Gas	174
	Neat	196
Bromobenzene	Gas	167
	Neat	180
Iodobenzene	Gas	—
	Neat	167

2. Totally symmetric vibrations with no change in the direction of the dipole moment:

$$\frac{\delta\theta}{\delta q_i} = 0$$

A blue shift or a red shift can be observed, depending on the sign of the

$$\frac{1}{\mu} \frac{\delta^2\mu}{\delta q_i^2}$$

quantity.

In summary and with the assumptions made, a red shift can only be observed for type-2 vibrations, such as C=O stretching vibrations; bending, rocking, scissoring (type-1) motions give rise to blue shifts only, in going from the gas phase to a solution. Examples are the 671 cm^{-1} γ_{C-H} benzene frequency [119] and the 829 cm^{-1} furfural frequency [86].

Semiempirical expressions have been devised for the red shifts observed in many cases, such as that of carbonyl bands. The Kirkwood-Bauer-Magat relation [19, 120]:

$$\frac{\nu-\nu_0}{\nu_0} = - A\frac{n^2-1}{2n^2+1} \quad A > 0$$

is sometimes improved [124] to include both the orientation polarization and the electronic polarization:

$$\frac{\nu-\nu_0}{\nu_0} = \frac{\Delta\nu}{\nu} = A\frac{n^2-1}{2n^2+1} + B(\frac{\epsilon-1}{2\epsilon+1} - \frac{n^2-1}{2n^2+1})$$

and is reasonably successful. The A term is determined experimentally in nonpolar solvents, for which the square of the refractive index n^2 approximates closely the dielectric constant ϵ. Information as to the nature of solvation can also be obtained from Hallam-Bellamy plots [27], in which $\Delta\nu/\nu$ for a vibrator in a solute 1 is plotted against $\Delta\nu/\nu$ for the same (or closely related) vibrator in another solute 2.

7   INFRARED ABSORPTION INTENSITIES

The influence of the medium on infrared (or Raman) vibrational modes can readily be understood; the transitions that are active depend on the exact geometry of the absorbing molecule. For nonvanishing of the transition integral,

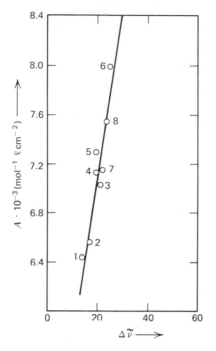

Fig. 4.11. Integrated intensities of carbonyl stretching frequency of acetone in different solvents. 1, n-Heptane; 2, cyclohexane; 3, dioxane; 4, benzene, 5, carbon tetrachloride; 6, chloroform; 7, chlorobenzene; 8, pyridine. [From J. Fruwert, G. Geiseler, and D. Luppa, *Z. Chem.*, **8**, 238 (1968).]

the normal mode must belong to one of the symmetry classes to which a representation of the dipole can be reduced.

The geometry in solution can and does differ from the gas-phase geometry. The pressure exerted by solvent molecules affects the intensities of many bands in the spectrum, by comparison to the gas phase; some transitions that are forbidden for the isolated molecule acquire finite probability and degeneracies, and are lifted as a result of changes in the local geometry.

In general, infrared absorption intensities are relatively little altered by a change in the nature of the solvent. A pronounced parallelism is often found between solvent effects on integrated intensities A and on absorption frequencies v. The example of the C=O stretch in acetone [79] is shown in Fig. 4.11.

These changes have been recently explained convincingly [220] by the coupling between the effective electric field E from the electromagnetic wave and the transition moment. It is in fact the square of the scalar product of the electric vector and of the transition moment that determines the spectral intensity.

The intensity Γ_i of an infrared absorption band is related to the square of the dipole gradient

$$\frac{\delta\mu}{\delta Q_i}$$

accompanying the transition:

$$\Gamma_i = \frac{1}{\nu_i}\frac{N\pi}{3c}\left(\frac{\delta\mu}{\delta Q_i}\right)^2$$

where Q_i is a normal vibrational coordinate, ν_i is the vibrational frequency, μ is the transition moment, and the other symbols have their usual meaning.

Another relation exists between Γ_i and the incident radiation density $\rho(\nu)$, in turn connected to the mean value of the electric field $E(\nu)$ acting on the molecule:

$$\Gamma_i \propto \rho(\nu)$$

$$4\pi\rho(\nu) = \left\langle E^2(\nu) \right\rangle$$

where the brackets imply time-averaging; even though the periodic function $E(\nu)$ has a zero time average:

$$\left\langle E(\nu) \right\rangle = 0$$

its square has a finite time-average:

$$\left\langle E^2(\nu) \right\rangle \neq 0$$

in the same way as the effective intensity of an alternating current equals its intensity maximum divided by $\sqrt{2}$.

The effective electric field E in the medium differs from the propagating electric field E_0 in the vacuum by a factor depending upon the characteristics of the solvent. In the absence of specific interactions, such as complex formation, the Debye-Onsager model leads to the following result:

$$\frac{\Gamma}{\Gamma_0} = \left(\frac{E}{E_0}\right)^2 = \frac{1}{\epsilon_\infty^{1/2}}\left(\frac{\epsilon_\infty+2}{3}\right) = \frac{1}{n}\left(\frac{n^2+2}{3}\right)^2$$

where the square n^2 of the refractive index n is identical to ϵ_∞, the dielectric constant at infinite frequency. It applies to a spherical cavity around the solute molecule. This equation has been applied with some success, but it fails to account for the observed changes whenever specific interactions occur. For instance, the stretching C—H vibration of chloroform has a highly solvent-

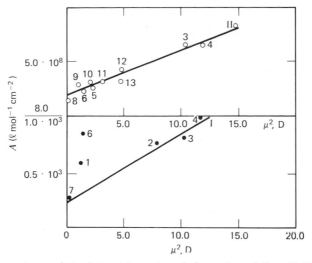

Fig. 4.12. Dependence of the integral intensity of absorption of the $\nu(C \equiv N)$ band on the square of the dipole moment of the solvent molecule (I) $\nu(C \equiv N)$ CH$_3$CN, (II) $\nu(C \equiv N)$ HCN; solvents: 1, (C$_2$H$_5$)$_2$O; 2, CH$_3$COCH$_3$; 3, C$_2$H$_5$NO$_2$; 4, CH$_3$NO$_2$; 5, CH$_2$Cl$_2$; 6, CHCl$_3$; 7, n-C$_6$H$_{14}$; 8, CCl$_4$; 9, CHBr$_3$; 10, CH$_2$Br$_2$; 11, 1,2-C$_2$H$_4$Cl$_2$; 12, t-C$_4$H$_9$Br; 13, t-C$_4$H$_9$Cl. [From A. E. Lutskii and S. N. Vragova, *Opt. Spectrosc.*, **31**, 113 (1971).]

dependent intensity; going from carbon tetrachloride to triethylamine solution, it increases thirtysix-fold [108].

By monitoring the intensity changes as a function of concentration, and assuming a 1:1 triethylamine-chloroform hydrogen-bonded complex, the formation constant K for this complex was determined [138]. Likewise, charge-transfer complexation has a pronounced effect on vibrational band intensities [67, 78].

In polar solvents the integrated intensity A increases with the square of the dipole moment of the solvent molecule [140] (Fig. 4.12).

The most general theory, devised by Buckingham [42], predicts that the function

$$F = \frac{2n^2+1}{3n^{3/2}} \left(\frac{A_{sol}}{A_{gas}}\right)^{1/2} - 1 - 0.6 \frac{n^2-1}{2n^2+1}$$

ought to be linear with respect to

$$\frac{(\epsilon-1)}{(2\epsilon+1)},$$

where ϵ is the dielectric constant of the solvent.

Experimental tests have been performed [53], also with limited success.

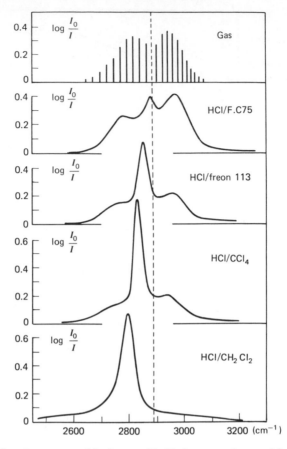

Fig. 4.13. Infrared spectrum of hydrogen chloride in vapor phase and in solution. [From M. Perrot, P. V. Houng, and J. Lascombe, *J. Chim. Phys.*, **68**, 614 (1971).]

8   THE SHAPES OF INFRARED ABSORPTION BANDS

For a given slit width, well below the width of an infrared absorption band, its aspect depends on the physical state of the sample; in the gaseous phase fine structure is noticeable, as a result of transitions between the rotational states. Three main spectral regions have been distinguished, the so-called P, Q, and R bands. Going into solution these spectral details [179] are blurred, and the absorption band becomes intermediate in shape between a pure gaussian, defined by an expression of the type:

$$I(\nu) = \frac{a}{b^2} \exp \left[-\frac{\ln 2}{b^2}(\nu - \nu_0)^2\right]$$

and a pure lorentzian, characterized by

$$I(\nu) = \frac{a}{b^2 + (\nu - \nu_0)^2}$$

where

$$\frac{a}{b^2} = I_{max}$$

(intensity maximum), and $2b = \Delta\nu_{1/2}$ (width at half-height), ν_0 being the position of the absorption maximum.

The contour of an infrared (or Raman) absorption band of a molecule dissolved in a liquid preserves precise information about its rotational behavior. We now examine this notion in somewhat more detail.

Correlation Functions

In the gaseous phase, between two successive collisions, molecules undergo translations and rotations. The fine structure observable in the vibrational spectrum allows determination of the energies of the rotational states. These depend on the geometric features of the molecule, acting through the various moments of inertia. It will be recalled that three principal axes of inertia, originating at the center of gravity for a molecule, can be defined. The corresponding moments of inertia dictate the angular frequency of the molecule undergoing independent rotations around each of these axes:

$$I_x = \sum_i m_i r_{ix}^2$$

$$I_y = \sum_i m_i r_{iy}^2$$

$$I_z = \sum_i m_i r_{iz}^2$$

A mean moment of inertia $\bar{I}$ is generally defined as:

$$\frac{1}{\bar{I}} = \frac{1}{3}\left(\frac{1}{I_x} + \frac{1}{I_y} + \frac{1}{I_z}\right)$$

In solution the separation between vibrational levels generally decreases (a red shift). The translation along the principal axes and the rotations around these axes are modified as a result of the presence of solvent molecules. There is a line broadening, with a concomitant loss of the rotational fine structure.

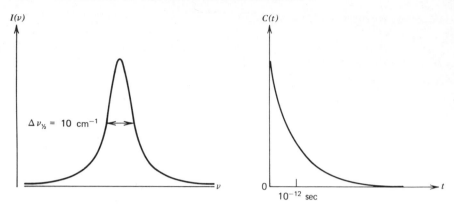

Fig. 4.14. Fourier transform of a pure lorentzian band.

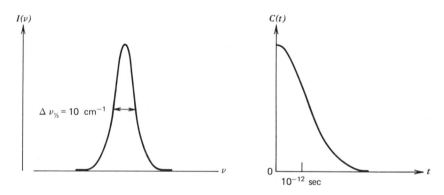

Fig. 4.15. Fourier transform of a pure gaussian band.

Gordon [88], Shimizu [215], and Bratos [39] have shown that it is possible to extract detailed information about translational and rotational motions from vibrational band profiles. Fourier inversion of the absorption band provides the correlation function $C(t)$. This function describes the time dependence of a unit vector μ colinear with the transition moment for the absorption band studied; $C(t)$ is obtained as a statistical average for all the molecules in the medium [64, 238]. It is obtained as the real part of the Fourier transform of the normalized frequency spectrum:

$$C(t) = \left\langle \mu(O) \cdot \mu(t) \right\rangle = \frac{\int_{\text{band}} I(\nu) \, \cos \, 2\pi c(\nu - \nu_0)t \, d\nu}{\int_{\text{band}} I(\nu) \, d\nu}$$

For a pure lorentzian band, $C(t)$ identifies to an exponential function of time (Fig. 4.14):

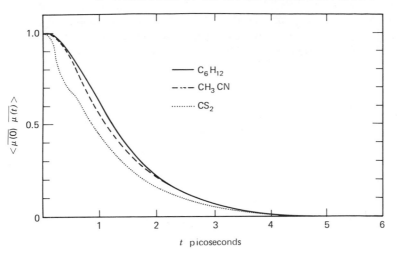

Fig. 4.16. Correlation functions for the camphor carbonyl stretching frequency in various solvents.

$$C(t) = \exp\,(-\beta t)$$

with

$$\beta = \pi c\,\Delta\nu_{\frac{1}{2}}$$

where $\Delta\nu_{\frac{1}{2}}$, the bandwidth at half-height, is expressed in reciprocal centimeters.

If the absorption band is a pure gaussian, $C(t)$ is also a gaussian function which decays more or less rapidly according to the bandwidth (Fig. 4.15), with

$$\frac{a}{b^2} = I_{max}, \quad \text{and} \quad 2b = \Delta\nu_{\frac{1}{2}}$$

The correlation functions $C(t)$ obtained for molecules in solution are generally observed to be gaussians at very short times and exponentials at very long times (thus corresponding to the hybrid character of the absorption bands which are found to be intermediate in shape between true gaussians and true lorentzians) [247] (Fig. 4.16).

What is the physical meaning of these correlation functions? Let us consider again the same unit vector μ along the direction of the transition moment. At some arbitrary time origin, it has a certain direction $\mu(0)$. At times t, its new direction is $\mu(t)$. Clearly, the magnitude of the scalar product:

$$\mu(0)\mu(t) = \mu(0)\,\mu(t)\,\cos\,[\mu(0),\,\mu(t)]$$

$$= \cos\,[\mu(0),\,\mu(t)]$$

is the measure of the angle $[\mu(0), \mu(t)]$ by which the molecule has rotated from its original orientation. The statistical average over all the identical molecules in the medium:

$$\langle \mu(0)\ \mu(t) \rangle$$

is the value assumed at time t by the correlation function $C(t)$.

At time zero the probability of finding a molecule with its original orientation is unity; at infinite time there is equal probability of finding the molecule with any orientation. The statistical average, over all molecules in the medium, gives zero as the value of the correlation function. Hence the decrease in $C(t)$ from 1 to 0 in a relatively short time, 10^{-9} to 10^{-13} sec depending on the molecular species (dependence on the molecular volume, competition of orientational and vibrational relaxation), and the viscosity and temperature of the medium.

By assuming that vibrational relaxation is negligible at very short times (which is not altogether warranted [112]), the behavior of the solute molecule as a free rotor, characterized by a mean moment of inertia $\overline{I}$, dictates the actual shape of the correlation function. $C(t)$ can be expressed as:

$$C(t) = 1 - \frac{kT}{\overline{I}}\ t^2 \qquad \text{(to first-order)}$$

where k is the Boltzmann constant, and T is the absolute temperature. This series development can be further exploited to account for the progressive slowdown of the decrease with time of the μ dipole correlation function as solute molecules encounter solvent molecules:

$$C(t) = 1 - \frac{kT}{\overline{I}}\ t^2 + [\frac{1}{3}(\frac{kT}{\overline{I}})^2 + (24\overline{I})^{-1}\ \langle (OV)^2 \rangle]\ t^4$$

where $\langle (OV)^2 \rangle$ is the mean-square torque exerted on the solute by the solvent molecules.

At longer times, because of the numerous collisions, the molecular reorientation becomes random and the correlation function is characterized by an exponential decay:

$$C(t) = \exp\ (-\beta t)$$

It can be related to the diffusion coefficients, provided some reliable model for the dynamic structure of the liquid is chosen. The Debye diffusion model [1, 54, 81], in which rotation occurs by infinitesimal steps, has been shown to be invalid in several cases [51]. More general theories are being evolved [9, 87, 121, 142].

As an indication of the accuracy possible with some of the best-studied examples, one is able to determine the isotope effect on the rotational diffusion coefficients for a symmetric top molecule such as ammonia or methyl iodide [80] (Table 4.8).

Table 4.8 Diffusion Constants (psec^{-1}) Obtained for Methyl Iodide and Methyl Iodide-d_3

	CH$_3$I	CD$_3$I
D_z	1.94	1.71
D_x	0.48	0.33

The diffusion constants about the symmetry axes are four to five times larger than those about the perpendicular (x or y) axes. The diffusion constants about the z axis have nearly the same values for methyl iodide and methyl iodide-d_3, while those about perpendicular axes differ substantially.

Thus the appearance of a vibrational band can be considerably modified according to the nature of its orientational behavior in a given solvent. By assuming that vibrational relaxation and the influence of the slit width can both be neglected, the following criteria should be useful:

1. The "free" rotation of a molecule is in fact a function of the nature of the solvent. When solute and solvent molecules are matched in size, it may occur for a smaller time and at a smaller frequency than when a small solute molecule tumbles in the free space between giant solvent molecules. This is reflected in the spectral wings of an absorption band; the time during which a solute is able to undergo free rotation is greater the less prominent the spectral wings.

2. The bandwidth, as measured by the $\Delta \nu_{1/2}$ parameter, is a useful characteristic of the motion. It is not even necessary to consider the correlation function [30]. The Heisenberg uncertainty principle dictates that the width be inversely proportional to the rotational lifetime. If it is assumed that the band shape is exclusively determined by rotational relaxation, the orientation

of the transition moment with respect to the inertial axes determines the bandwidth. For a spectral band with a transition moment parallel to one of the inertial axes,

$$\Delta\nu_{\frac{1}{2}} = \frac{1}{\pi c} (D_2 + D_3)$$

where D_2 and D_3 are the rotational diffusion constants relative to the other two axes of inertia [66]. Take for instance the case of nitrous oxide, a linear molecule [225]. In labeling as 1 the internuclear axis, $D_1 = 0$ and $D_2 = D_3 = D$. Thus

$$(\Delta\nu_{\frac{1}{2}})_{\parallel} = \frac{2D}{\pi c} \qquad (\Delta\nu_{\frac{1}{2}})_{\perp} = \frac{D}{\pi c}$$

Equal bandwidths are expected for ν_1, ν_3, and their harmonics, with half the value of the bandwidth for ν_2:

$$
\begin{array}{lll}
\leftarrow N - \leftarrow N - O \rightarrow & \nu_1 \\
\qquad\quad \uparrow \\
N - N - O & \nu_2 \\
\downarrow \qquad\quad \downarrow \\
\leftarrow N - N \rightarrow\leftarrow O & \nu_3
\end{array}
$$

Experimentally, this expectation is rather well fulfilled (Table 4.9).

Table 4.9 Solvent Effect on Vibrations of Nitrous Oxide

	$\Delta\nu_{\frac{1}{2}}$		
Solvent	ν_1	ν_3	ν_2
$n\text{-}C_6H_{14}$	17.0	17.4	7.3
C_6H_{12}	14.5	11.5	6.0
CCl_4	11.2	8.1	4.3
CS_2	12.5	10.5	6.0
C_6H_6	14.3	9.6	6.5
CH_3CN	12.9	12.5	6.6

In the presence of complexation at a site distant from the vibrational band under examination, any increase in the moment of inertia is displayed as band narrowing. Conversely, the bandwidth increases with temperature with a pseudo-Arrhenius behavior [187, 219]:

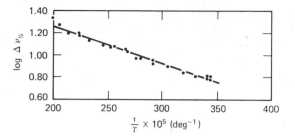

Fig. 4.17. Dependence of the logarithm of the halfwidth of the 2252 cm^{-1} absorption band of acetonitrile on $1/T$. The dots represent the points obtained in an experiment with a 30 mol % solution of acetonitrile in diethyl ether. [From A. I. Sidorova, M. G. Batishcheva, and E. N. Shermatov, *Opt. Spectrosc., Supp.,* **1-3,** 96 (1963).]

$$\Delta\nu_{\frac{1}{2}} = C \exp\left(- U/RT\right)$$

An example (Fig. 4.17) of such temperature broadening is provided by the 2252 cm^{-1} absorption band of acetonitrile (30 mol % in diethyl ether) [216].

Whenever vibrational relaxation has a characteristic time of the same order as that for rotational relaxation, the absorption bandwidth is determined by both factors, and likewise the correlation function is affected by both types of relaxation [112, 159].

Rather similar considerations apply to the solvent dependence of Raman line shapes [38, 147]. Actually, fruitful comparisons can be made between infrared and Raman line shapes [165]; they yield pure rotational or vibrational correlation functions.

9   SOLVENT EFFECTS ON NMR SPECTRA

Solvent effects on NMR spectra have been extensively discussed in an earlier review [127]. We confine our remarks here to solvent effects on chemical shifts, to the exclusion of solvent changes of linewidths or line shapes [127], or coupling parameters [8, 49, 50, 72, 109, 111, 114, 127, 143, 217, 218]. Similar arguments (and similar mistakes) have been published for solvent changes of chemical shifts and of coupling parameters. Since the data on the former are more abundant, and more easily and perhaps more accurately obtained, we restrict our consideration to these changes in the shielding constant of a nucleus.

In proton NMR, solvent shifts are generally measured by use of an internal reference because of ease of operation and because use of an external reference requires rather inaccurate susceptibility corrections, unless one has access to two spectrometers of different design. (See below under bulk magnetic susceptibility.)

In ^{13}C medium shift studies, in which shifts are approximately one order of magnitude greater than for protons, external referencing and the accompanying susceptibility correction are preferred [248].

We attempt to answer here the questions: Can NMR solvent shifts be factored into contributions from independent physical mechanisms? Is it legitimate to view each of these separate factors as a product function of solute and solvent parameters, as initially proposed long ago by Bothner-By [32]? To what possible uses can they be put?

The shielding constant σ for a nucleus within a solute in a fluid differs from its value σ_0 for the isolated molecule. The usual interpretations are founded on the factorization [43]:

$$\sigma = \sigma_0 + \sigma_b + \sigma_a + \sigma_E + \sigma_W$$

in which σ_b originates with the bulk susceptibility of the medium submitted to the uniform magnetic field H_0, σ_a derives from the contribution of magnetically anisotropic solvent molecules, σ_E represents the contribution from polarization of the electron density by local electric fields, and σ_W accounts for the van der Waals pressure experienced by the solute. That such axiomatic factorization can provide a working scheme for the experimental results has been tested, using factor analysis, by Weiner, Malinowski, and Levinstone [230, 232].

Factor analysis is a mathematical technique for defining the number of different factors in a given property. It is postulated that this property can be expressed as a linear sum of terms or factors. Factor analysis yields the minimum number of independent factors necessary to describe the parameter within, for example, its experimental uncertainty. When applied to the chemical shift of a polar solute in solution, measured with respect to a reference substance codissolved (a so-called internal standard), it points to the presence of fundamental factors. The minimum basis necessary to account for the chemical shift of solute nuclei thus includes:

1. The gas-phase shift of the solute.
2. The solvent anisotropy contribution.
3. The van der Waals term.
4. The electric field (reaction field) dependence.
5. The bulk susceptibility contribution, if it has not been already corrected for.

Let us illustrate this dissection of the solvent shift into various terms with the rather elementary example of methane, going from the gas phase into benzene solution [205] (Fig. 4.18).

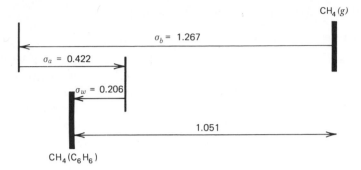

Fig. 4.18. Gas to solution shift for methane (in parts per million).

NMR: Magnetic Susceptibility Correction

If and when an external reference is used—in a container separate from the sample—a correction should be applied to compensate for the difference in magnetic susceptibility between the reference and the sample [126]. Coaxial cylindrical tubes are generally used. When their axis of rotation is parallel to the uniform external field, the correction is:

$$\sigma_b(\text{solution 1}) - \sigma_b(\text{solution 2}) = \frac{4\pi}{3} \left[\chi_m(\text{solution 2}) - \chi_m(\text{solution 1}) \right]$$

When the axis of the cylindrical sample is perpendicular to the applied field, the correction is:

$$\sigma_b(\text{solution 1}) - \sigma_b(\text{solution 2}) = \frac{2\pi}{3} \left[\chi_m(\text{solution 1}) - \chi_m(\text{solution 2}) \right]$$

The first possibility is achieved in the design of spectrometers with super-conducting solenoid magnets. The second possibility is embodied in spectrometers with more conventional magnets. Actually, an elegant method, both accurate and rapid, for determining magnetic susceptibilities and obtaining reference-independent solvent shifts [25, 105] consists of applying both types of measurements to the sample. Values of the molar susceptibilities χ_M for the more important spectroscopic solvents are given in Table 4.9. The magnetic susceptibility χ_m (sometimes written χ_v, for volume susceptibility) is related to the molar susceptibility χ_M:

$$\chi_M = \chi_m \frac{M}{\rho}$$

where M is the molar mass, and ρ is the density.

The Van der Waals Term

When the solute is a nonpolar, isotropic molecule, perturbation of the electron distribution by surrounding magnetically isotropic, nonpolar molecules in the medium reduces to the van der Waals interaction. The solvent shift is defined as the difference between the chemical shift for the isolated molecule, obtained in the gas phase by extrapolating to zero pressure, and the chemical shift in solution at infinite dilution. For a given nucleus the observed solvent shift depends somewhat on its situation within the molecule. For instance, in the series benzene-toluene-p-xylene-mesitylene, the solvent shift for the aromatic protons is progressively reduced with substitution, whereas the methyl protons are less affected (Table 4.10); the degree of exposure to the solvent influences the magnitude of the solvent shift, predominantly because of van der Waals interactions.

Table 4.10   Solvent Shifts (^{1}H) for Aromatic Molecules in Carbon Tetrachloride or Tetraethylsilane Solution at 60 MHz (Hz)[a]

	Solvent	
Solute	CCl_4	$Si(C_2H_5)_4$
Benzene	26.6	14.4
Toluene (aryl)	23.3	13.3
p-Xylene (aryl)	20.4	12.0
Mesitylene (aryl)	17.5	11.0
Toluene (CH_3)	27.5	16.8
p-Xylene (CH_3)	25.4	14.5
Mesitylene (CH_3)	25.0	15.0

[a]M. A. Raza and W. T. Raynes, *Mol. Phys.*, **19**, 199 (1970).

For protons downfield shifts are observed on going from the gas phase to the liquid phase. Likewise, gas-to-liquid chemical shifts of the ^{31}P nucleus in the phosphorus tribromide molecule are downfield shifts, ranging between 1.5 and 4.5 ppm in various solvents [97].

Van der Waals shifts depend on the nature of the solvent (Table 4.10). Their magnitude increases with increasing molecular polarizability of the solvent molecules, which is larger for carbon tetrachloride than for tetraethylsilane. Going through the series methylene chloride-chloroform-carbon tetrachloride, the gas-to-solution downfield shift of the ^{19}F nuclei in hexafluorobenzene also gradually increases in magnitude (Table 4.11).

Table 4.11   Solvent Shifts (^{19}F) for C_6F_6
in Halogenated Solvents (ppm)[a]

Solvent	$\Delta\sigma$[b]
CH_2Cl_2	7.75
$CHCl_3$	8.67
CCl_4	9.09

[a]N. Cyr, M. A. Raza, and L. W. Reeves,
Mol. Phys., **24**, 459 (1972).
[b]Corrected for bulk diamagnetic susceptibility in a cylindrical sample.

A simplified theory [123, 158] of the dispersion interaction between a pair
of nonpolar isotropic molecules yields an expression of the type:

$$\sigma_W = -B \left\langle F^2 \right\rangle$$

where the mean-square effective electric field from all solvent molecules acting
on the solute (molecule 1) is given by:

$$\left\langle F^2 \right\rangle = 3a_2 I_2 / R^6 \left[I_1 / (I_1 + I_2) \right]$$

Here a_2 and I_2 are the polarizability and ionization potential of the solvent
molecules, I_1 is the ionization potential of the solute molecule, and R is the
distance between the centers of solute and solvent molecules. The B coefficient
is expressed as:

$$B = \frac{1}{2} \frac{3I_1 + 2I_2}{(I_1 + I_2)^2} \sigma_0 \, a_1$$

where σ_0 is the shielding for an isolated unperturbed molecule, and a_1 is the
solute polarizability.

The polarizability a_2 of a solvent molecule can be expressed [173] as a
function of the refractive index n and of the radius a of the spherical cavity:

$$a_2 = \frac{n^2 - 1}{n^2 + 2} a^3$$

For hydrogen in C—H bonds, the parameter B has values [206] in the range
0.4 to 1.5 $\times$ 10^{-18} esu, which compare well with the theoretical value of
0.74 $\times$ 10^{-18} esu for a hydrogen atom [148].

Use of such expressions yields order-of-magnitude agreement with observed solvent shifts for rare gas atoms dissolved in various gaseous solvents (Table 4.12). For nuclei such as ^{129}Xe, shifts are very large and relatively less affected by experimental uncertainty than are ^{19}F and, worse yet, ^{1}H shifts.

Table 4.12   Dispersion Shifts for ^{129}Xe in Various Gases[a]

	$\sigma_W \times 10^3$	
Solvent	*Calculated*	*Experimental*
Ar	6.3	3.080
Kr	9.9	6.009
Xe	13.4	12.188
CO_2	7.6	3.779
CF_4	8.2	4.262
CH_4	8.7	6.165

[a] R. A. Kromhout and B. Linder, *J. Chem. Phys.*, **54**, 1834 (1971).

Solute-solvent interaction can also be written as a product of two functions. One function characterizes the solute, and one function characterizes the solvent. They have been expressed [231] as:

$$x_1(\text{solute}) = \frac{8\pi^2 N^2 e^2 B}{9V_1}$$

$$y_2(\text{solvent}) = \frac{n^2 + 2}{2n^2 + 1} \frac{\sum\limits_{j}\langle r_j^2 \rangle}{V_2}$$

Here n is the refractive index of the medium, V_1 and V_2 are the molar volumes of the solute and of the solvent, respectively, and r_j is the radius of the electron cloud distribution in the solvent molecules. The

$$\sum\limits_{j}\langle r_j^2 \rangle$$

term can be estimated from the volume diamagnetic susceptibility [106], or from semiempirical estimates [150]. The

$$\frac{(n^2 + 2)}{(2n^2 + 1)}$$

term varies relatively little from solvent to solvent.

More often, the solvent refractive index function is represented by the McRae term [131] $(n^2 - 1)/(2n^2 + 1)$. Correlations between magnetic resonance parameters (chemical shifts [130], coupling constants [131]), and the McRae term have been observed for centrosymmetric nonpolar solutes in nonpolar solvents. The influence of dispersion forces on the chemical shift of methane protons [131] parallels the increase in the refractive index n (Table 4.13). For the 11 different halocarbon solvents of Table 4.13, there is a remarkably successful correlation between the solvent shift $\Delta\sigma$ and the McRae term $(n^2 - 1)/(2n^2 + 1)$; the correlation coefficient is better than 0.98. The origin of the empirical additivity rules of Raynes was thus adequately explained [188].

Table 4.13 Methane Gas-to-Solution Shift in Relation to Refractive Index of the Solvent [131]

Solvent	Refractive index n	Downfield shift $\Delta\sigma$ at 60 MHz (Hz)
CH_3Br	1.4234	23.8
CH_2Cl_2	1.4237	24.4
$CHCl_3$	1.4455	25.2
CCl_4	1.4630	26.9
CH_2ClBr	1.48	28.5
CCl_3Br	1.5061	32.5
CH_3I	1.5317	30.9
CH_2Br_2	1.5419	31.9
$CHClBr_2$	1.5482	34.1
$CHBr_3$	1.5976	39.1
CH_2I_2	1.7425	46.0

The Reaction Field

The chemical shift of a proton in an X–H bond is affected by the local electric field E at the nucleus [41, 164]:

$$\Delta\sigma = -\,AE_z - B\mathbf{E}^2$$

where E_z is the component of $\mathbf{E}$ along the X–H bond, and A and B are estimated [94, 180] as 3×10^{-12} esu and 10^{-18} esu, respectively. The E_z component of the electric field draws electrons away from the hydrogen nucleus, thus reducing its shielding.

Such electric fields are being produced by the permanent electric moments in the polar solvent molecules, or by the Onsager *reaction field* $\mathbf{R}$ from polarizable solvent molecules. The reaction field is a local field created by polarization of the solvent, conceived as a continuum of dielectric constant ϵ in which the polar solute is immersed [174].

The polar solute is, as a first approximation, reduced to a point dipole μ at the center of a spherical cavity whose radius a is essentially the molecular radius:

$$R = \frac{\mu}{3a}\ \frac{2(\epsilon-1)(n^2-1)}{2\epsilon + n^2}$$

where the isotropic molecular polarizability a of the solute is obtained [174] from the refractive index n:

$$a = \frac{n^2-1}{n^2+2}\ a^3$$

Neglecting the quadratic term $-BE^2$, the contribution of the reaction field to the shielding change can be expressed as the following product:

$$\Delta\sigma = -\frac{A_1\mu_1\ \cos\ \theta_1}{3a_1}\ \frac{2(\epsilon_2-1)(n_1^2-1)}{2\epsilon_2 + n_1^2}$$

where subscripts 1 and 2 refer to solute and solvent, respectively.

Several studies have attempted to correlate solvent shifts for polar solutes in media of varying permittivity or dielectric constant ϵ. They have met with some success. For instance, the solvent shifts of acetonitrile appear to incorporate a substantial reaction field contribution [189] (Table 4.14), after allowance for bulk susceptibility and van der Waals contributions.

Table 4.14 Gas-to-Infinite Dilution Proton Chemical Shifts at 60 MHz of CH_3CN in Various Solvents (Hz) [189]

	Solvent					
	n-C_6H_{14}	C_6H_{12}	C_5H_{10}	$Si(C_2H_5)_4$	$Sn(C_2H_5)_4$	$SiCl_4$
$\sigma_E(R)$, obs.	-3.3	-4.1	-4.4	-4.1	-4.9	-4.9
$\sigma_E(R)$, calc.	-3.65	-4.01	-4.08	-4.25	-4.66	-4.94

A smoothly varying range of dielectric constants is provided by changing the concentrations of the components in binary solvent mixtures, such as carbon tetrachloride-acetone, cyclohexane-cyclohexanone, or carbon tetrachloride-nitromethane [128]. Apparently, there is no unique reaction field acting on a polar molecule in a liquid. No firm correspondence between the above expressions and the actual solvent shift at a given molecular position was found in studies of bromocyclohexanones [128], pyridine [115], or dibenzo-thiophene [18]. These findings are not too surprising. The Onsager continuum approach is inherently incapable of revealing the dependence of the dielectric constant on the configuration of solvent molecules around the polar solute [224]. It thus ignores the fluctuations of local order in the liquid. Likewise, alignment of polar molecules such as nitrobenzene or *p*-nitrotoluene in liquids by *external* electric fields gives shifts reduced in size by almost an order of magnitude with respect to predictions of the Onsager local-field model [211].

The Solvent Anisotropy Term

Some solvents, such as benzene, are characterized by large values of their susceptibility anisotropy Δ_χ, as measured by the difference between the axial χ_{ax} and the tranverse χ_{tr} molecular susceptibilities:

$$\Delta_\chi = \chi_{ax} - \chi_{tr}$$

Consider a spherical solute molecule moving over the surface of a benzene solvent molecule, which is disc-like. The lines of force of the magnetic field produced by the circulating π electrons in the presence of the applied magnetic field are well known. If the center of the solute is on the benzene sixfold symmetry axis, there will be maximum shielding. Conversely, if the solute is at van der Waals contact, with its center in the plane of the benzene ring, it will experience maximum deshielding.

On the average, the solute spends more time over the face of a benzene ring than around its edge, thus experiencing a net shielding. The opposite is true for a rodlike molecule such as carbon disulfide.

The shielding contribution from surrounding solvent molecules having a large susceptibility anisotropy Δ_χ can be evaluated for nonpolar centrosym-

Table 4.15 Solvent Anisotropy Term and Molecular Volume

Solute	V (cm^3)	σ_a (ppm)
CH_4	38.65	0.422
$C(CH_3)_4$	119.9	0.481
$C(C_2H_5)_4$	170.3	CH_3, 0.454
		CH_2, 0.479
$Si(CH_3)_4$	136.1	0.488
C_6H_6	88.86	0.521
C_6H_{12}	108.0	0.491

metric solutes, subtracting from the observed gas-to-solution shift the calculated bulk susceptibility and van der Waals terms. Values of σ_a obtained [104, 205] for various solutes in benzene solution are given in Table 4.15, together with their molar volumes V. It is seen that the values of σ_a depend slightly on the molecular volume V [24, 26, 241].

Calculations of σ_a have given order-of-magnitude agreement [24, 26, 104, 205, 241]; it is assumed that:

1. The magnetic field originating from a solvent molecule is approximated by that of a point dipole located at its center.
2. Solvent molecules can assume all possible orientations with respect to to solute.
3. Solute and solvent molecules are in van der Waals contact.

Some investigators have entertained the mistaken notion that "even the benzene molecules surrounding a quasi-spherical non polar molecule have a preferred geometry relative to such a solute which in these cases is the sole reason for a σ_a effect in the first place" [207]. It is to be very strongly emphasized that a nonzero σ_a must inevitably be found for spherical solutes whenever the solvent molecules do not have spherical symmetry, even if the solute-solvent encounters are totally at random.

The resulting equations [24, 26, 104, 207, 241] show that σ_a is proportional to Δ_χ, in rough accord with observation (Table 4.16), and that there is indeed an inverse dependence on $V^{1/6}$.

Table 4.16 Effect of the Solvent Susceptibility Anisotropy [104]

Solvent	$-\Delta_\chi$ (10^{-30} emu)	$\sigma_a(CH_4)$ (ppm)	$\sigma_a(C_6H_{12})$ (ppm)
CS_2	5.00	-0.14	-0.13
C_6H_6	9.91	$+0.42$	$+0.49$

For benzene solutions the following semiempirical relationship [207] appears to be quite satisfactory:

$$\sigma_a \text{ (ppm)} = + 1.08 \ V^{-1/6}$$

Note that σ_a is independent of the position within the solute of the nucleus under observation, as is indeed found experimentally; compare the very similar values for the methyl and methylenic protons of tetraethylmethane (Table 4.15). The site factor is thus unity for all nuclei in the solute, irrespective of their distance from the center of the molecule. This has been explained by Becconsall [24, 26]; for a nucleus whose distance from the molecular center is a, the average magnetic field **B** is produced by external sources (solvent molecules). It thus satisfies the double condition; **curl B** = 0 and **div B** = 0. Since **B** can be expressed as the gradient of a scalar potential, and applying Gauss' theorem to the spherical surface of radius a, it can be shown that **B** has the same value as at the center of the molecule.

Aromatic Solvent-Induced Shifts

The phenomenon of aromatic solvent-induced shifts (ASIS) [127] has been investigated in numerous cases of a polar solute dissolved in an aromatic solvent. By definition [125], ASIS are measured as the difference in the chemical shift of a given nucleus between a suitably inert reference solvent (a hydrocarbon or fluorocarbon; carbon tetrachloride; sometimes an even less fortunate choice [125], chloroform) and an aromatic solvent. Why an aromatic solvent? Because its pronounced magnetic anisotropy helps to reveal subtle solute-solvent interactions which otherwise would be too weak to be measurable with any accuracy.

The ASIS phenomenon was rapidly found to be useful. The shifts appear to be characteristic of the position of the nucleus under examination with respect to the various (polar) chemical groups in the solute molecule. Furthermore, these shifts are additive to a very good approximation. Hence they have been applied to numerous problems of structure, configuration, and conformation. Table 4.17 is a selected list, by compound type, of the most successful applications.

Table 4.17 Main Structural Types for which ASIS have been Applied

Compound Type	Ref.
Acetanilides	186
Acetophenones	11
Alkene sulfonic acids and chlorides	156
Adamantanes (2-substituted)	93
Erythromycin aglycones	56
Alcohols	57, 58
Enamino aldehydes and ketones	122
a,β-Unsaturated aldoximes and ketoximes	242
Aliphatics	100, 101
Alkenes	2
Amidoximes	6
Amines	16
Anhydrides	152
Aromatic hydrocarbons	17, 65, 135, 178
Aziridines	246
t-Butylbenzene, t-butoxybenzene, nitro-t-butylbenzene	135

Table 4.17 Main Structural Types (continued)

Compound Type	Ref.
Benzene derivatives	35, 36, 172, 178
Borane adducts	48
Bornane derivatives	13
Camphor and thiocamphor	55
Carane derivatives	40
Cyclopropyl- and epoxycarbinols	181
Three-membered ring carbonyl compounds	37
Metal carbonyls	161
Chloroethylenes	99, 102
Cyclopropane derivatives	62
Cyclopropanols	182
Cyclopropyl carboxamides	214
Chromanones	170
Substituted cyclohexanones	61, 71
N,N-Dialkylamides	91
a-Diketones	34
Dioxanes	183
Diquat, paraquat	96
Epoxides	237
Methylbenzoic esters	4
Monoethyl, diethyl, and cyanoethyl esters	117
Ethers	166, 237
Cyclic ethers	166
Five-membered rings containing heteroatoms	221
1-Chloro-1-fluoroethylene	109
Syn- and *anti*-2-furanaldoximes	228
Methyl halides	102
Halogenopropenes	107
Helicenes	149
Henokiflavone	177
Heterocyclic compounds	208
Hydroxy and methoxy compounds	239
Imides	152
Imines	246
Indoles	199
Alicyclic ketones	46, 47
Cyclic ketones	110, 133
Cyclopropyl ketones	3, 213
Enamino ketones	122
Epoxy ketones	213
Steroidal ketones	85
a,β-Unsaturated ketones	14, 198, 223
N-Methyl lactams	160
γ-Lactones	167

Table 4.17 Main Structural Types (continued)

Compound Type	Ref.
Cyclic lactones	110
Methoxybenzenes	35, 36
Aromatic methoxy compounds	65
Monosubstituted (poly)methylbenzenes	172
Acetylated monosaccharides	75, 76
Natural products	11, 83, 191, 209
Nitriles	153, 223
Nitroalkanes	5
Nitroform	103
N-Alkyl nitrones	7
Nitrosamines	44
Norbornene derivatives	129
Bicyclo[2.2.2] octenes	89
Organoantimony and organotin compounds	141
Oxathianes and oxathiolanes	184
Oxazoles	33
Oximes	246
O-Alkyloximes	7
Oxymercurials	192
Pseudoguaianolides	98
Pyrazines	176
Pyridines	199
Alkyl-substituted pyridines	45
Pyrimidines	176
Pyrimidine bases	202
Quinolines	199
Pyrroles	199
Steroidal sapogenins	236
Silylamines	197
Steroids	69, 70, 92, 95
Sulfoxides	169, 171, 222
Benzene chromium tricarbonyls	145
N,N-Disubstituted trifluoroacetamides	90
Triterpenes	240
Ubiquinone analogs	235
a,β-Unsaturated compounds	200
Vinyl metallic compounds	29

It has been recently estimated [207] that approximately 400 papers appear every year that make a more or less extensive use of ASIS. A classic example is the case of compounds containing a carbonyl moiety dissolved in benzene

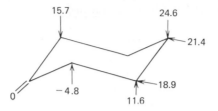

Fig. 4.19. CCl$_4 \rightarrow$ C$_6$H$_6$ solvent shifts (in Hz at 60 MHz) for methyl protons at the indicated positions of a cyclohexanone.

[12, 28, 31, 71, 83, 110, 132, 198, 209, 240]; characteristic ASIS are obtained for the methyl substituents in the α, β, and γ positions of cyclohexanones (Fig. 4.19). These can be used, for instance, to assess the position of the conformational equilibrium in methyl-2-cyclohexanone.

A second example of the detailed structural information accessible through ASIS is in the aromatic series. A linear relationship has been found between the ASIS in the meta and para positions and the dipole moments of monosubstituted benzenes, C$_6$H$_5$X; for instance,

$$\mu = 7.14(\text{ASIS})_{\text{meta}}$$

where (ASIS)$_{\text{meta}}$ is expressed in parts per million [196].

It is even possible to obtain from the ASIS the magnitude and the direction of the dipole moment projection along the C$-$X bond in C$_6$H$_5$X compounds for which the C$-$X bond is not colinear with the dipole moment. For aniline, this projection is 0.9 ± 0.2 D, corresponding to a moment angle of between 110 and 140°. For phenol, the moment angle is smaller (ca. 60°), as inferred from the component of μ along the C$-$O bond, 0.8 D.

Likewise, Hammett substituent constants can be determined with great ease from ASIS measurements [196, 229]. For 1-X-3,5-dichlorobenzenes, for instance, the resonance frequency of proton 4 is obtained from simple spectral inspection as that of a single line, the antisymmetric transition in the AB_2 system for the ring protons. Measurements of its position for 3 mol % solutions in cyclohexane and in benzene yield the corresponding ASIS $\Delta_p{}^X$, which is related [229] to $\sigma_p{}^0$ [193] through:

$$\Delta_p{}^X = 35.8 \; \sigma_p{}^0 + 5.4 \quad \text{Hz at 60 MHz}$$

The molecular dynamics of the system formed by a polar solute with an aromatic solvent are very informative with respect to the physical nature of the ASIS phenomenon. Whereas the component molecules in the (hydrogen-bonded) chloroform-dimethyl sulfoxide system indeed move as a joint entity, which can

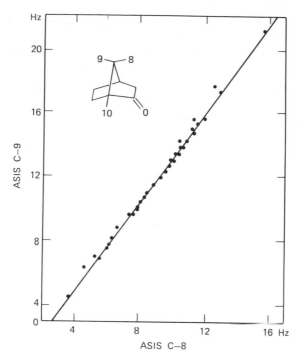

Fig. 4.20. Plot of the C-9 methyl ASIS versus the C-8 methyl ASIS of camphor for a large number of aromatic solvents. Correlation coefficient = 0.995. [From P. Laszlo and E. M. Engler, *J. Am. Chem. Soc.*, **93**, 1317 (1972).]

legitimately be termed a complex, the molecular partners in the looser acetone-benzene system undergo independent translational and rotational motions [10]. Likewise, the molecular motion of chloroform [203] or camphor [112] in benzene solution is that of an unhindered rotor with respect to any of the three principal axes of inertia.

More indirect arguments can also be adduced to rule out the description of such weak interactions in terms of long-lived, well-defined complexes. These are characterized by equilibrium constants K whose values reflect the strength of the complex—they are always found in the range of 0 to 0.51 l mol^{-1}—and whose logarithms should vary linearly with reciprocal temperature, the slope being the heat of formation of the postulated complex.

In fact, relationships of the type:

$$K = \frac{B}{T} - 1$$

are found to best represent the dependence of the apparent equilibrium constant on temperature for steroidal ketones [132] or 1,3-dioxanes in toluene solution [63].

Of even more interest, the ASIS are linear functions of the *density* of toluene at various temperatures [63, 183]. Another line of argument serves to disprove the notion [33, 134] of complexes between polar solutes and aromatic solvent molecules, with a well-defined orientation of one solvent molecule with respect to the solute dipole. This refutation is based on comparisons of the ASIS in various aromatic solvents. Such comparisons can be *intra*molecular, when the ASIS for two different groups of nuclei within the same molecule are compared; or they can be *inter*molecular, when the ASIS for groups of nuclei in different molecules are compared.

An example of an *intra*molecular comparison is provided by the outstanding linear correlation obtained when the ASIS are plotted for the 9-methyl group in camphor versus those for the 8-methyl group in the same molecule [63] (Fig. 4.20).

The existence of such a correlation cannot be reconciled easily with the notion of 1:1 camphor-benzene complex. For example, in going from benzene to nitrobenzene solvent, both the geometry and the equilibrium constant for the 1:1 complex would have to change drastically in order to comprise the dipolar interaction between the carbonyl group of the solute and the nitro substituent in the solvent molecule. It could not be expected that the change in the C-9 ASIS would be linear in that for the C-8 ASIS irrespective of the solvent used. *Intra*molecular comparisons of the same sort have been extremely successful for a highly polar bicyclic solute, 1,4,7,7-tetrachloronorbornane, for the ASIS of the endo and of the exo protons [146].

Moreover, *inter*molecular comparisons are also possible. The ASIS for 1,4,7,7-tetrachloronorbornane and those for camphor are linearly correlated in 11 different aromatic solvents, with correlation coefficients of 0.96 to 0.97 [146]. Several further intermolecular comparisons can be gleaned from the literature, some of which are presented graphically in Fig. 4.21 [63, 146, 162]. The numbers under the chemical structures are the ratios of the differences in chemical shifts between the indicated nuclei, in benzene and in fluorobenzene solutions. The constancy of the benzene/fluorobenzene ratio at 1.60 ± 0.15 is a rather spectacular indication that the ASIS should be expressible as product functions of solute and solvent parameters.

Provided that internal chemical shifts are used, that is, that the solute resonances are referred to one of their lot (the Δ-difference method [207]), the ASIS can indeed be expressed as product functions of solute and solvent parameters [63].

In an investigation of the origin of the ASIS phenomenon, Engler and Laszlo [63] studied camphor in a variety of aromatic solvents. Reference-independent

1.62

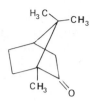

1.55

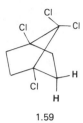

1.59

1.61

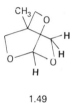

1.49

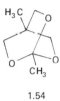

1.54

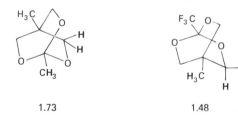

1.73   1.48

Fig. 4.21. Ratios of the differential ASIS between the indicated protons in benzene and in fluorobenzene solution.

ASIS were determined by what was to be later called the Δ-difference method [207] for the C-8 and the C-9 methyl groups relative to the C-10 methyl proton resonance. It was then found that there is an outstanding linear correlation of the C-8 and the C-9 ASIS for more than 50 different aromatic solvents, with zero intercept, in accord with a formulation of the type (Fig. (4.20):

$$ASIS = (solute\ property)\ (solvent\ parameter)$$

The solute property was identified [63] with a site factor depending only on the geometry of the solute. As a first approximation, localizing the main source for the ASIS as the polar carbonyl group of the camphor molecule, the *apparent* change in the magnetic anisotropy of the carbonyl brought about by the proximity of magnetically anisotropic solvent molecules can be expressed by the dipolar term $(3 \cos^2 \theta - 1)/r^3$. The dipole moment of the solute, or of its polar part(s), should also enter the solute parameter. For benzenoid hydrocarbons the solvent parameter is simply proportional to the concentration of benzene rings in the medium.

What is the nature of the local ordering* implicit in the ASIS phenomenon? Certainly, it cannot be pictured as a relative *orientation* of solute and solvent molecules, if one has in mind the angle formed by the two molecular axes. Since relaxation of a solvent cage occurs on the picosecond time scale, any *angular* information about the encounter is necessarily lost on the NMR time scale.

Notion of a preferred geometry should be carefully avoided. We visualize the ASIS phenomenon as the time-averaged change in the *radial* part of the solute-solvent interaction, as modulated by the electronic distribution in the solute molecule. The solvent internal pressure is greater at certain points on the solute surface, depending on their (geometric) relationship with respect to the polar centers or bonds in the molecule.

Note added in proof:

Since submission of our manuscript, the relaxation time (~ 40 psec) for the reorganization of ethanol solvent molecules around a highly dipolar (~ 23 D) solute has been measured directly through picosecond spectroscopy [W. S. Struve, P. M. Rentzepis, and J. Jortner, *J. Chem. Phys.*, **59**, 5014 (1973)]. Energy-wise, excited state solvation is commensurate with ground state solvation for the $n \rightarrow \pi^*$ blue shift incurred by carbonyl compounds in polar solvents [P. Haberfield, *J. Am. Chem. Soc.*, **96**, 6526 (1974)]. The lack of a further red shift on raising the pressure in water solvent when observing weak electronic transitions is ascribed to dipole-quadrupole interactions between the solvent molecules and the chromophore; repulsive interactions, leading to greater destabilization of the excited state, can be important [A. Zipp and W. Kauzmann, *J. Chem. Phys.*, **59**, 4215 (1973)].

---

*Abuse of terms such as *collision complex* or *contact complex* has become widespread. The former was coined by Ketelaar [118]. It was originally meant to explain the appearance of new transitions in the vibrational spectrum. These are due to the fleeting intermolecular interactions occurring on a collision, whose lifetimes are short with respect to the angular momentum correlation time τ_j. Likewise, a contact complex serves to label the charge transfer between colliding molecules in a *random* encounter [23, 175]. Both terms are singularly inappropriate in reference to NMR.

Finally, we mourn the brutal death on March 18, 1973, of Dr. Michel Jauquet shortly after completion of this chapter. This is the last contribution of an exceptionally gifted young scientist.

References

1. A. Abragam, *The Principles of Nuclear Magnetism*, Clarendon, Oxford, 1961.
2. C. Agami and S. Combrisson, *Bull. Soc. Chim. Fr.*, **1968**, 2138.
3. C. Agami and J. L. Pierre, *Bull. Soc. Chim. Fr.*, **1969**, 1963.
4. N. E. Alexandrou, P. M. Hadjimihalakis, and E. G. Pavlidou, *Org. Magn. Resonance*, **3**, 299 (1971).
5. N. E. Alexandrou and D. Jannakoudakis, *Tetrahedron Lett.*, **1968**, 3841.
6. N. E. Alexandrou and D. N. Nicolaides, *Org. Magn. Resonance*, **4**, 229 (1972).
7. N. E. Alexandrou and A. G. Varvoglis, *Org. Magn. Resonance*, **3**, 293 (1971).
8. T. D. Alger and H. S. Gutowsky, *J. Chem. Phys.*, **48**, 4625 (1968).
9. J. E. Anderson, *J. Chem. Phys.*, **47**, 4879 (1967).
10. J. E. Anderson, *J. Chem. Phys.*, **51**, 3578 (1969).
11. H. T. Anthonsen, *Acta Chem. Scand.*, **22**, 352 (1968).
12. K. M. Baker and B. R. Davis, *J. Chem. Soc., B*, **1968**, 261.
13. K. M. Baker and B. R. Davis, *Tetrahedron*, **24**, 1663 (1968).
14. J. E. Baldwin, *J. Org. Chem.*, **30**, 2423 (1965).
15. P. Baraldi, P. Mirone, and E. S. Guidetti, *Z. Naturforsch.*, **A26**, 1852 (1971).
16. D. J. Barraclough, P. W. Hickmott, and O. Meth-Cohn, *Tetrahedron Lett.*, **1967**, 4289.
17. K. D. Bartle, D. W. Jones, and R. S. Matthews, *J. Chem. Soc., A*, **1969**, 876.
18. K. D. Bartle and T. T. Mokoena, private communication, June 7, 1972.
19. E. Bauer and M. Magat, *J. Phys. Radium*, **9**, 319 (1938).
20. M. E. Baur and M. Nicol, *J. Chem. Phys.*, **44**, 3337 (1966).
21. N. S. Bayliss, *J. Chem. Phys.*, **18**, 292 (1950).
22. N. S. Bayliss, *J. Mol. Spectrosc.*, **31**, 406 (1969).
23. N. S. Bayliss and E. G. McRae, *J. Phys. Chem.*, **58**, 1002 (1954).
24. J. K. Becconsall, *Mol. Phys.*, **15**, 129 (1968).
25. J. K. Becconsall, G. D. Daves, Jr., and W. R. Anderson Jr., *J. Am. Chem. Soc.*, **92**, 430 (1970).
26. J. K. Becconsall, T. Winkler, and W. von Philipsborn, *J. Chem. Soc., D*, **1969**, 430.
27. L. F. Bellamy, H. E. Hallam, and R. L. Williams, *Trans. Faraday Soc.*, **54**, 1120 (1958).
28. S. Bien and U. Michael, *Chem. Ind.* (London), **1967**, 664.

29. D. J. Blears, S. Cawley, and S. S. Danyluk, *J. Mol. Spectrosc.*, **26**, 524 (1968).

30. A. E. Boldeskul and V. E. Pogorelov, *Opt. Spectrosc.*, **28**, 248 (1970).

31. S. Bory, M. Fetizon, P. Laszlo, and D. H. Williams, *Bull. Soc. Chim. Fr.*, **1965**, 2541.

32. A. A. Bothner-By, *J. Mol. Spectrosc.*, **5**, 52 (1960).

33. J. H. Bowie, P. F. Donaghue, and H. J. Rodda, *J. Chem. Soc., B,* **1969**, 1122.

34. J. H. Bowie, G. E. Gream, and M. H. Laffer, *Aust. J. Chem.*, **21**, 1799 (1968).

35. J. H. Bowie, J. Ronayne, and D. H. Williams, *J. Chem. Soc., B,* **1966**, 785.

36. J. H. Bowie, J. Ronayne, and D. H. Williams, *J. Chem. Soc., B,* **1967**, 535.

37. D. W. Boykin, Jr., A. B. Turner, and R. E. Lutz, *Tetrahedron Lett.*, **1967**, 817.

38. S. Bratos and E. Maréchal, *Phys. Rev.*, **A4**, 1078 (1971).

39. S. Bratos, J. Rios, and Y. Guissani, *J. Chem. Phys.*, **52**, 439 (1970).

40. H. C. Brown and A. Suzuki, *J. Am. Chem. Soc.*, **89**, 1933 (1967).

41. A. D. Buckingham, *Can. J. Chem.*, **38**, 300 (1960).

42. A. D. Buckingham, *Proc. Roy. Soc.*, **A248**, 169 (1958); **255**, 32 (1960); *Trans. Faraday Soc.*, **56**, 753 (1960).

43. A. D. Buckingham, T. Schaefer, and W. G. Schneider, *J. Chem. Phys.*, **32**, 1227 (1960).

44. Y. L. Chow and M. M. Feser, *Chem. Commun.*, **1967**, 239.

45. R. J. Chuck and E. W. Randall, *J. Chem. Soc., B,* **1967**, 261.

46. J. D. Connolly and R. McCrindle, *Chem. Ind.*, (London), **1965**, 379.

47. J. D. Connolly and R. McCrindle, *J. Chem. Soc., C,* **1966**, 1613.

48. A. H. Cowley, M. C. Damasco, J. A. Mosbo, and J. G. Verkade, *J. Am. Chem. Soc.*, **94**, 6715 (1972).

49. R. H. Cox and L. W. Harrison, *J. Magn. Resonance.*, **6**, 84 (1972).

50. R. H. Cox and S. L. Smith, *J. Magn. Resonance*, **1**, 432 (1969).

51. R. I. Cukier and K. Lakatos-Lindenberg, *J. Chem. Phys.*, **57**, 3427 (1972).

52. E. A. Cutmore, M. Sc. Thesis, University of Wales, 1961. Numerous such plots have been reported by these workers.

53. J. G. David and H. E. Hallam, *Trans. Faraday Soc.*, **65**, 2838, 2843 (1969).

54. P. Debye, *Polar Molecules,* Chemical Catalog, New York, 1929.

55. P. V. Demarco, *Chem. Commun.*, **1969**, 1418.

56. P. V. Demarco, *Tetrahedron Lett.*, **1969**, 383.

57. P. V. Demarco, E. Farkas, D. Doddrel, B. L. Mylari, and E. Wenkert, *J. Am. Chem. Soc.*, **90**, 5480 (1968).

58. P. V. Demarco and L. A. Spangle, *J. Org. Chem.*, **34**, 3205 (1969).

59. P. Diehl and C. L. Khetrapal, *N.M.R.—Basic Principles and Progress,* Vol. 1, Springer-Verlag, Berlin, 1969.

60. J. E. Dubois and A. Bienvenüe, *J. Chim. Phys.*, **65**, 1259 (1968).

61. P. Dufey, J. Delmau, and J. C. Duplan, *Bull. Soc. Chim. Fr.*, **1967**, 1336.

62. H. Dürr, *Justus Liebigs Ann. Chem.*, **703**, 109 (1967).

63. E. M. Engler and P. Laszlo, *J. Am. Chem. Soc.*, **93**, 1317 (1971).

64. G. E. Ewing, *Accounts Chem. Res.*, **2**, 168 (1969).
65. H. M. Fales and K. S. Warren, *J. Org. Chem.*, **32**, 501 (1967).
66. L. D. Favro, *Phys. Rev.*, **119**, 53 (1960).
67. E. E. Ferguson and F. A. Matsen, *J. Am. Chem. Soc.*, **82**, 3268 (1960).
68. M. Fétizon and P. Foy, *Bull. Soc. Chim. Fr.*, **1967**, 2653.
69. M. Fétizon, M. Golfier, and J. C. Gramain, *Bull. Soc. Chim. Fr.*, **1968**, 275.
70. M. Fétizon, M. Golfier, and P. Laszlo, *Bull. Soc. Chim. Fr.*, **1965**, 3205.
71. M. Fétizon, J. Goré, P. Laszlo, and B. Waegell, *J. Org. Chem.*, **31**, 4047 (1966).
72. H. Finegold, *J. Phys. Chem.*, **72**, 3244 (1968).
73. R. Foster, *Organic Charge-Transfer Complexes,* Academic, New York, 1969.
74. R. Freeman and N. S. Bhacca, *J. Chem. Phys.*, **45**, 3795 (1966).
75. M. H. Freemantle and W. G. Overend, *Chem. Commun.*, **1968**, 503.
76. M. H. Freemantle and W. G. Overend, *J. Chem. Soc., B*, **1969**, 547.
77. Y. I. Frenkel, *Sobr. Izbr. Trudov, Akad. Nauk SSR*, 3 (1959).
78. H. B. Friedrich and W. B. Person, *J. Chem. Phys.*, **44**, 2161 (1966).
79. J. Fruwert, G. Geiseler, and D. Luppa, *Z. Chem.*, **8**, 238 (1968).
80. T. Fujiyama and B. Crawford, Jr., *J. Phys. Chem.*, **73**, 4040 (1969).
81. W. H. Furry, *Phys. Rev.*, **107**, 7 (1957).
82. R. Gallardo-Herrero, G. Torri, J. F. Gal, and M. Azzaro, *C. R. Acad. Sci.*, **C275**, 1319 (1972).
83. J. P. Garratt, F. Scheinmann, and F. Sondheimer, *Tetrahedron*, **23**, 2413 (1967).
84. R. Gleiter, D. Schmidt, and J. Streith, *Helv. Chim. Acta*, **54**, 1645 (1971).
85. E. Glotter and D. Lavie, *J. Chem. Soc., C*, **1967**, 2298.
86. D. L. Glusker and H. W. Thompson, *Spectrochim. Acta*, **6**, 434 (1954).
87. R. G. Gordon, *J. Chem. Phys.*, **44**, 1830 (1966).
88. R. G. Gordon, *J. Chem. Phys.*, **41**, 1819 (1964); **43**, 1307 (1965); **44**, 1830 (1966); *Advan. Magn. Resonance*, **3**, 1 (1968).
89. M. Gordon, Q. O. Gurudata, and J. B. Stothers, *Can. J. Chem.*, **48**, 1098 (1970).
90. L. L. Graham, *Org. Magn. Resonance*, **4**, 335 (1972).
91. L. L. Graham and M. R. Miller, *Org. Magn. Resonance*, **4**, 327 (1972).
92. J. C. Gramain, H. P. Husson, and P. Potier, *Bull. Soc. Chim. Fr.*, **1969**, 3585.
93. J. W. Greidanus, *Can. J. Chem.*, **48**, 3593 (1970).
94. G. K. Hamer and W. F. Reynolds, *Can. J. Chem.*, **46**, 3813 (1968).
95. V. B. Hampel and J. M. Kraemer, *Tetrahedron*, **22**, 1601 (1966).
96. R. Haque, W. R. Coshow, and L. F. Johnson, *J. Am. Chem. Soc.*, **91**, 3822 (1969).
97. G. Heckmann, *Mol. Phys.*, **23**, 627 (1972).
98. W. Herz, P. S. Subramaniam, and N. Dennis, *J. Org. Chem.*, **34**, 3691 (1969).
99. J. Homer and M. C. Cooke, *J. Chem. Soc., A*, **1969**, 773.

100. J. Homer and M. C. Cooke, *J. Chem. Soc., A,* **1969**, 777.
101. J. Homer and M. C. Cooke, *J. Chem. Soc., A,* **1969**, 1984.
102. J. Homer and M. C. Cooke, *J. Chem. Soc., A,* **1969**, 2862.
103. J. Homer and P. J. Hulk, *J. Chem. Soc., A,* **1968**, 277.
104. J. Homer and D. L. Redhead, *J. Chem. Soc. Faraday Trans. II,* **68**, 793 (1972).
105. J. Homer and P. M. Whitney, *Chem. Commun.,* **1970**, 153.
106. B. B. Howard, B. Linder, and M. T. Emerson, *J. Chem. Phys.,* **36**, 485 (1962).
107. F. Hruska, D. W. McBride, and T. Schaefer, *Can. J. Chem.,* **45**, 1081 (1967).
108. C. M. Huggins and G. C. Pimentel, *J. Chem. Phys.,* **23**, 896 (1955).
109. H. M. Hutton and T. Schaefer, *Can. J. Chem.,* **45**, 1111 (1967).
110. Y. Ichikawa and T. Matsuo, *Bull. Chem. Soc. Jap.,* **40**, 2030 (1967).
111. A. M. Ihrig and S. L. Smith, *J. Am. Chem. Soc.,* **94**, 34 (1972).
112. M. Jauquet and P. Laszlo, *Chem. Phys. Lett.,* **15**, 600 (1972).
113. P. R. Jefferies and T. G. Payne, *Tetrahedron Lett.* **1967**, 4777.
114. M. D. Johnston, Jr., and M. Barfield, *J. Chem. Phys.,* **54**, 3083 (1971); **55**, 3483 (1971); *Mol. Phys.,* **22**, 831 (1971).
115. V. I. P. Jones and J. A. Ladd, *Mol. Phys.,* **19**, 233 (1970).
116. M. L. Josien, *Pure Appl. Chem.,* **4**, 33 (1962).
117. A. Kemula and R. T. Iwamoto, *J. Phys. Chem.,* **72**, 2764 (1968).
118. J. A. Ketelaar, *Rec. Trav. Chim. Pays-Bas,* **75**, 857 (1956).
119. H. Kienitz, *Naturwissenchaften,* **42**, 511 (1955).
120. J. G. Kirkwood, *J. Chem. Phys.,* **5**, 14 (1937).
121. D. Kivelson and T. Keyes, *J. Chem. Phys.,* **57**, 4599 (1972).
122. L. Kozerski and J. Dabrowski, *Org. Magn. Resonance,* **4**, 253 (1972).
123. R. A. Kromhout and B. Linder, *J. Magn. Resonance,* **1**, 450 (1969).
124. J. Lascombe, Dissertation, University of Bordeaux, 1960.
125. P. Laszlo, *Bull. Soc. Chim. Fr.,* **1964**, 2658.
126. P. Laszlo, in *Progress in Nuclear Magnetic Resonance Spectrometry,* J. W. Emsley, J. Feeney, and L. H. Sutcliffe, Eds., Vol. 3, Pergamon, Oxford, 1967, pp. 237-240.
127. P. Laszlo, in *Progress in Nuclear Magnetic Resonance Spectroscopy,* J. W. Emsley, J. Feeney, and L. H. Sutcliffe, Eds., Vol. 3, Pergamon, Oxford, 1967, Chap. 6.
128. P. Laszlo and J. I. Musher, *J. Chem. Phys.,* **41**, 3906 (1964).
129. P. Laszlo and P. V. R. Schleyer, *J. Am. Chem. Soc.,* **86**, 1171 (1964).
130. P. Laszlo and A. Speert, *J. Magn. Resonance,* **1**, 291 (1969).
131. P. Laszlo, A. Speert, and W. T. Raynes, *J. Chem. Phys.,* **51**, 1677 (1969).
132. P. Laszlo and D. H. Williams, *J. Am. Chem. Soc.,* **88**, 2799 (1966).
133. T. Ledaal, *Tetrahedron Lett.,* **1968**, 651.
134. T. Ledaal, *Tetrahedron Lett.,* **1968**, 1683.
135. I. Leupold, H. Musso, and J. Vicar, *Chem. Ber.,* **104**, 40 (1971).
136. R. A. Levenson, H. B. Gray, and G. P. Caesar, *J. Am. Chem. Soc.,* **92**, 3653 (1970).

137. W. Liptay, *Angew. Chem., Int. Ed.,* **8**, 177 (1969); *Z. Naturforsch.,* **21A**, 1605 (1966).

138. R. C. Lord, D. Nolin, and H. D. Stidham, *J. Am. Chem. Soc.,* **77**, 1365 (1955).

139. G. R. Luckhurst, *Quart. Rev.* (London), **22**, 179 (1968).

140. A. E. Lutskii and S. N. Vragova, *Opt. and Spectrosc.,* **31**, 113 (1971).

141. A. Mackor and H. A. Meinema, *Rec. Trav. Chim. Pays-Bas,* **91**, 911 (1972).

142. R. E. D. McClung, *J. Chem. Phys.,* **51**, 3842 (1969).

143. C. J. McDonald and T. Schaefer, *Can. J. Chem.,* **45**, 3157 (1967).

144. E. G. McRae, *J. Phys. Chem.,* **61**, 562 (1957).

145. A. Mangini and F. Taddei, *Inorg. Chem. Acta,* **2**, 8 (1968).

146. A. P. Marchand, W. R. Weimar, Jr., A. L. Segre, and A. M. Ihrig, *J. Magn. Resonance,* **6**, 316 (1972).

147. E. Maréchal, *Ber. Bunsenges. Phys. Chem.,* **75**, 343 (1971).

148. T. W. Marshall and J. A. Pople, *Mol. Phys.,* **1**, 199 (1958).

149. R. H. Martin, N. Defay, H. P. Figeys, M. Flammang-Barbieux, J. P. Cosyn, M. Gelbcke, and J. J. Schurter, *Tetrahedron,* **25**, 4985 (1969).

150. P. G. Maslov, *J. Phys. Chem.,* **72**, 1414 (1968).

151. N. Mataga and T. Kubota, *Molecular Interactions and Electronic Spectra,* Dekker, New York, 1970.

152. T. Matsuo, *Can. J. Chem.,* **45**, 1829 (1967).

153. T. Matsuo and Y. Kodera *J. Phys. Chem.,* **70**, 4087 (1966).

154. Y. T. Mazurenko, *Opt. and Spectrosc.,* **33**, 22 (1972).

155. Y. T. Mazurenko and N. G. Bakhshiev, *Opt. and Spectrosc.,* **28**, 490 (1970).

156. C. Y. Meyers and I. Sataty, *Tetrahedron Lett.,* **42**, 4323 (1972).

157. J. Michl, E. W. Thulstrup, and J. H. Eggers, *J. Phys. Chem.,* **74**, 3878 (1970).

158. S. Mohanty and H. J. Bernstein, *J. Chem. Phys.,* **54**, 2254 (1971).

159. H. Morawitz and K. B. Eisenthal, *J. Chem. Phys.,* **55**, 887 (1971).

160. R. M. Moriarty and J. M. Kliegman, *Tetrahedron Lett.,* 891 (1966).

161. J. A. Mosbo, J. R. Pipal, and J. G. Verkade, *J. Magn. Resonance,* **8**, 243 (1972).

162. J. Mosbo and J. G. Verkade, *J. Magn. Resonance,* **8**, 250 (1972).

163. R. S. Mulliken and W. B. Person, *Molecular Complexes. A Lecture and Reprint Volume,* Wiley-Interscience, New York, 1969.

164. J. I. Musher, *J. Chem. Phys.,* **37**, 34 (1962).

165. L. A. Nafie and W. L. Peticolas, *J. Chem. Phys.,* **57**, 3145 (1972).

166. C. R. Narayanan and N. R. Bhadane, *Tetrahedron Lett.,* **1968**, 1557.

167. C. R. Narayanan and N. K. Venkatasubramanian, *Tetrahedron Lett.,* **1966**, 5865.

168. G. W. Nederbragt and J. Pelle, *Mol. Phys.,* **1**, 97 (1958).

169. M. Nishio, *Chem. Commun.,* **1969**, 51.

170. M. Nishio, *Chem. Pharm. Bull.,* **17**, 274 (1969).

171. M. Nishio, *Chem. Pharm. Bull.,* **17**, 262 (1969).

172. Y. Nomura and Y. Takeuchi, *Tetrahedron,* **25**, 3825 (1969).

173. L. Onsager, *J. Am. Chem. Soc.*, **58**, 1486 (1936).
174. L. Onsager, *J. Am. Chem. Soc.*, **58**, 1485 (1936).
175. B. E. Orgel and R. S. Mulliken, *J. Am. Chem. Soc.*, **79**, 4839 (1958).
176. W. W. Paudler and S. A. Humphrey, *Org. Magn. Resonance*, **3**, 217 (1971).
177. A. Pelters, R. Warren, J. N. Usmani, M. Ilyas, and W. Rahman, *Tetrahedron Lett.*, **1969**, 4259.
178. H. H. Perkampus, U. Krüger, and W. Krüger, *Z. Naturforsch.*, **B24**, 1365 (1969).
179. M. Perrot, P. V. Huong, and J. Lascombe, *J. Chim. Phys.*, **68**, 614 (1971).
180. L. Petrakis and H. J. Bernstein, *Can. J. Chem.*, **43**, 81 (1965).
181. J. L. Pierre, R. Perraud, and P. Arnaud, *C. R. Acad. Sci., Paris*, **270**, 1663 (1970).
182. J. L. Pierre, R. Perraud, and G. Bouvard, *C. R. Acad. Sci., Paris*, **273**, 429 (1971).
183. K. Pihlaja and M. Ala-Tuori, *Acta Chem. Scand.*, **26**, 1891 (1972).
184. K. Pihlaja and M. Ala-Tuori, *Acta Chem. Scand.*, **26**, 1904 (1972).
185. C. Pimentel and A. L. McClellan, *The Hydrogen Bond*, W. H. Freeman San Francisco, 1960.
186. I. D. Rae, *Can. J. Chem.*, **46**, 2589 (1968).
187. A. V. Rakov, Dissertation, *Proc. P. N. Lebedev Phys. Inst.*, **27**, (1965); *Opt. Spectrosc.*, **13**, 203 (1962).
188. W. T. Raynes, *J. Chem. Phys.*, **51**, 3138 (1969).
189. W. T. Raynes and M. A. Raza, *Mol. Phys.*, **6**, 1241 (1972).
190. J. Reisse and G. Chiurdoglu, *Spectrochim. Acta*, **20**, 441 (1964).
191. G. S. Ricca and B. Danieli, *Gazzetta*, **99**, 133 (1969).
192. R. F. Richter, J. C. Philips, and R. D. Bach, *Tetrahedron Lett.*, **1972**, 4327.
193. C. D. Ritchie and W. F. Sager, in *Progress in Physical Organic Chemistry*, Vol. 2, S. G. Cohen, A. Streitwieser, and R. W. Taft, Eds., Interscience, New York, 1964.
194. W. W. Robertson, A. D. King, Jr., and O. E. Weigang, Jr., *J. Chem. Phys.*, **35**, 464 (1961).
195. G. W. Robinson, *J. Chem. Phys.*, **46**, 572 (1967).
196. L. G. Robinson, J. Jacobus, and G. B. Savitsky, *J. Magn. Resonance*, **5**, 328 (1971).
197. G. Rocktäschel, E. A. V. Ebsworth, D. W. H. Rankin, and J. C. Thompson, *Z. Naturforsch.*, **B23**, 598 (1968).
198. A. Roger, J. J. Godfroid, and J. Wiemann, *Bull. Soc. Chim. Fr.*, 3030 (1967).
199. J. Ronayne and D. H. Williams, *J. Chem. Soc., B*, **1967**, 805.
200. J. Ronayne and D. H. Williams, *J. Chem. Soc., C*, **1967**, 2642.
201. H. Rosen, Y. R. Shen, and F. Stenman, *Mol. Phys.*, **22**, 33 (1971).
202. I. Rosenthal, *Tetrahedron Lett.*, **1969**, 3333.
203. W. G. Rothschild, *J. Chem. Phys.*, **51**, 5187 (1969).
204. W. G. Rothschild, *J. Chem. Phys.*, **57**, 991 (1972).

205. F. H. A. Rummens, *J. Am. Chem. Soc.*, **92**, 3214 (1970).
206. F. H. A. Rummens, *Mol. Phys.*, **21**, 535 (1971).
207. F. H. A. Rummens and R. H. Krystynak, *J. Am. Chem. Soc.*, **94**, 6914 (1972).
208. T. Schaefer and W. G. Schneider, *J. Chem. Phys.*, **32**, 1224 (1960).
209. F. Scheinmann, *Chem. Commun.*, **1967**, 1015.
210. D. Schmidt, Ph.D. Thesis, Eidgenössiche Technische Hochschule, Zürich, 1970.
211. R. E. J. Sears, and E. L. Hahn, *J. Chem. Phys.*, **45**, 2753 (1966).
212. A. V. Sechkaryov, *Opt. and Spectrosc.*, **19**, 401 (1965).
213. J. Seyden-Penne, P. Arnaud, J. L. Pierre, and M. Plat, *Tetrahedron Lett.*, **1967**, 3719.
214. J. Seyden-Penne, T. Strzalko, and M. Plat, *Tetrahedron Lett.*, **1965**, 4597.
215. H. Shimizu, *J. Chem. Phys.*, **43**, 2453 (1965).
216. A. I. Sidorova, M. G. Batishcheva, and E. N. Shermatov, *Opt. Spectrosc.*, *Suppl.*, **1-3**, 96 (1963).
217. S. L. Smith and R. H. Cox, *J. Phys. Chem.*, **72**, 198 (1968).
218. S. L. Smith and A. M. Ihrig, *J. Chem. Phys.*, **46**, 1181 (1967).
219. J. Soussen-Jacob, J. Vincent-Geisse, D. Beaulieu, and T. Tsakiris, *J. Chim. Phys.*, **67**, 1118 (1970).
220. D. Steele, *Theory of Vibrational Spectroscopy*, W. B. Saunders, Philadelphia, 1971.
221. E. T. Strom, B. S. Snowden, Jr., H. C. Custard, D. E. Woessner, and J. R. Norton, *J. Org. Chem.*, **33**, 2555 (1968).
222. E. T. Strom, B. S. Snowden, and P. A. Toldan, *Chem. Commun.*, **1969**, 50.
223. J. Timmons, *Chem. Commun.*, **22**, 576 (1965).
224. J. H. Van Vleck, *Mol. Phys.*, **24**, 341 (1972).
225. J. Vincent-Geisse, J. Soussen-Jacob, N. T. Tai, and D. Descout, *Can. J. Chem.*, **48**, 3918 (1970).
226. S. N. Vinogradov and R. H. Linnell, *Hydrogen Bonding*, Van Nostrand Reinhold, New York, 1971.
227. W. R. Ware, S. K. Lee, G. J. Brant, and P. P. Chow, *J. Chem. Phys.*, **54**, 4729 (1971).
228. R. Wasylishen and T. Schaefer, *Can. J. Chem.*, **50**, 274 (1972).
229. R. Wasylishen, T. Schaefer, and R. Schwenk, *Can. J. Chem.*, **48**, 2885 (1970).
230. P. H. Weiner and E. R. Malinowski, *J. Phys. Chem.*, **75**, 1207 (1971); **75**, 3160 (1971); **75**, 3971 (1971).
231. P. H. Weiner and E. R. Malinowski, *J. Phys. Chem.*, **75**, 3160 (1971).
232. P. H. Weiner, E. R. Malinowski, and A. R. Levinstone, *J. Phys. Chem.*, **74**, 4537 (1970).
233. D. Wendish, *Z. Naturforsch.*, **B23**, 616 (1968).
234. D. H. Whiffen, *Quart. Rev.*, (London), **4**, 131 (1950).
235. J. J. Wilczynski, G. D. Daves, and K. Folkers, *J. Am. Chem. Soc.*, **90**, 5593 (1968).

236. D. H. Williams and N. S. Bhacca, *Tetrahedron,* **21,** 1641 (1965).
237. D. H. Williams, J. Ronayne, H. W. Moore, and H. R. Shelden, *J. Org. Chem.,* **33,** 998 (1968).
238. G. Williams, *Chem. Rev.,* **72,** 55 (1972).
239. R. G. Wilson, D. E. A. Rivett, and D. H. Williams, *Chem. Ind.* (London), **1969,** 109.
240. R. G. Wilson and D. H. Williams, *Tetrahedron,* **25,** 155 (1969).
241. T. Winkler and W. von Philipsborn, *Helv. Chim. Acta,* **51,** 183 (1968).
242. Z. W. Wolkowski, N. Thoai, and J. Wiemann, *Tetrahedron Lett.,* **1970,** 93.
243. A. Yogev, L. Margulies, D. Amar, and Y. Mazur, *J. Am. Chem. Soc.,* **91,** 4558 (1969).
244. A. Yogev, L. Margulies, and Y. Mazur, *J. Am. Chem. Soc.,* **93,** 249 (1971).
245. A. Yogev, J. Riboid, J. Marero, and Y. Mazur, *J. Am. Chem. Soc.,* **91,** 4559 (1969).
246. T. Yonezawa, I. Morishima, and K. Fukuta, *Bull. Chem. Soc. Jap.,* **41,** 2297 (1968).
247. R. P. Young and R. N. Jones, *Chem. Rev.,* **71,** 219 (1971).
248. D. Ziessow and M. Carroll, *Ber. Bunsenges Phys. Chem.,* **76,** 61 (1972).

Chapter **V**

SOLVENTS IN CHEMICAL TECHNOLOGY

Romesh Kumar and J. M. Prausnitz

Chemical industry is concerned with chemical and physical changes of materials, and many of these are in the liquid state. Rational design of chemical plants therefore requires some quantitative understanding of liquid-phase properties, especially for liquid mixtures. This chapter provides a brief survey and selected overview of equilibrium properties of mixtures from the standpoint of industrial application. The overview is necessarily selected, because of the finite number of pages budgeted for this chapter and because of the limited experience of the authors. Emphasis is directed toward applications in industries concerned with large-scale production of organic materials (e.g., petrochemicals and polymers) and only little attention is given to applications in industries processing inorganic materials or biological products.

Even though attention is focused on equilibrium properties of fluids found in the production of organic chemicals, it has been necessary to restrict the discussion to those topics which in our judgment are of primary interest to the readers of this volume. Because of space limitations, many specialized topics could not be included.

The chapter begins with a summary of the basic thermodynamic relations for describing phase equilibria, followed by a discussion on how these relations may be used to calculate solubilities. Special attention is then given to a variety of semiempirical methods for estimating solubility behavior of amorphous polymers. The chapter continues with a discussion of solvent-aided separation processes and closes with some general remarks concerning the use of solvents in chemical engineering.

1 BASIC THERMODYNAMIC CONSIDERATIONS

The solubility of one substance in another is determined by the equations of phase equilibrium. As shown in Fig. 5.1, we consider two phases α and β, at the same temperature, such that the components in one phase can travel freely across the interface to the other. The net transfer across the interface ceases when equilibrium is attained; this condition is given by the Gibbs-Lewis relation. For every component i,

$$f_i^\alpha = f_i^\beta \tag{5.1}$$

where f stands for fugacity.

In a gaseous phase it is convenient to relate f_i to the total pressure P and to the composition (here expressed by the gas-phase mole fraction y_i) through the fugacity coefficient ϕ_i:

$$f_i = \phi_i y_i P \tag{5.2}$$

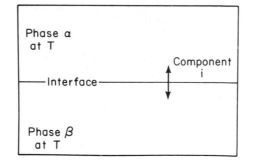

Fig. 5.1. Component i travels freely between phases a and β. At equilibrium the fugacity of i is the same in both phases.

For a mixture of ideal gases, $\phi_i = 1$. In general, ϕ_i depends on temperature, pressure, and gas-phase composition; it can be evaluated from experimental gas-phase volumetric data, or from an equation of state as discussed elsewhere [54]. At modest pressures (of the order of a few atmospheres), it is often satisfactory to assume that ϕ_i is close to unity.

In a condensed phase the fugacity of component i is conveniently related to its concentration (here expressed by mole fraction x_i) through the activity coefficient γ_i:

$$f_i = \gamma_i x_i f_i^0 \tag{5.3}$$

where f_i^0 is the fugacity of i in an arbitrarily chosen standard state. In ordinary liquid mixtures at moderate pressures, it is customary to choose for the standard state pure liquid i at system temperature and pressure. In this event γ_i is normalized such that $\gamma_i \to 1$ as $x_i \to 1$. If component i is supercritical and therefore cannot exist as a pure liquid at the system temperature, it is sometimes convenient to set f_i^0 equal to Henry's constant for solute i in the solution at system temperature and pressure. In this event γ_i is normalized such that $\gamma_i \to 1$ as $x_i \to 0$. It is customary to designate activity coefficients normalized in this manner with an asterisk: γ_i^*. (For more details see Refs. 54 and 60).

In principle it is possible to calculate fugacities from volumetric properties alone, but extensive data are required and, except for limited cases (e.g., fluid mixtures of light hydrocarbons), such a calculation is not practical for fugacities in condensed mixtures. It is more convenient to relate activity coefficients to a useful function g^E (the molar excess Gibbs energy) and to construct semi-empirical models which relate g^E to the temperature and composition of the condensed phase. Two important equations relate g^E to γ:

$$\frac{g^E}{RT} = \sum_i x_i \ln \gamma_i \qquad (5.4)\dagger$$

$$\ln \gamma_i = \frac{1}{RT} \left(\frac{\partial n_T g^E}{\partial n_i}\right)_{T,P,n_j} \qquad (5.5)\dagger$$

where n_i is the number of moles of i, and n_T is the total number of moles. The derivative in (5.5) is taken with all mole numbers n_j constant, except n_i.

Many equations have been proposed for g^E. The best known, perhaps, are those of Margules and Van Laar and, more recently, those of Wilson and Renon [54]. To illustrate, we consider a particularly simple case: a binary mixture of ordinary liquids which are chemically similar (e.g., benzene-cyclohexane at $30°C$). For such a mixture we assume

$$g^E = Ax_1x_2 \qquad (5.6)$$

where A is a constant, dependent on temperature but not on composition. This constant must be found empirically, from binary data. Only in few cases is it possible to estimate A from pure-component data alone.

Substitution in (5.5) gives the activity coefficients as functions of mole fraction:

$$\ln \gamma_1 = \frac{A}{RT}x_2^2 \quad \text{and} \quad \ln \gamma_2 = \frac{A}{RT}x_1^2 \qquad (5.7)$$

While A depends on temperature, this dependence is usually weak, and to a first approximation it is often possible to neglect it. The advantage of (5.7) is that, while only limited binary data (in principle, only one experimental datum) are needed to determine the constant A, the equations enable us to calculate activity coefficients for the entire composition range $x_1 = 0$ to $x_1 = 1$.

For more complex mixtures it is usually necessary to use for g^E an expression with more than one empirical constant. For most typical liquid mixtures of nonelectrolytes, an expression for g^E with two or three adjustable constants is sufficient.

One of the great advantages of the excess Gibbs energy concept follows from its application to mixtures containing more than two components. It is often possible to predict activity coefficients in multicomponent mixtures using only experimental data for the constituent binaries [48, 56, 62].

†These equations are also valid for γ_i^*. For applications of γ_i^* at high pressures, see the monograph by Prausnitz and Chueh [55].

While the excess Gibbs energy is commonly used to represent experimental data for liquid-phase mixtures, it can also be used to represent data for solid-phase mixtures.

Having briefly summarized the major parts of the thermodynamic framework for describing phase equilibrium, we can now present some techniques for estimating solubilities.

2 SOLUBILITY OF SOLIDS IN LIQUIDS

Consider a two-phase system as shown in Fig. 5.1, and let phase a stand for a liquid and phase β for a solid. To express the fugacity of component i in the liquid phase [see (5.3)], it is convenient to use for the standard state the pure liquid at system temperature and pressure. If the system temperature T is below T_t, the triple-point temperature of pure component i, pure liquid i at T is in a hypothetical (subcooled) state. As shown in many standard thermodynamics texts, $f^L_{\text{pure } i}$, the fugacity of subcooled liquid i at temperature T, is related to $f^S_{\text{pure } i}$, the fugacity of pure solid i at T, by

$$\ln f^L_{\text{pure } i} = \ln f^S_{\text{pure } i} + \frac{\Delta h^f}{RT}\left(1 - \frac{T}{T_t}\right) - \frac{\Delta c_p}{R}\left(\frac{T_t - T}{T}\right) + \frac{\Delta c_p}{R}\ln\frac{T_t}{T} \quad (5.8)$$

where Δh^f is the enthalpy of fusion at T_t, and Δc_p is the molar heat capacity at constant pressure of pure liquid minus that of pure solid. Equation (5.8) is based on the assumption that Δc_p is independent of temperature, and that the effect of pressure on condensed-phase properties is negligible. For practical purposes it is common practice to neglect the last two terms in (5.8), since they are of opposite sign and tend to cancel one another; further, since the melting temperature is usually a weak function of pressure, it is customary to replace the triple-point temperature by the normal melting temperature T_m. A simpler and more useful form of (5.8), therefore is

$$\log_{10}\left(\frac{f^L}{f^S}\right)_{\text{pure } i} = \frac{\Delta S^f}{2.3R}\left(\frac{T_m}{T} - 1\right) \quad (5.9)$$

where ΔS^f is the entropy of fusion at T_m. (The entropy of fusion at T_m is the enthalpy of fusion at T_m divided by T_m.)

Let the solid component be designated by subscript 2, and assume that the solid phase in equilibrium with the saturated solution is pure (negligible solubility of solvent 1 in the solid). Then

$$\log_{10}\gamma_2 x_2 = -\frac{\Delta S_2^f}{2.3R}\left(\frac{T_{m_2}}{T} - 1\right) \quad (5.10)$$

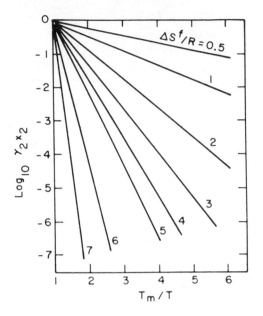

Fig. 5.2. Activities of solid solutes as a function of melting temperature and entropy of fusion.

A plot of (5.10) is shown in Fig. 5.2. If we also assume that $\gamma_2 = 1$, we can then calculate the so-called ideal solubility.

Equation (5.10) shows that, for a system in which γ_2 is not temperature-dependent, the solubility increases with rising temperature. The rate of increase is approximately proportional to the enthalpy of fusion and, to a first approximation, does not depend on the melting temperature. Equation (5.10) also shows that, for a fixed solvent and fixed temperature, if two solids have similar enthalpies of fusion, the solid with the lower melting temperature has the higher solubility; similarly, if the two solids have nearly the same melting temperature, the one with the lower enthalpy of fusion has the higher solubility.

For typical nonpolar systems, in the absence of specific interactions, γ_2 is likely to be larger than unity; the ideal solubility is therefore likely to represent an upper limit. For example, the solubilities of various aromatic hydrocarbons in carbon tetrachloride have been studied by McLaughlin and Zainal [41]. For these hydrocarbons the entropy of fusion varies little; the average is 13.0 cal mol^{-1} K^{-1}. The ideal solubility curve using this value is shown in Fig. 5.3. Also shown are the experimental results that fall below those calculated with $\gamma_2 = 1$.

No general method is available for predicting activity coefficients for subcooled liquids dissolved in solvents. For simple mixtures, however, it is sometimes

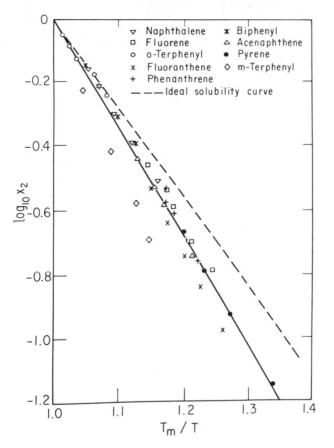

Fig. 5.3. Solubility of aromatic solids in carbon tetrachloride.

possible to make good estimates using the Scatchard-Hildebrand theory of regular solutions as discussed by Preston [57, 58].

Equation (5.10) is of interest in solvent selection, because it enables us to estimate the freezing point of a binary solvent mixture, which in many cases is lower than that of either pure solvent. In low-temperature processing it is often desirable to conduct a separation operation at as low a temperature as possible. To illustrate, consider a mixture of two solvents designated by subscripts 1 and 3. For simplicity assume that these solvents form an ideal liquid mixture ($\gamma_1 = \gamma_3 = 1$); further assume that these solvents are completely immiscible in the solid phase. The solubility of (solid) solvent 3 in (liquid) solvent 1 is given by

$$\ln x_3 = -\frac{\Delta S_3^f}{R}\left(\frac{T_{m_3}}{T} - 1\right) \qquad (5.11)$$

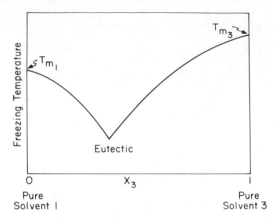

Fig. 5.4. Freezing-point diagram for a simple mixture of solvents 1 and 3. (For most compositions, the freezing temperature is below T_{m_1} and T_{m_3}.)

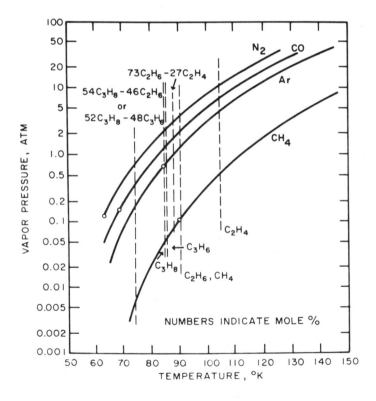

Fig. 5.5. Purification of hydrogen prior to liquefaction. Vapor pressures of solutes and solvent freezing points.

and similarly, the solubility of (solid) solvent 1 in (liquid) solvent 3 is given by

$$\ln x_1 = - \frac{\Delta S_1^f}{R} \left(\frac{T_{m_1}}{T} - 1 \right) \tag{5.12}$$

Equations (5.11) and (5.12) give a freezing-point diagram as shown in Fig. 5.4; the two equations intersect at the eutectic point. We see that in the mid-composition region the freezing point of a simple mixture of two solvents is lower than that of either pure solvent.

The operating advantages of a low-freezing solvent for absorption have been pointed out by Cheung and Wang [12], who presented the results shown in Fig. 5.5. These investigators were concerned with the optimum design of a liquid-hydrogen plant; prior to liquefaction, it is necessary to remove impurities (nitrogen, carbon monoxide, argon, methane) by absorption and, as shown in Fig. 5.5, it is advantageous to absorb at as low a temperature as possible, since the vapor pressures of the impurities are then small, providing a large solubility. The vertical dashed lines show the freezing temperatures of pure solvents (ethylene, methane, propylene, ethane, and propane) and of several solvent mixtures. Absorption with a solvent mixture is much more efficient than that with a pure solvent, because of the significantly larger solubilities of the condensable impurities that can be obtained on lowering the operating temperature.

3 SOLUBILITY OF LIQUIDS IN LIQUIDS

Consider a two-phase system as shown in Fig. 5.1, and let both α and β represent liquid phases. On substituting (5.3) into (5.1), we obtain the equations of equilibrium; for every component i,

$$(\gamma_i x_i)^\alpha = (\gamma_i x_i)^\beta \tag{5.13}$$

If we have information giving γ as a function of x, we can find the equilibrium compositions x_i^α and x_i^β. To illustrate, we consider a simple case: a binary mixture in which the excess Gibbs energy is given by (5.6); the activity coefficients are then given by (5.7). If A/RT is known, there are two unknowns: x_1^α and x_1^β. These can be found by solution of the two coupled equations

$$x_1^\alpha \exp \frac{A}{RT}(1 - x_1^\alpha)^2 = x_1^\beta \exp \frac{A}{RT}(1 - x_1^\beta)^2 \tag{5.14}$$

$$(1 - x_1^\alpha) \exp \frac{A}{RT}(x_1^\alpha)^2 = (1 - x_1^\beta) \exp \frac{A}{RT}(x_1^\beta)^2 \tag{5.15}$$

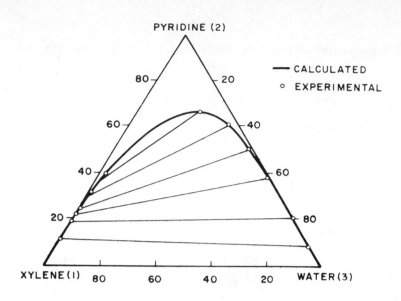

Fig. 5.6. Ternary liquid-liquid diagram for xylene-pyridine-water at 25°C. (From H. Renon, *Calcul sur Ordinateur des Equilibres Liquide-Vapeur et Liquide-Liquide,* p. 78, Editions Technip, Paris, 1971.)

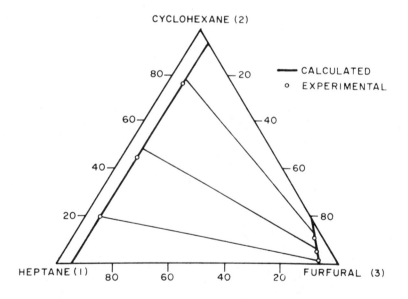

Fig. 5.7. Ternary liquid-liquid diagram for *n*-heptane-cyclohexane-furfural. (From H. Renon, *Calcul sur Ordinateur des Equilibres Liquide-Vapeur et Liquide-Liquide,* p. 80, Editions Technip, Paris, 1971.)

To be physically meaningful, x_1^a and x_1^β must satisfy

$$0 < x_1^a < 1 \quad \text{and} \quad 0 < x_1^\beta < 1 \tag{5.16}$$

Equations (5.14) to (5.16) are satisfied, provided $A/RT > 2$.

While experimental data for g^E can be used to calculate mutual solubilities, it is also possible to reverse the procedure and to obtain information on g^E from experimental mutual solubility data (see, e.g., Ref. 8).

Ternary (and higher) liquid-liquid phase diagrams can be calculated in a manner analogous to that indicated by (5.14) and (5.15). Computer programs for performing such calculations have been described by Renon et al. [62].

Calculated results for two ternary systems, presented by Renon, are shown in Figs. (5.6) and (5.7). These calculations were performed with the nonrandom, two-liquid (NRTL) equation for g^F (see Table 5.3). Although this calculation can be made using only binary data, such a calculation does not consistently produce results of sufficient accuracy. For high accuracy in calculating ternary liquid-liquid equilibria, it is often necessary to have a few experimental ternary data points for adjusting the NRTL constants. The constants used by Renon are given in Table 5.1.

Table 5.1 NRTL Constants Used by Renon to Calculate Ternary Liquid Diagrams[a]

i	j	a_{ij}	$g_{ij} - g_{jj}$ (cal mol^{-1})	$g_{ji} - g_{ii}$ (cal mol^{-1})
Xylene(1)-pyridine(2)-water(3)				
1	2	0.30	−63	282
1	3	0.18	1934	4314
2	3	0.59	−90	1615
n-Heptane(1)-cyclohexane(2)-furfural(3)				
1	2	0.30	−535	611
1	3	0.35	1401	1418
2	3	0.35	1443	992

[a] The NRTL equation is given in Table 5.3.

Liquid-liquid diagrams are described in detail by Treybal [75] and by Francis [20], who review the extensive experimental literature. Since gases at high pressure have a solvent power similar to that of liquids, in recent years attention has also been given to "fluid-fluid" diagrams in binary and multicomponent systems in which one or more of the components is above its critical temperature

[20, 64]. Such systems offer a variety of possibilities for new separation operations.

4 SOLUBILITY OF GASES IN LIQUIDS

Consider a two-phase system as shown in Fig. 5.1; let β stand for a liquid phase and α for a gaseous phase. Let subscript 2 stand for the light component whose mole fraction in the gas phase is y_2.

If Raoult's law holds, the solubility of (gaseous) solute 2 is given by

$$x_2 = \frac{y_2 P}{P_2{}^s} \tag{5.17}$$

where P is the total pressure, and $y_2 P$ is the partial pressure of component 2 in the gas phase; $P_2{}^s$ is the saturation (vapor) pressure of liquid 2 at system temperature. Since the system temperature T may exceed the critical temperature T_{c_2}, it is necessary to find $P_2{}^s$ by extrapolation; it is customary to do so by using a straight line on a plot of $\ln P_2{}^s$ versus $1/T$. When calculated in this manner, the solubility x_2 (called the ideal solubility) provides only a rough, order-of-magnitude estimate. Contrary to observation, (5.17) gives a solubility that is independent of the nature of the solvent and that always falls with rising temperature. Table 5.2 compares some observed gas solubilities with those calculated from (5.17). Qualitative agreement is reasonable for nonpolar solvents but, in most cases, the ideal solubility tends to be too large.

Table 5.2 Ideal and Observed Gas Solubilities at $25°C$ and 1 atm Partial Pressure [31][a]

| | Solute | | | |
Solvent	H_2	N_2	CH_4	C_2H_6
Ideal	8	10	35	250
n-Hexane	6.5	12.5	42	177
Carbon tetrachloride	3.3	6.4	29	213
Methyl acetate	3.1	6.0	20	108
Benzene	2.6	4.4	21	151
Chlorobenzene	2.7	4.3	21	148
Water	0.15	0.12	0.24	0.33

[a] Solubilities are in mole fraction $\times 10^4$.

If x_2 is small compared to unity, it is convenient to use Henry's law:

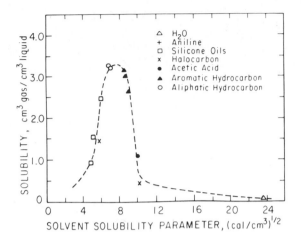

Fig. 5.8. Solubility of xenon at ambient temperature and low pressure.

$$x_2 = \frac{y_2 P}{H} \tag{5.18}$$

where Henry's constant H depends on temperature and on the chemical nature of both solute and solvent. Equation (5.18) follows from the general equation of equilibrium on assuming ideal gas behavior in the gas phase ($\phi_2 = 1$) and ideal dilute solution behavior in the liquid phase ($\gamma_2^* = 1$).

When solute and solvent are significantly different in chemical properties, H is large and the solubility is small. To illustrate, Fig. 5.8 shows experimental results reported by Steinberg and Manowitz [69], who were interested in separating radioactive xenon from a nuclear reactor by absorption in a liquid solvent. The solubility of xenon is plotted against the solubility parameter of the solvent (see Section 6). A sharp maximum is obtained in the region where $\delta \approx 7.5$; this value is close to that of "liquid" xenon at $25°C$ as estimated from critical properties. Figure 5.8 shows that the solubility is largest when the similarity between solute and solvent (as measured by δ) is a maximum, that is, when δ (solute) = δ (solvent). Several authors, notably Shair and Prausnitz [65] and Yen and McKetta [81], have expressed this relationship for nonpolar systems by writing

$$\ln H = \text{constant} + \frac{V_2^L (\delta_1 - \delta_2)^2}{RT} \tag{5.19}$$

where V_2^L and δ_2 are, respectively, the liquid volume and solubility parameter

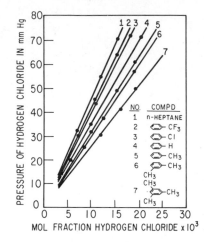

Fig. 5.9. Solubility of hydrogen chloride at −78.51°C in *n*-heptane and in 5 mol % solutions of aromatics in *n*-heptane. (Data of Brown and Brady.)

of the solute, and δ_1 is the solubility parameter of the solvent. The constant is independent of the solvent and, for a fixed solute, depends only on the temperature. Gas solubility correlations based on the solubility parameter concept are reviewed by Hildebrand, Prausnitz, and Scott [32].

Gas solubility is enhanced when there is a specific interaction between solute and solvent. For example, in liquid acetone, the solubility of acetylene is very much larger than that of ethylene, as a result of hydrogen bonding between the hydrogen atom in acetylene and the oxygen atom in acetone. Another example is provided by the measurements of Brown and Brady [9], shown in Fig. 5.9; the solubility of hydrogen chloride in heptane is significantly enhanced by the addition of small amounts of aromatic hydrocarbons, because of inter-action between the polar gaseous solute and the highly polarizable π bonds of the aromatic additive. Methyl substitution on the benzene ring increases the electron-donating ability of the hydrocarbon, and therefore the solubility of hydrogen chloride rises with methyl substitution as shown.

Gas solubility data are reviewed by Battino and Clever [3] and in Chapter VII of this volume.

5   LIQUID–PHASE ACTIVITY COEFFICIENTS

Many workers have attempted to construct methods of predicting activity coefficients in liquid mixtures from pure-component data alone. Despite much effort, such methods have not as yet been established, but a few useful correlations of limited scope and applicability have been presented, as shown in later sections of this chapter.

Table 5.3 Some Common Expressions for Molar Excess Gibbs Energy g^E for Binary Systems

Name	Number of Constants	Equation
van Laar	Two, A and B	$g^E = \dfrac{A x_1 x_2}{(x_1 A/B + x_2)}$
Three-suffix Margules	Two, A and B	$g^E = x_1 x_2 [A + B(x_1 - x_2)]$
Wilson	Two, Λ_{12} and Λ_{21}	$\dfrac{g^E}{RT} = -x_1 \ln (x_1 + \Lambda_{12} x_2) - x_2 \ln (x_2 + \Lambda_{21} x_1)$
		$\Lambda_{12} = (v_2/v_1) \exp - (\lambda_{12} - \lambda_{11})/RT$
		$\Lambda_{21} = (v_1/v_2) \exp - (\lambda_{12} - \lambda_{22})/RT$
		v_i = Molar liquid volume of pure liquid i
		λ_{ij} = Energy of interaction between molecules i and j
NRTL	Three, τ_{12}, τ_{21}, and α	$\dfrac{g^E}{RT} = x_1 x_2 \left(\dfrac{\tau_{21} G_{21}}{x_1 + x_2 G_{21}} + \dfrac{\tau_{12} G_{12}}{x_2 + x_1 G_{12}} \right)$
		$\tau_{12} = \dfrac{(g_{12} - g_{22})}{RT}$; $\tau_{21} = \dfrac{(g_{12} - g_{11})}{RT}$
		$G_{12} = \exp (-\alpha \tau_{12})$; $G_{21} = \exp (-\alpha \tau_{21})$
		g_{ij} = Free-energy parameter for the ij interaction
		α = Nonrandomness parameter

While we are not able to predict the properties of any liquid mixture using pure-component information alone, we can frequently calculate such properties over a wide range of composition using only very little binary experimental data in addition to pure-component data. To calculate activity coefficients for the entire composition range in a binary system, it is often sufficient to obtain one or two experimental measurements, taken at one or two compositions. The number of measurements required depends on the complexity of the mixture and on the desired degree of accuracy. For most typical liquid mixtures, activity coefficients [calculated from (5.5)] require an expression for g^E with two or three adjustable constants; for very simple mixtures one constant is often sufficient. Only a few reliable experimental measurements are needed to determine these constants.

Many algebraic expressions for g^E have been proposed; some of the more common ones are given in Table 5.3. In each case the constants appearing in the equations must be determined from some experimental measurements on the mixture itself. In some of the equations, an attempt is made to introduce an explicit temperature dependence; this attempt usually weakens, but does not eliminate, the temperature dependence of the constants.

Technical literature abounds with experimental studies of binary liquid mixtures; an excellent review and compilation is given by Hala et al. [24]. However, despite the wealth of data that has been reported, the number of possible binary mixtures is so large that there is still a continuing need for more (and better) experimental information.

To determine the constants in the binary expression for g^E, a few experimental measurements on the binary mixture are required. For example, consider a binary liquid mixture of known mole fraction x_1 at known temperature T. Let us measure the pressure P and the mole fraction y_1 of the vapor that is in equilibrium with this mixture. From these measurements we can calculate two activity coefficients:

$$\gamma_1 = \frac{y_1 P}{x_1 P_1{}^s} \tag{5.20}*$$

$$\gamma_2 = \frac{y_2 P}{x_2 P_2{}^s} \tag{5.21}*$$

where the pure-component saturation pressures $P_1{}^s$ and $P_2{}^s$ are known functions of temperature. Now let us assume that the molar excess Gibbs energy g^E is some arbitrary function of x which contains two adjustable parameters A and B:

$$g^E = f(x,A,B) \tag{5.22}**$$

*For simplicity, we have here assumed that the pressure is sufficiently low to set the fugacity coefficient $\phi = 1$ and to neglect the effect of pressure on liquid-phase properties.

**However, the arbitrary function must satisfy the boundary conditions $g^E = 0$ when $x_1 = 0$, and $g^E = 0$ when $x_2 = 0$.

Since g^E is related to γ_1 and γ_2 as shown in (5.4) and (5.5), we can use the two experimentally obtained activity coefficients to determine the constants A and B; once these have been found, we can use them to calculate activity coefficients over the entire range $0 \leqslant x_1 \leqslant 1$. The constants A and B do not depend on x but, unfortunately, they depend on T. In many practical cases, however, we can neglect the temperature dependence without serious error.

In principle it is sufficient to measure only one equilibrium T, P, x, and y in order to calculate two characteristic constants. In practice, however, one experimental point is often not enough, especially if that point corresponds to a composition in which x_1 is close to unity or to zero. For reliable results it is desirable to make at least two measurements, preferably at different ends of the composition range.

One common source of experimental information is azeotropic data; at the azeotropic composition $x_i = y_i$ and therefore, to calculate γ_1 and γ_2 from (5.20) and (5.21), it is necessary only to know the azeotrope's temperature and pressure in addition to the pure-component saturation pressures. Since azeotropic data are plentiful [33], it is often possible to estimate activity coefficients over the entire range of composition from such data. However, such estimates provide only rough approximations, especially if the azeotropic composition is not near the center of the composition range.

Useful information on liquid-phase activity coefficients can sometimes be obtained from the solubility of a solid in a liquid, or from mutual solubilities of two liquids [39, 54]. Unfortunately, these data sources are frequently of insufficient accuracy to give reliable results for the variation of g^E with composition. For good results it is best to measure the partial pressures of the volatile components in the liquid mixture at selected compositions, or to measure the total pressure (at constant temperature) over the entire composition range.

While Table 5.3 gives some expressions for g^E for binary systems, these and other expressions are readily extended to systems containing any number of components [24, 54]. On using a multicomponent expression for g^E, coupled with (5.5), it is often possible to predict with good accuracy activity coefficients in multicomponent mixtures using only experimental data for the pure components and for the binary systems that constitute the multicomponent mixture [43].

6 SOLUBILITY OF AMORPHOUS POLYMERS

The following discussion is concerned with the dissolution of polymeric materials in organic solvents. The discussion is directed at common problems such as those encountered in the paint and coatings industry, where it is often necessary to devise a solvent system for a particular film-forming polymer or resin such that the resulting solution has specified physical properties, including viscosity, evaporation rate, and blush resistance. Several solvents and solvent mixtures can provide the desired characteristics. However, not all of them can dissolve the polymer to any appreciable extent. Since experimental solubility data are not available for many polymer-solvent systems of possible interest, it is desirable to predict the solubility behavior of such systems without resorting to experiment. Various methods have been proposed for providing such estimates.

The Flory-Huggins Interaction Parameter

Subject to well-defined assumptions, it was shown by Flory [19] that a polymer and solvent are completely miscible provided the Flory parameter χ meets the condition

$$\chi \leqslant \tfrac{1}{2} \, (1 \, + \, 1/\sqrt{m})^2 \qquad (5.23)$$

where m is the ratio of the molar volume of the polymer to that of the solvent. For polymers of large molecular weight, $m \gg 1$, and thus the critical Flory parameter $\chi_c \simeq 0.5$.

Experimental values of χ can be obtained from a variety of measurements, including solution viscosity, vapor pressure, and osmotic pressure. (A compilation of available data up to 1966 is given by Sheehan and Bisio [66]). Some attempts have been made to estimate χ from pure-component polymer and solvent properties. Since χ is a Gibbs-energy parameter, it consists of an entropy term and an enthalpy term:

$$\chi \, = \, \chi_S \, + \, \chi_H \qquad (5.24)$$

Experimental studies show that χ_S varies from about 0.1 to 0.5; for many systems it is equal to about 0.34. For nonpolar systems the enthalpy term may be expressed in terms of the solubility parameter δ (see below):

$$\chi_H \, = \, V_2 (\delta_1 - \delta_2)^2 / RT \qquad (5.25)$$

where subscript 1 refers to polymer, and 2 refers to solvent. While the solubility parameter of the solvent can be obtained from pure-component data alone, the polymer solubility parameter must be determined indirectly, since polymers cannot readily be vaporized. Sheehan and Bisio combined (5.24) and (5.25) to obtain

$$\delta_2 \, = \, \delta_1 \, \pm \, [RT(\chi - \chi_S)/V_2]^{\tfrac{1}{2}} \qquad (5.26)$$

Assuming χ_S equal to 0.34, they found that, for homologous series of solvents, plots of $[RT \, (\chi - \chi_S)/V_2]^{\tfrac{1}{2}}$ versus δ_2 were straight lines for each polymer.

Sheehan and Bisio proposed using such plots to obtain χ for those binaries for which the solvent solubility parameter is known, and they presented plots for some polymers in alkanes, halogenated hydrocarbons, ketones, and esters as solvents. However, since many systems of practical interest are polar, many data points deviate considerably from the straight lines, and this method therefore is not generally useful.

As indicated by Burrell [11], the interaction parameter χ is only of limited use:

1. It has little theoretical significance, and can only be considered an empirical constant. Its entropy and enthalpy components differ from what might be predicted, with the differences, however, tending to be compensatory.

2. It is not a constant, since it is often a function of composition and temperature.

3. It cannot be calculated accurately and must be determined experimentally for each polymer-solvent system. The necessary experiments tend to be laborious.

4. For multicomponent systems all binary interactions must be determined.

5. It cannot handle systems in which there are significant polar, hydrogen-bonding, or acid-base interactions between the polymer and the solvent.

Burrell strongly suggests that the Flory χ parameter be abandoned for practical polymer solubility work. This suggestion, however, may require reconsideration, in view of recent work by Flory and others [50] which extends polymer solution theory along lines introduced by Prigogine [59].

The Solubility Parameter

Over 50 years ago, Hildebrand introduced the concept of cohesive energy density to explain solubility relationships. The cohesive energy density of a liquid is the ratio of the isothermal energy of vaporization of that liquid to the ideal gas state divided by the liquid volume. The square root of the cohesive energy density is called the solubility parameter δ. It is commonly expressed in Hildebrand units: 1 hildebrand is equal to 1 $(\text{cal cm}^{-3})^{\frac{1}{2}}$.

Subject to several well-defined simplifying assumptions which are reasonable for nonpolar fluids, Hildebrand showed that the enthalpy of mixing for a binary mixture can be expressed by

$$\Delta h(\text{mixing}) = (x_1 V_1 + x_2 V_2)\phi_1\phi_2 A_{12} \qquad (5.27)$$

where x is the mole fraction, V is the molar liquid volume, and ϕ is the volume fraction:

$$\phi_1 = \frac{x_1 V_1}{x_1 V_1 + x_2 V_2} \qquad \phi_2 = \frac{x_2 V_2}{x_1 V_1 + x_2 V_2} \qquad (5.28)$$

An estimate for the interchange energy density is given by

$$A_{12} = (\delta_1 - \delta_2)^2 \tag{5.29}$$

Equations (5.27) and (5.29) can be used to estimate solubilities, provided an estimate of the entropy of mixing can also be obtained. For mixtures of nonpolar molecules of similar size, the ideal entropy of mixing is often used; for a binary mixture,

$$\Delta s(\text{mixing}) = -R(x_1 \ln x_1 + x_2 \ln x_2) \tag{5.30}$$

For mixtures of nonpolar molecules that differ appreciably in size, the entropy of mixing is often calculated from Flory's equation; for a binary mixture,

$$\Delta s(\text{mixing}) = -R(x_1 \ln \phi_1 + x_2 \ln \phi_2) \tag{5.31}$$

The Gibbs energy of mixing is related exactly to Δh and to Δs by

$$\Delta g(\text{mixing}) = \Delta h(\text{mixing}) - T\Delta s(\text{mixing}) \tag{5.32}$$

Prior to the use of the solubility parameter concept by paint technologists, there was no practical theory of solubility that could be used in the formulation of solvent systems for resins and polymers. Based on experience with individual polymers, various substances were categorized as active solvents, latent solvents or cosolvents, and nonsolvents or diluents for the particular polymer. An active solvent was one that could dissolve the polymer at all concentrations. An active solvent could be mixed partially with a nonsolvent and still maintain solubility of the polymer in the solvent mixture. By the addition of selected other substances, the cosolvents, even more of the diluents could be added before the polymer solubility in the solvent mixture was impaired.

During the early 1950s extensive work was done by Burrell and co-workers on the determination of polymer solubilities and their correlation with the solubility parameters of polymers and various solvents [10]. It was found that, if the liquids were classified according to their hydrogen-bonding characteristics, a good correlation of solubility relationships was obtained using polymer and solvent solubility parameters. The liquids were divided into three classes: strongly hydrogen-bonded (alcohols, acids, and amines), moderately hydrogen-bonded (esters, ketones, and ethers), and weakly hydrogen-bonded (most nonpolar compounds such as hydrocarbons). Within each class of liquid, polymer solubility could be predicted with a 95% probability or better. The polymer solubility parameters were determined empirically. The polymers were tested for solubility in a series of solvents about 0.5 hildebrands apart. The range of solvent δ values that dissolved the polymer in each class of liquids was then assigned to that polymer. Thus a polymer dissolves in a liquid (or mixture of liquids) whose δ value lies within the range of polymer values for that class of liquids.

Several other methods have been proposed for estimating the solubility parameters of liquids, including polymers; one of these is a group contribution method suggested by Small [67], and later extended by Konstam and Feairheller [35]. Small calculated a molar attraction constant F from the equation

$$F = (\Delta U^v V)^{0.5} \tag{5.33}$$

for different functional groups. The energy of vaporization ΔU^v was calculated from vapor pressure and heat-of-vaporization data. According to the group contribution hypothesis, the F values of the various groups in a substance can be added together to obtain the F value of the substance. From its molar volume the solubility parameter of the substance can then be calculated by

$$\delta = F/V \tag{5.34}$$

Small showed F to be additive for the methylene group. For other groups, however, steric, ring closure, conjugation, polar, and hydrogen-bonding effects lead to some deviations from the additivity of F.

Konstam and Feairheller found that if the solubility parameters of a family of monofunctional compounds are plotted against the reciprocal of the molar volumes, straight lines are obtained (except for hydrocarbons). Thus, for substances other than hydrocarbons, δ could be expressed as a linear function of $1/V$:

Table 5.4 Molar Attraction Constants and Molar Volumes of Functional Groups at 25°C

Group		F, $(cal\ ml)^{1/2}\ mol^{-1}$		Molar Volume $(ml\ mol^{-1})$
		Small	Konstam & Feairheller	
CH₃	Single-bonded	214	—	—
CH₂		133	—	—
−CH		28	—	—
−C−		−93	—	—
CH₂=	Double bonded	190	—	—
−CH=		111	—	—
>C=		19	—	—

Table 5.4 Molar Attraction Constants and Molar Volumes of Functional Groups at 25°C

	F, (cal ml)$^{1/2}$ mol^{-1}		Molar Volume
Group	Small	Konstam & Feairheller	(ml mol^{-1})
CH≡C−	285	−	−
−C≡C−	222	−	−
Phenyl	735	−	−
Phenylene (o, m, p)	658	−	−
Naphthyl	1,146	−	−
Ring, five-membered	105-115	−	−
Ring, six-membered	95-105	−	−
Conjugation	20-30	−	−
S sulfides	225	−	−
SH thiols	315	−	−
ONO$_2$ nitrates	440	−	−
NO$_2$ (aliphatic nitrocompounds)	440	−	−
PO$_2$ (organic phosphates)	500	−	−
H (variable)	80-100	−	−
CHO	−	403	37.22
O ethers	70	281	27.96
CO ketones	275	333	32.52
COO esters	310	−	−
−COOH	−	525	36.28
CN	410	464	31.23
Cl single	270	311	34.98
Cl as in >CCl$_2$	260	257	−
Cl as in −CCl$_3$	250	241	−
Cl as in −C(Cl)−C(Cl)−	−	243	−
Br single	340	360	33.07
Br as in >CBr$_2$	−	331	−
Br as in −CBr$_3$	−	284	−
Br as in −C(Br)−C(Br)−	−	308	−
F single	−	199	30.30
CF$_2$ ⎱ n-fluorocarbons	150	152	−
CF$_3$ ⎰	274	−	−
I single	425	410	42.30
OH single	170	399	19.66
OH in diols	−	310	−
NH$_2$	−	347	29.49

$$\delta = A + B/V \tag{5.35}$$

where A and B are constants. Using (5.35) for a particular family of fluids and substituting the molar volume of the functional group, they obtained what they called the solubility parameter of the particular functional group.

Table 5.4 shows F values and molar volumes for functional groups presented by Small and by Konstam and Feairheller. These may be used for rough estimates of polymer solubility parameters. However, such estimates are often not useful for practical purposes, since they do not account for the hydrogen-bonding characteristics of the polymer-solvent system. As shown by Burrell [10], solubility predictions obtained by matching polymer and solvent solubility parameters are reliable only within a given hydrogen-bonding class of solvents.

Although some experimental data must be obtained for accurate results, an estimate of the maximum allowable difference in the solubility parameters of the polymer and solvent may be obtained from the Flory interaction parameter. From (5.23) and (5.26),

$$(\delta_1 - \delta_2)^2_{max} - -[\tfrac{1}{2} (1 + 1/\sqrt{m})^2 - \chi_S] RT/V_2 \tag{5.36}$$

To illustrate, consider polystyrene with a molecular weight of 1.4×10^6. If the solvent molar volume is 100 ml mol^{-1}, and χ_S is taken as 0.34, then at $30°C$ (5.36) yields $(\delta_1 - \delta_2)_{max} = 1.005$ hildebrands which compares reasonably well with 1.25 obtained experimentally for weakly hydrogen-bonded solvents. (Agreement, however, is poor for solvents with significant hydrogen bonding). Equation (5.36) also shows that $(\delta_1 - \delta_2)_{max}$ increases with rising temperature, falling solvent molar volume, and falling χ_S.

Burrell's work was instrumental in initiating widespread use of the solubility parameter in the paint and coatings industry. It explained why active solvents, cosolvents, and nonsolvents behaved as they did. Its use permitted solvent system formulation with much greater ease. A specific solvent could be added to the mixture to impart certain desired characteristics; the mixture solubility parameter could then be adjusted by the addition of a suitable other solvent, in order that it be in the desired range. Hansen and Beerbower [27] have presented a comprehensive review of the use of solubility parameters in the paint industry.

By itself the solubility parameter can explain the behavior of only a relatively small group of solvents, those with little or no polarity and with weak or no hydrogen-bonding forces. To extend its use to different kinds of solvents, it is necessary to consider specific forces and interactions. Quantitative or semiquantitative measures of the effects of different intermolecular forces are required. Toward that end, several investigators have classified these forces and estimated the contributions of each class to the cohesive energy, as briefly discussed in the next section

Contributions to the Cohesive Energy

The potential energy between a pair of molecules is due to one or more of the following.

1. Dispersion forces, which result from transient charge separations in all, including nonpolar, molecules. The magnitude of the interaction depends on the ionization potentials and the polarizabilities of the molecules.

2. Dipole-dipole forces, which depend on the dipole moments of polar molecules.

3. Dipole-induced dipole forces, which are a function of the dipole moments and the polarizabilities of the molecules.

4. Specific ("chemical") forces such as hydrogen bonding or charge-transfer complex formation.

The relative effects of these intermolecular forces for any given solvent-solute system cannot be determined in a straightforward manner. Several approximations are discussed below.

Polarity

As suggested by van Arkel [76] and developed by Blanks and Prausnitz [6], the total energy of vaporization can be split into a nonpolar and a polar part:

$$\frac{\Delta U^v}{V} = \frac{\Delta U^v_{np}}{V} + \frac{\Delta U^v_p}{V} = \lambda^2 + \tau^2 \tag{5.37}$$

where λ and τ are, respectively, the nonpolar and polar solubility parameters. Blanks and Prausnitz used the homomorph concept (see Section 7, p. 295) along with the group contribution method of Small [67] to determine the λ and τ parameters for several polymers and solvents. These parameters were then used to obtain the interchange energy density A_{12} to estimate the heat of mixing by (5.27). For a nonpolar solute in a nonpolar solvent,

$$A_{12} = (\delta_1 - \delta_2)^2 = (\lambda_1 - \lambda_2)^2 \tag{5.38}$$

and for a nonpolar solute in a polar solvent,

$$A_{12} = (\lambda_1 - \lambda_2)^2 + (\tau_2^2 - 2\psi) \tag{5.39}$$

where ψ is an empirical induction parameter. For a polar solute in a polar solvent,

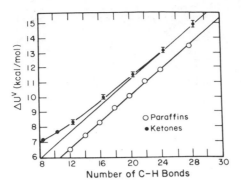

Fig. 5.10. Vaporization energies of ketones versus the number of C−H bonds in the molecule.

$$A_{12} = (\lambda_1 - \lambda_2)^2 + (\tau_1 - \tau_2)^2 \tag{5.40}$$

From (5.27), (5.31), and (5.32), the Gibbs energy of mixing can then be determined. This method, however, is difficult to apply to multicomponent systems, and even for binary systems it requires an independent estimate of the entropy of mixing. The concept of splitting the solubility parameter into nonpolar and polar parts was later extended by Hansen [28] to include a hydrogen-bonding contribution (see below).

Meyer and Wagner [44] have presented another method for estimating the relative magnitudes of dispersion, polar, and induction energies. They plotted the energy of vaporization versus the number of C−H bonds for a homologous series of fluids. A similar curve for paraffins was also drawn on the same plot. Fig. 5.10 shows such a plot for ketones. With increasing size of the alkyl chain, the dipole-dipole interactions decrease, the induction forces remain unchanged, and the dispersion forces increase at the same rate as those for paraffins. Thus for a ketone of a given number of C−H bonds, the paraffin curve gives the dispersion contribution to ΔU^v; the difference between the paraffin curve and the asymptote for the higher ketones is due to the induction forces, and that between the ketone curve and the asymptote is due to the polar forces. This method appears to be useful for obtaining the various contributions to the cohesive energy of solvents. However, before this concept of Meyer and Wagner can be used in practical applications, extensive polymer solubility testing needs to be done along the lines used by Burrell.

In the development of the Hildebrand-Scatchard regular-solution theory for nonpolar systems, it was assumed that the cohesive energy density between dissimilar nonpolar molecules is the geometric mean of the individual cohesive energy densities of the two components. For polar systems, Gardon [21] proposes instead:

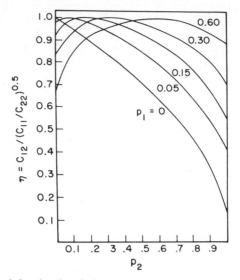

Fig. 5.11. Effect of fractional polarities on the cohesive energy density between dissimilar molecules.

$$A_{12} = \delta_1^2 + \delta_2^2 - 2\eta\delta_1\delta_2 \qquad (5.41)$$

When both components 1 and 2 are nonpolar, $\eta = 1$. Subject to certain assumptions, Gardon writes η in terms of p_1, p_2, d_1, and d_2:

$$\eta = (p_1 p_2)^{0.5} + (d_1 d_2)^{0.5} + 0.0721 \, [(d_1 p_2)^{0.5} + (d_2 p_1)^{0.5}] \qquad (5.42)$$

Here p denotes the fraction of the total potential energy in a fluid due to dipole-dipole forces, and d the fraction due to dispersion forces. The dispersion fraction may also be expressed in terms of p:

$$d = 1 - 0.9898p - 0.142(p-p^2)^{0.5} \qquad (5.43)$$

so that η may be calculated from p_1 and p_2 only, by combining (5.42) and (5.43). Figure 5.11 shows the value of η as a function of p_1 and p_2. Gardon concludes that,

1. $\eta = 1$ only if $p_1 = p_2$, and that in general $\eta < 1$.
2. A "good" solvent matches both the δ and p of the solute.
3. Even if δ_1 and δ_2 are equal, an endothermic heat of mixing can be obtained if p_1 is not equal to p_2.

Gardon calculated the dispersion, polar, and induction forces between two identical molecules from the potential energy equations of London, Keesom, and Debye; he then calculated p as the polar fraction of the total potential energy and called it the fractional polarity of the substance. The fractional polarity has been used along with the solubility parameter to characterize polymers and solvents. (See below.)

More recently, Bagley et al. [2] have proposed using internal pressure measurements in liquids to determine the contribution of polarity as well as hydrogen bonding to the total cohesive energy. However, few data are given by these investigators, and the practical usefulness of their technique cannot be determined with the limited information currently available.

Hydrogen Bonding

Hydrogen bonding strongly influences polymer solubility in solvents. Pimentel and McClellan [53] classify liquids into four groups:

1. Proton donors, such as substantially halogenated hydrocarbons, for example, chloroform.
2. Proton acceptors, such as ketones, aldehydes, ethers, and esters.
3. Substances that are simultaneously proton donors and acceptors, such as water, alcohols, acids, and primary and secondary amines.
4. Liquids that are neither proton donors nor acceptors, and therefore do not participate in hydrogen-bond formation. Paraffins, for example, belong to this category.

Most of the early work regarding the effect of hydrogen bonding on solubilities was qualitative and at best semiquantitative. Small [67] proposed that the contribution to the heat of mixing due to hydrogen bonding may be expressed as:

$$\Delta h(\text{due to hydrogen bonding}) = \phi_1 \phi_2 (A_1 - A_2)(D_1 - D_2) \qquad (5.44)$$

where ϕ_1 and ϕ_2 are the component volume fractions, and A and D are, respectively, (positive) proton-accepting and proton-donating capacities of the particular liquids. No numerical values were proposed for A and D, but (5.44) can be used qualitatively. For example, if component 2 belongs to class 4, and component 1 to any other class, both A_2 and D_2 are zero but A_1 and/or D_1 are finite, resulting in a positive contribution to the heat of mixing. The miscibility of two such components therefore tends to be limited. If component 1 belongs to class 1, and component 2 to class 2, then A_1 and D_2 are zero and the hydrogen-bonding contribution to the heat of mixing is always negative. When both liquids are hydrogen-bonded, all four, A_1, A_2, D_1, and D_2, are nonzero, and the net effect cannot easily be predicted.

The qualitative discussion can be summarized by saying that miscibility is enhanced if hydrogen bonds are formed on mixing, (e.g., chloroform and acetone), while it is reduced if hydrogen bonds tend to be broken (e.g., water and a paraffin). The effect of hydrogen bonding is small for mixing of fluids belonging to the same class.

In an effort to establish a quantitative scale for hydrogen-bond formation, Gordy [23] compared 1 M solutions of deuterated methanol in various solvents with 0.1 M deuterated methanol solutions in benzene; he observed the effect of hydrogen bonding on the O–D absorption band in the wavelength range of 3.6 to 4.35 μ. The wavelength of the absorption peak shifts, depending on the test solvent. Gordy tabulated the shifts produced by several solvents and proposed that this shift be used as a quantitative measure of the hydrogen-bond energy.

Small [67] determined the "iodine-bonding number" of a few solvents and suggested that it be used as a "hydrogen-bonding number". These numbers were obtained by a visual matching of the color produced by dissolving iodine in different solvents. However, Small did not list iodine-bonding numbers for any solvents; no other workers appear to have used these numbers.

Hansen and Skaarup [29] used a value of 5 kcal mol^{-1} (based on infrared spectroscopy) as the enthalpy for an O–H $\cdot \cdot \cdot$ O hydrogen bond; they then considered this enthalpy to be additive for each additional O–H $\cdot \cdot \cdot$ O bond. However, numerical values for the enthalpies of other types of hydrogen bonds were not assigned.

Refinements of Gordy's work have been presented by Nelson, Hemwall, and Edwards [46], who made measurements using benzene and carbon tetrachloride as the reference solvents. Nelson, Hemwall, and Edwards also proposed a net hydrogen-bonding index θ. Since the effect on component solubility depends on whether hydrogen bonds are broken or formed on mixing, a weighting factor k_i was introduced:

$$\theta = \sum_{i=1}^{n} k_i \phi_i \Gamma_i \tag{5.45}$$

where ϕ_i is the volume fraction of the component, and Γ_i is the hydrogen-bonding parameter taken as 10% of the wave number shift produced by the solvent. The weighting factor k_i is taken as -1 for donor-acceptor fluids such as alcohols, zero for ether alcohols, and $+1$ for all other substances. The sign and magnitude of θ reflect the hydrogen-bonding contribution to solubility; systems having a large positive value of θ are better solvents than those with a negative or a small positive value of θ.

Gordy's scale has been used in some industrial laboratories for characterizing solvents. Nelson's extension of this work seems to be a useful one; it was developed in an industrial laboratory and has been used for commercial formulation of solvents for paints and coatings. Some applications are discussed below.

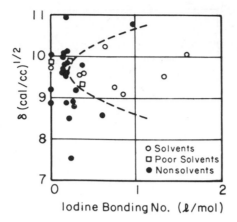

Fig. 5.12. Two-dimensional solubility region for polyvinyl chloride.

Multidimensional Solubility Correlation

To obtain reliable predictions of polymer solubility behavior, various workers have attempted to correlate experimental data with specific intermolecular forces. These correlations first characterize the different polymers and solvents on the basis of two or three parameters, and then establish two- or three-dimensional regions of solubility for each polymer.

One of the earliest two-dimensional methods was proposed by Small [67], who used the solubility parameter and the iodine-bonding number. Figure 5.12 shows the solubility of polyvinyl chloride in various solvents. A general region of solubility was obtained, as shown by the area to the right of the dotted curve in Fig. 5.12. Although no subsequent use appears to have been made of these two parameters, Small's work is nevertheless significant, as it pioneered the concept of the multidimensional solubility correlation.

The method of Burrell [10] is essentially a method for delineating a two-dimensional region of solubility. The second dimension is not quantitative; it is a qualitative grouping according to the hydrogen-bonding tendency of the solvent. Lieberman [40] later assigned arbitrary numerical values for the hydrogen-bonding tendency of solvents, based on Burrell's work. These were then used for preparing solubility maps similar to Fig. 5.12.

Wyart and Dante [80] give an example of the use of fractional polarity for solvent system formulations. A region of solubility is experimentally determined for a given polymer on a two-dimensional plot of δ and p of the solvents, as shown in Fig. 5.13. If the δ and p are known for each potential component in the solvent mixture, various mixtures can be formed which dissolve the polymer. The only criterion for solubility is that the mixture parameters (volume fraction average) be such that the mixture lies inside the solubility region.

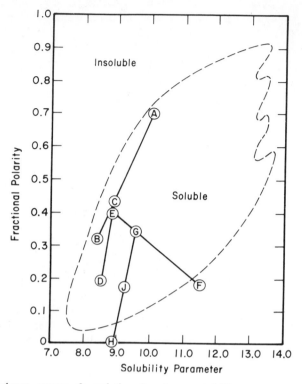

Fig. 5.13. Solvent system formulation based on solubility parameter and fractional polarity.

For example, it may be desired that a solvent blend, suitable for a lacquer formulation, posses the following characteristics:

1. A ratio of 3:7 of low- to medium-boiling active solvent.
2. A ratio of 13:87 of high- to low- and medium-boiling active solvents.
3. A 3:1 ratio of active to latent solvents.
4. A 1:1 ratio of hydrocarbon to oxygenated solvents.

The polymer-dissolving characteristics of the solvent blend may be predicted by a graphical construction illustrated in Fig. 5.13. Points A and B represent the low- and medium-boiling active solvents, respectively, the point C being determined by criterion 1. Point D represents the high-boiling active solvent, and E the fulfillment of requirement 2. The latent solvent point F lies outside the solubility region but, when F is mixed with E, the resultant point G is inside the region of solubility. Finally, enough aromatic hydrocarbon, point H, is added to meet the last criterion. The final solvent blend is then represented by point J. Since J lies inside the solubility region, the polymer dissolves in this solvent blend.

The same parameters are also used by Nelson et al. [45] who, in addition, use the hydrogen-bonding index θ. The δ, p, and θ values for many common coatings solvents are listed by these workers and are shown in Table 5.5. These investigators also describe a computerized solvent-formulation procedure which can be used to obtain the least expensive blend that satisfies a given set of constraints.

Table 5.5 Physical Chemical Parameters of Solvents

Solvent[a]	Solubility Parameter, δ	Fractional Polarity p	Hydrogen-Bonding, θ
Hydrocarbons			
Shell Sol B	7.25	0	0
Hexane	7.30	0	0
Shell Sol B-8	7.40	0	0.4
Tolu Sol 5	7.20	0	0
Tolu Sol 10	7.40	0	0.5
Tolu Sol 25	7.60	0	3.5
Shell Sol 260	7.20	0	0.07
Toluene	8.90	0.001	4.2
Special VM & P naphtha	7.40	0	0.32
Super VM & P naphtha 66	7.45	0	0
Chlorobenzene	9.50	0.058	2.7
Ethylbenzene	8.80	0.001	4.2
Xylene	8.85	0.001	4.5
Tolu Sol 28	8.50	0.001	3.4
Shell mineral spirits 135W	7.60	0	4.2
Cyclo Sol 53	8.75	0.001	5.0
Shell Sol 70	7.15	0	0
Shell Sol 360	7.45	0	0.5
Shell Sol 340	7.50	0	0.13
Cyclo Sol 63	8.70	0.001	5.3
Shell Sol 140	7.40	0	0.5
Ketones			
Acetone	10.0	0.695	12.5
Methyl ethyl ketone	9.3	0.510	10.5
Methyl isobutyl ketone	8.4	0.315	10.5
Methyl isoamyl ketone	8.3	0.255	10.9
Cyclohexanone	9.9	0.380	13.7
Ethyl amyl ketone	8.2	0.223	8.5

Table 5.5 Physical Chemical Parameters of Solvents (Continued)

Solvent[a]	Solubility Parameter, δ	Fractional Polarity p	Hydrogen-Bonding, θ
Pentoxone solvent	8.5	0.190	12.5
Diisobutyl ketone	7.8	0.123	9.8
Diacetone alcohol	9.2	0.312	0
Isophorone	9.1	0.190	14.9
Mesityl oxide	9.0	0.332	12.0
Esters			
Ethyl acetate (85-88%)	9.6	0.182	4.9
Ethyl acetate (99%)	9.1	0.167	8.4
Isopropyl acetate	8.6	0.100	8.5
n-Propyl acetate	8.75	0.129	8.5
s-Butyl acetate	8.2	0.082	8.3
Isobutyl acetate	8.3	0.097	8.7
n-Butyl acetate (90-92%)	9.0	0.118	5.7
n-Butyl acetate (99%)	8.6	0.120	8.0
Amyl acetate	8.45	0.067	8.2
Methyl amyl acetate	8.2	0.050	8.3
Isobutyl isobutyrate	7.7	0.042	8.0
Ethylene glycol monomethyl ether acetate	9.2	0.095	10.5
Ethylene glycol monoethyl ether acetate	8.7	0.073	10.1
Ethylene glycol monobutyl ether acetate	8.2	0.060	10.3
Ethers and glycol ethers			
Tetrahydrofuran	9.9	0.075	12.0
Dioxane	10.0	0.006	14.6
Methyl Oxitol glycol ether	10.8	0.126	0
Oxitol glycol ether	9.9	0.086	0
Butyl Oxitol glycol ether	8.9	0.048	0
Methyl Dioxitol glycol ether	10.2	0.058	0
Dioxitol glycol ether	9.6	0.043	0
Butyl Dioxitol glycol ether	8.9	0.028	0
Alcohols			
Methyl alcohol	14.5	0.388	-19.8
Neosol proprietary solvent	13.2	0.296	-17.7
Isopropyl alcohol	11.5	0.178	-16.7

Table 5.5 Physical Chemical Parameters of Solvents (Continued)

Solvent[a]	Solubility Parameter, δ	Fractional Polarity p	Hydrogen-Bonding, θ
n-Propyl alcohol	11.9	0.152	-16.5
s-Butyl alcohol	10.8	0.123	-17.5
Isobutyl alcohol	10.7	0.111	-17.9
n-Butyl alcohol	11.4	0.096	-18.0
Methyl isobutyl carbinol	10.0	0.066	-18.7
n-Amyl alcohol	10.9	0.074	-18.2
Cyclohexanol	11.4	0.075	-16.5
2-Ethyl hexanol	9.5	0.042	-18.7
Miscellaneous			
Dichloromethane	9.7	0.12	1.5
Chloroform	9.3	0.017	1.5
1,2-Dichloroethane	9.8	0.043	1.5
1,1,2,2-Tetrachloroethane	10.4	0.092	1.5
2-Nitropropane	9.9	0.72	4.0
Dimethylformamide	12.1	0.772	18.9

[a]Shell Sol, Tolu Sol, Cyclo Sol, Pentoxone, Oxitol, Dioxitol, and Neosol are registered trademarks of Shell Chemical Company.

As an illustration of their computer program, Nelson et al. present the reformulation of a solvent blend for a nitrocellulose lacquer which complies with Rule 66 of the Los Angeles County Air Pollution Control District (see Section 9). The original blend, which does not meet the air pollution standards, is used as the starting point. The first part of the computer program calculates its properties, such as solubility parameter, fractional polarity, hydrogen-bonding index, viscosity, specific gravity, evaporation rate, and cost. Constraints for the new formulation are then set to yield a solvent blend, with similar properties, that meets the air pollution requirements. The second part of the program calculates the optimum blend from a given list of candidate solvents by means of linear programming. The formulator may wish to modify or add further constraints to obtain different optimal blends. Solvent blends selected by this technique then serve as starting points for more extensive laboratory testing.

For the example given, the compositions, costs, and some properties of the original and reformulated blends are shown in Table 5.6. The replacement was found to cost 4.4% more than the original blend. A cost increase of this

Table 5.6 Compositions and Properties of a Nitrocellulose Lacquer Solvent Blend and Its Replacement to Meet Air Pollution Code Requirements

	Original Blend	*Replacement*
Solvent Composition Vol %		
Isobutyl acetate	26.0	—
n-Butyl acetate	—	34
Methyl isobutyl ketone	—	7
Methyl amyl acetate	—	5
n-Butyl alcohol	4.8	—
Neosol proprietary solvent	—	15
Xylene	27.6	6
Toluene	27.6	4
Super VM & P naphtha 66	—	29
	100.0	100
Solvent properties		
Cost, cents/gal (in 1972)	60.9	63.6
Specific gravity, 25/25°C	0.864	0.817
Estimated viscosity, cps, 25°C	0.65	0.69
Determined viscosity, cps, 25°C	0.62	0.66

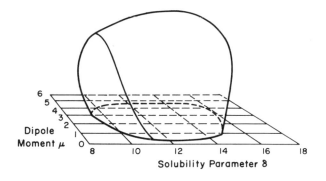

Fig. 5.14. Solubility region of cellulose acetate butyrate polymer.

magnitude is typical for nitrocellulose lacquer system reformulation to meet Rule 66 requirements.

Other workers have also used three-parameter correlations for predicting polymer solubilities in solvents. The three parameters used by Crowley, Teague, and Lowe [13] are the solubility parameter δ, the dipole moment μ, and the hydrogen-bonding parameter Γ from Gordy's work. The values of the three

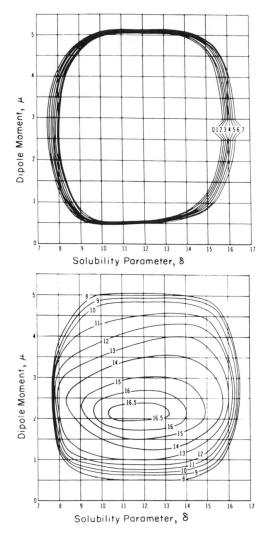

Fig. 5.15. Two-dimensional representation of the solubility region of cellulose acetate butyrate.

parameters have been tabulated by these investigators for several common solvents. Three-dimensional models of solubility behavior were constructed, as shown in Fig. 5.14. Such models were prepared for cellulose nitrate, cellulose acetate, cellulose acetate butyrate, vinyl chloride acetate copolymer, and polymethylmethacrylate. The three-dimensional models were also converted to two-dimensional contour maps, as illustrated in Fig. 5.15, for ease of graphical formulation of solvent systems.

Hansen [28] also used a three-dimensional representation of polymer solubility behavior. His three parameters were derived from the solubility parameter δ. Dividing δ into dispersion, polar, and hydrogen-bonding contributions, he obtained

$$\delta^2 = \delta_d^2 + \delta_p^2 + \delta_h^2 \tag{5.46}$$

The parameter δ_d was obtained from the homomorph concept. Initially, the parameters δ_p and δ_h were assigned values empirically, (5.46) being satisfied at all times. δ_p and δ_h were determined by trial and error from experimental solubility data; the values of the two parameters were adjusted so that *all* solvents for a given polymer were represented by points inside a three-dimensional region and all nonsolvents lay outside that region. Later, Hansen and Skaarup [29] presented independent calculation procedures for δ_p and δ_h. It was found that, if the unit distance along δ_d was set at twice that along the other two axes, spherical regions of solubility were obtained. Each polymer could then be characterized by the center and radius of this sphere. However, no theoretical or physical reasoning was given to justify spherical solubility regions. Hansen studied 33 polymers and resins, 25 pigments, 10 plasticizers, and about 90 solvents.

To alleviate the complexity of working in three-dimensional space, Teas [72] has suggested using an empirical form of Hansen's parameters in a triangular diagram. The coordinates for such a plot were defined by

$$f_d = 100\delta_d/\Delta \quad f_p = 100\delta_p/\Delta \quad f_h = 100\delta_h/\Delta \tag{5.47}$$

where

$$\Delta = \delta_d + \delta_p + \delta_h \tag{5.48}$$

Using Hansen's data, Teas plotted the solubility behavior of certain polymers on triangular diagrams; again, more or less well-defined regions of solubility were obtained, as shown in Fig. 5.16. However, the reliability of these solubility predictions was less than that using Hansen's procedure.

In all the multidimensional correlations of polymer solubility, the respective regions of miscibility are obtained only after considerable experimentation. Hansen's work, for example, entailed over 10,000 individual solubility tests. These were followed by a time-consuming process of parameter manipulation to obtain coherent volumes of solubility in three-dimensional space. Further, solubility data on one polymer cannot be used to predict the solubility behavior of another. No attempts have so far been reported on correlating the characteristics of the solubility regions with those of the respective polymers. However,

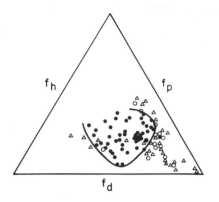

Fig. 5.16. Triangular diagram for the solubility of Pentalyn 225 resin. •, soluble; ○, partially soluble; ∧, insoluble.

multidimensional representations do yield highly reliable empirical solubility predictions, particularly for solvent blends, and are therefore widely used in the paint and coatings industry.

7 SOLVENT-AIDED SEPARATION PROCESSES

Separation of a mixture into its components is frequently achieved with the aid of an additional chemical component; if a liquid, such a separating agent is generally called the solvent. The purpose of the separating agent is either to create a second phase or to alter the equilibrium compositions of the two phases created from the feed mixture by some other means, such as by partial evaporation. Liquid-liquid extraction, extractive distillation, azeotropic distillation, and absorption are examples of solvent-aided separation processes. This section deals briefly with the selection of suitable solvents for such processes.

Solvent-aided separation processes, also called indirect separation processes, are used when more direct methods, such as distillation, are either impossible or impractical. For example, simple distillation cannot be used to obtain nearly pure ethyl alcohol from a mixture with water, since the two form an azeotrope at 95.6 wt % alcohol. However, if benzene is added to the mixture, the binary azeotrope is eliminated and alcohol of any desired purity may be obtained. Paraffinic and olefinic hydrocarbons are frequently separated by extractive distillation with a suitable solvent, such as furfural.

An example of liquid-liquid extraction is provided by the separation of tantalum from niobium. The solubility of metal fluorides in methyl isobutyl ketone depends on the acidity of the aqueous solution of the salts; only tantalum fluoride is soluble in the organic solvent at low acidity of the feed solution, whereas both fluorides are soluble in the ketone at high acidity of the aqueous solution. Thus both metal fluorides may be dissolved in the organic solvent from a highly acid aqueous phase, and the niobium may then be stripped from the organic solution by contact with a low-acidity aqueous phase.

Separation Factors and Selectivity

Mixture separation with the aid of a solvent is based primarily on the preferential affinity of the solvent for some of the components of the feed mixture relative to the others. As a result, the concentrations of the components in one phase are different from those in the other phase. The separation factor a_{AC} for separating components A and C from a mixture is defined by [34]:

$$a_{AC} = \frac{x_A^\beta}{x_A^a} \left(\frac{x_C^\beta}{x_C^a} \right)^{-1} \tag{5.49}$$

where x^a and x^β are the equilibrium compositions in the two phases. For the solvent to be effective as a separating agent, a_{AC} must be different from unity; the further removed a_{AC} is from unity, the more selective the solvent. From the equations of equilibrium it follows that, for a liquid-liquid system,

$$a_{AC} = \frac{\gamma_A^a}{\gamma_A^\beta} \left(\frac{\gamma_C^a}{\gamma_C^\beta} \right)^{-1} \tag{5.50}$$

and for a vapor-liquid system,

$$a_{AC} = \frac{\gamma_A \phi_C \, f_A^0}{\gamma_C \phi_A \, f_C^0} \tag{5.51a}$$

where γ is the liquid-phase activity coefficient, f^0 is the standard-state fugacity, and ϕ is the vapor-phase fugacity coefficient. At low or moderate pressures, we set $\phi = 1$ and replace the standard-state fugacities by the pure-component saturation pressures P_A^s and P_C^s, evaluated at system temperature. Thus (for low or moderate pressure):

$$a_{AC} = \frac{\gamma_A P_A^s}{\gamma_C P_C^s} \tag{5.51b}$$

In vapor-liquid systems a_{AC} is often called the relative volatility.

In the presence of a solvent, the ratio γ_A/γ_C depends on the temperature, the nature of the solvent, and the solvent concentration.

References for liquid-liquid equilibrium data have been compiled by Francis [20], and those for vapor-liquid equilibrium are given by Hala et al. [24].

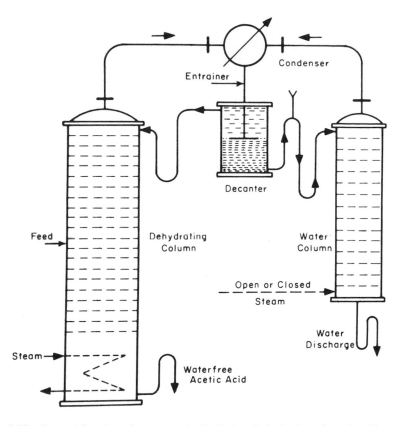

Fig. 5.17. General flowsheet for azeotropic distillation. Dehydration of acetic acid.

References for a few common systems are also given by Perry, Chilton, and Kirkpatrick [51]. Timmermans [74] lists data on many binary systems. More recent data are referenced in the annual reviews "Distillation" and "Thermodynamics" in *Industrial and Engineering Chemistry* (published near the end of each year in the journal through 1970).

While much experimental work has been reported on specific systems, there have been relatively few investigations directed toward systematic determination of the most selective or otherwise optimal solvent. The few studies that have considered this topic are primarily directed at separations in the petroleum and petrochemical industry.

Azeotropic and Extractive Distillation

A general flowsheet for azeotropic distillation with only binary azeotropes is shown in Fig. 5.17. The separation of acetic acid from water is carried out in

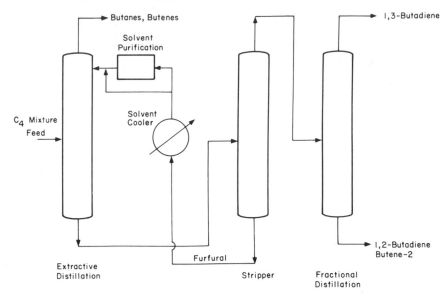

Fig. 5.18. 1,3-Butadiene recovery by extractive distillation with furfural.

such a system using ethylene dichloride, *n*-propyl acetate, or *n*-butyl acetate as the solvent (entrainer). The acid-water mixture is fed to the dehydrating column; the water forms an azeotrope with the entrainer and leaves as the overhead from the column. The vapor overhead, when condensed, forms two liquid phases. The entrainer-rich phase is returned as reflux to the dehydrating column. The entrainer in the aqueous layer in the decanter is distilled out in the water column and returns to the condenser azeotroped with water. Since this azeotrope composition is the same as that of the overhead from the dehydrating column, a common condenser for the two overhead streams is frequently used. Water-free acetic acid and acid-free water are obtained as the bottom products of the dehydrating column and the water column, respectively.

Extractive distillation is used for the recovery of 1,3-butadiene from a mixture with other C_4 hydrocarbons. A simplified flowsheet of the process using furfural as the solvent is shown in Fig. 5.18. The more polarizable components (butadienes and 2-butene) appear with furfural in the bottom product of the extractive distillation column. After removal of furfural in the stripper, purified 1,3-butadiene is separated by ordinary distillation.

For economic reasons a solvent-aided separation process is used in many cases in which simple distillation can also effect the separation. For example, acetic acid and water do not form an azeotrope, but the relative volatility between the two is low and a large number of equilibrium stages is required to separate the two by fractional distillation. To be economically attractive,

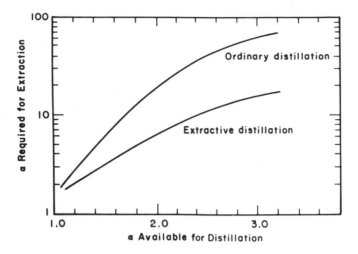

Fig. 5.19. Separation factors required for comparative costs as calculated by Souders [68] ; a is the separation factor.

the separation factor in a solvent-aided separation process must be greater than that for ordinary distillation. Souders [68] has made an approximate comparison between the separation factors required to give equivalent process costs for distillation, extractive distillation, and extraction. The results of Souders' calculations are shown in Fig. 5.19. For extractive distillation, Souders assumed a 2:1 molar ratio of solvent to hydrocarbon in the main column. He also assumed liquid flow rates in extraction and in extractive distillation to be four times those in ordinary distillation. As shown in Fig. 5.19, the separation factors required in extraction are an order of magnitude greater than those for distillation for the same separation costs; separation factors for extractive distillation need to be two to three times the ones available for ordinary distillation. Such a comparison, however, is useful only for making a crude estimate of the applicability of a solvent-aided separation process. A detailed economic evaluation is almost always needed for making the final choice.

Azeotropic distillation has been used for the separation of benzene from similar-boiling nonaromatics using acetone as the solvent, for separation of toluene from acetic acid using methanol or methyl ethyl ketone, and for separating ethylbenzene from styrene using isobutanol or 1-nitropropane. These processes were established in the early 1940s; in later years little further use of azeotropic distillation seems to have developed for separating hydrocarbon mixtures [33a].

Mair, Glasgow, and Rossini [42] considered the separation of various hydrocarbon mixtures by azeotropic distillation. Potential entrainers were selected to meet the following requirements:

1. It must have a boiling point within 30 to 40°C of that of the mixture to be separated.

2. It must be completely soluble in water and only partially soluble in the hydrocarbons at room temperature, in order that it can be removed by water-washing the condensate.

3. It must be completely soluble in the hydrocarbons at the temperature of the distillation column, and a few degrees below this temperature, to avoid the occurrence of two liquid phases in the column.

4. It must be available at low cost and in a fairly pure state.

5. It must be chemically inert with respect to the hydrocarbons, as well as to the materials of construction of the still.

These investigators worked with methanol, ethanol, acetonitrile, acetic acid, and some glycol ethers. They found that separation of aromatics from naphthenes and paraffins, of monoolefins from aromatics, paraffins, and diolefins, and of diolefins from naphthenes could be accomplished easily; however, separation of naphthenes from paraffins, of cyclic hydrocarbons with a different number of rings from one another, and of monoolefins from naphthenes was not facilitated by these entrainers.

Ewell, Harrison, and Berg [18] have given a qualitative discussion of the effect of hydrogen bonding on the suitability of a substance as an entrainer for azeotropic distillation. They classified substances according to their hydrogen-bonding capabilities (see Section 6); hydrocarbons have no ability to form hydrogen bonds. Therefore hydrocarbons tend to form heterogeneous (condensate forms two liquid phases), minimum-boiling azeotropes with strong hydrogen-bonding liquids such as water, alcohols, glycols, and organic acids.

A large number of possible entrainers was investigated by Berg et al. [5] for the separation of naphthenes from paraffins, olefins from paraffins, and C_8 aromatic hydrocarbons from one another. Table 5.7 shows the separations studied, along with the best entrainer and the accompanying improvement in relative volatility. Entrainer performance could not be correlated with the various physical characteristics of the entrainer and the mixture to be separated. The only general conclusion appears to be that the best entrainers are all capable of hydrogen bonding.

Gerster, Gorton, and Eklund [22], considered the separation of n-pentane from 1-pentene by extractive distillation. Thirty-three polar solvents were studied at four different temperatures between 0 and 45°C. The activity coefficients of each hydrocarbon at infinite dilution in the various solvents were determined experimentally. It was found that for every solvent the activity coefficient of the pentene was always smaller than that of pentane. This result is physically reasonable, since the polar solvent induces a larger dipole in the more polarizable pentene than it does in pentane. As shown by (5.51b), the

Table 5.7 Entrainers and Relative Volatilities for Some Hydrocarbon
Separations by Azeotropic Distillation [5]

Mixture	Best Entrainer	Relative Volatility	
		Without Entrainer	With Best Entrainer
Naphthene-paraffin			
Neohexane-cyclopentane	n-Propylamine	1.006	0.986
2,4-Dimethylpentane-cyclohexane	Acetone	1.007	1.026
2,2,4-Trimethylpentane-methylcyclohexane	Ethanol, 1,4-Dioxane	1.047	1.083
Paraffin-olefin			
3-Methylpentane-1-hexene	Methylene chloride	1.009	1.159
3-Methylpentane-2-ethyl-1-butene	Ethyl formate	1.037	1.156
n-Hexane-2-ethyl-1-butene	Chloroform	1.056	1.094
2,2,4-Trimethylpentane-2,2,4-trimethyl pentene-1	Isopropyl acetate	1.040	1.129
n-Heptane-2,2,4-trimethyl-pentene-1	Isopropyl acetate	1.045	1.129
C_8 Aromatics			
Ethylbenzene-p-xylene	2-Methyl butanol	1.037	1.079
p-Xylene-m-xylene	2-Methyl butanol	1.020	1.029
m-Xylene-o-xylene	Formic acid	1.105	1.154

relative volatility between the two hydrocarbons in the presence of a solvent is proportional to the ratio of their activity coefficients. This ratio in turn is a function of the two infinite-dilution activity coefficients and the solvent concentration:

$$\frac{\gamma_A}{\gamma_C} = \frac{\gamma_A^\infty}{\gamma_C^\infty} \, f(x_s) = S^\infty f(x_s) \qquad (5.52)$$

At constant temperature $f(x_s) \leqslant 1$ with few exceptions, and $f(x_s) \rightarrow 1$ as $x_s \rightarrow 1$. Thus at a fixed temperature the relative volatility is largest when there is an excess of solvent. The relative volatility is then given by

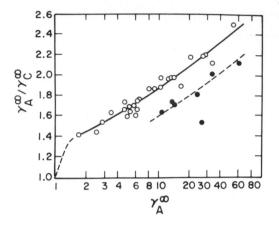

Fig. 5.20. Solvent selectivity for *n*-pentane-1-pentene separation versus the pentane activity coefficient at infinite dilution ($25°C$). [Reprinted with permission from *J. Chem. Eng. Data*, **5**, 423 (1960). Copyright by the American Chemical Society.]

$$a_{AC} = S^{\infty}\left(\frac{P_A^{\ s}}{P_C^{\ s}}\right)f(x_s) \tag{5.53}$$

that is, at a given solvent concentration the relative volatility is proportional to the ratio of the infinite-dilution activity coefficients of the two hydrocarbons. This ratio is called the selectivity.

Gerster, Gorton, and Eklund plotted the ratio $\gamma_A^{\infty}/\gamma_C^{\infty}$ as a function of γ_A^{∞}, where subscript A refers to paraffin and subscript C to olefin; the results are shown in Fig. 5.20. It was found that for 26 nonhydrogen-bonding solvents the data points could be represented by the continuous curve. The dotted curve shows the results for 7 solvents that form strong hydrogen bonds.

Two main conclusions can be drawn from Gerster's study. One is that the selectivity of a solvent increases with the extent of nonideality of a solution formed by that solvent and either of the hydrocarbons. The other is that (contrary to Berg's conclusions for azeotropic distillation) hydrogen-bonded solvents are not suitable, since they give lower selectivities than nonhydrogen-bonded solvents at the same extent of nonideality. However, even though S^{∞} increases with rising γ_A^{∞}, the two quantities are not directly proportional; a twofold increase in γ_A^{∞} produces only a modest increase in S^{∞}. Further, an increase in nonideality lowers the mutual solubility of the solvent and the hydrocarbons. There is thus a practical limit to the highest selectivity for an efficient separation.

The preceding discussion applies only to systems in which no significant chemical bonding is present between the solvent and one or more of the feed components. In the presence of such bonds (e.g., the bond formed between cuprous ammonium acetate and 1,3-butadiene), a much higher value of S^{∞} may be obtained than that predicted from a plot such as Fig. 5.20.

A systematic study of solvents for use in liquid-liquid extraction does not seem to have been reported in the literature. For various applications of extraction, however, Treybal [75] provides information on solvents and industrial processes. Some of the major applications are discussed below.

Liquid-liquid extraction is widely used in the petroleum industry. A typical application is in solvent refining of lubricating oils to remove aromatics, naphthenics, and other components of the oils that form sludges or varnishes in service. Although a large number of solvents has been proposed for this application, relatively few are in common use. These include nitrobenzene, furfural, phenol, propane, and sulfur dioxide (either by itself or as a mixture with benzene). Liquid sulfur dioxide was used as the solvent in the first commercial-scale use of extraction in petroleum refining in the Edeleanu process; aromatic hydrocarbons were removed from kerosene by extraction into sulfur dioxide to improve the burning qualities of kerosene in lamps. Sulfur dioxide is also used for the recovery of aromatics from naphtha to obtain nitration-grade benzene and toluene and high-grade xylene. In the Udex process diethylene glycol, containing small amounts of water, is used as an alternative solvent to sulfur dioxide for aromatics recovery by extraction. Aqueous sodium hydroxide solution is often used in the desulfurization of petroleum oils by extraction.

The pharmaceutical industry uses liquid-liquid extraction in a large number of separation and purification processes. For example, penicillin is recovered from an acidic growth medium by extraction into butyl or amyl acetate; the extract is then stripped with an aqueous phosphate buffer solution at a pH of 6.8 to 7.5 to transfer the antibiotic to the aqueous phase. Repeated alternate extraction and stripping are generally used to concentrate as well as to purify the product.

Other fairly large-scale uses of liquid-liquid extraction include the recovery of many metals, separation and purification of rare earth metals, purification of nuclear reactor fuels, and reprocessing of spent fuels [4, 26]. Extraction processes in the metals industry often depend on the formation of chemical complexes, such as chelates, coordinate compounds, and ion associates, between the metal salts and the extracting solvents. Chelate formation is used extensively in the analytical chemistry of metals, while ion associates are most important for industrial metal separations.

When considering the separation of a given mixture by a solvent-aided separation process, it is frequently necessary to include solvent systems for which phase equilibrium data are not available in the literature. It is then desirable to estimate the activity coefficients of various components in the system from correlations or from pure-component properties. For many systems

Table 5.8 Correlation Constants for Activity Coefficients [52]

Solute Series	Solvent Series	Temp. (°C)	A Constant	B Term	B Constant	C Term	C Constant	D Term	D Constant	F Term	F Constant
n-Acids	Water	25	-1.00	Bn_1	0.622	$C(1/n_1)$	0.490	None	—	None	—
n-Primary alcohols	Water	25	-0.995	Bn_1	0.622	$C(1/n_1)$	0.558	None	—	None	—
n-s-Alcohols	Water	25	-1.220	Bn_1	0.622	$C(1/n_1' + 1/n_1'')$	0.170	None	—	None	—
n-t-Alcohols	Water	25	-1.740	Bn_1	0.622	$C\left[1/n_1' + 1/n_1'' + 1/n_1'''\right]$	0.170	None	—	None	—
Alcohols, general	Water	25	-0.525	Bn_1	0.622	$C\left[(1/n_1' - 1) + (1/n_1'' - 1) + (1/n_1''' - 1)\right]$	0.475	None	—	None	—
n-Allyl alcohols	Water	25	-1.180	Bn_1	0.622	$C(1/n_1)$	0.558	None	—	None	—
n-Aldehydes	Water	25	-0.780	Bn_1	0.622	$C(1/n_1)$	0.320	None	—	None	—
n-Alkene aldehydes	Water	25	-0.720	Bn_1	0.622	$C(1/n_1)$	0.320	None	—	None	—
n-Ketones	Water	25	-1.475	Bn_1	0.622	$C(1/n_1' + 1/n_1'')$	0.500	None	—	None	—
n-Acetals	Water	25	-2.556	Bn_1	0.622	$C\left[1/n_1' + 1/n_1'' + 2/n_1'''\right]$	0.486	None	—	None	—
n-Ethers	Water	20	-0.770	Bn_1	0.640	$C(1/n_1' + 1/n_1'')$	0.195	None	—	None	—
n-Nitriles	Water	25	-0.587	Bn_1	0.622	$C(1/n_1)$	0.760	None	—	None	—
n-Alkene nitriles	Water	25	-0.520	Bn_1	0.622	$C(1/n_1)$	0.760	None	—	None	—
n-Esters	Water	20	-0.930	Bn_1	0.640	$C(1/n_1' + 1/n_1'')$	0.260	None	—	None	—
n-Formates	Water	20	-0.585	Bn_1	0.640	$C(1/n_1)$	0.260	None	—	None	—

n-Monoalkyl chlorides	Water	20	1.265	Bn_1	0.640	$C(1/n_1)$	0.073	None	—	None	—
n-Paraffins	Water	16	0.688	Bn_1	0.642	None	—	None	—	None	—
n-Alkyl benzenes	Water	25	3.554	Bn_1	0.622	$C[(1/n_1 - 4)]$	−0.466	None	—	None	—
n-Alcohols	Paraffins	25	1.960	None	—	$C\left[\begin{array}{l}(1/n_1'-1)+\\(1/n_1''-1)+\\(1/n_1'''-1)\end{array}\right]$	0.475	$D(n_1 - n_2)^2$	−0.00049	None	—
n-Ketones	Paraffins	25	0.0877	None	—	$C(1/n_1' + 1/n_1'')$	0.757	$D(n_1 - n_2)^2$	−0.00049	None	—
n-Alcohols	Water	25	0.760	None	—	None	—	None	—	$F(1/n_2)$	−0.630
s-Alcohols	Water	80	1.208	None	—	None	—	None	—	$F(1/n_2' + 1/n_2'')$	−0.690
n-Ketones	Water	25	1.857	None	—	None	—	None	—	$F(1/n_2' + 1/n_2'')$	−1.019
Ketones	n-Alcohols	25	−0.088	$B(n_1/n_2)$	0.176	$C(1/n_1' + 1/n_1'')$	0.50	$D(n_1 - n_2)^2$	−0.00049	$F(1/n_2)$	−0.630
Aldehydes	n-Alcohols	25	−0.701	$B(n_1/n_2)$	0.176	$C(1/n_1)$	0.320	$D(n_1 - n_2)^2$	−0.00049	$F(1/n_2)$	−0.630
Esters	n-Alcohols	25	0.212	$B(n_1/n_2)$	0.176	$C(1/n_1' + 1/n_1'')$	0.260	$D(n_1 - n_2)^2$	−0.00049	$F(1/n_2)$	−0.630
Acetals	n-Alcohols	60	−1.10	$B(n_1/n_2)$	0.138	$C(1/n_1' + 1/n_1'' + 2/n_1'')$	0.451	$D(n_1 - n_2)^2$	−0.00057	$F(1/n_2)$	−0.440
Paraffins	Ketones	25	None	$B(n_1/n_2)$	0.1821	None	—	$D(n_1 - n_2)^2$	−0.00049	$F(1/n_2' + 1/n_2'')$	0.402

no suitable correlation is available. However, for typical mixtures of organic liquids, three partially successful techniques have been suggested for making such estimates.

Correlation of Activity Coefficients at Infinite Dilution

A method for estimating infinite dilution activity coefficients for binary liquid solutions is discussed by Pierotti, Deal, and Derr [52]. Experimental data for a large number of systems were correlated with the structure of the solute and solvent. The average deviation in the infinite dilution activity coefficients obtained from these expressions was 8%.

The natural logarithm of the infinite-dilution activity coefficient was assumed to be the sum of contributions from various interactions. The data were correlated with the general expression:

$$\ln \gamma_1^{\infty} = A_{12} + B_2 n_1/n_2 + C_1 n_1^{-1} + D(n_1 - n_2)^2 + F_2 n_2^{-1} \quad (5.54)$$

where γ_1^{∞} is activity coefficient of component 1 at infinite dilution in component 2, n_1 and n_2 are number of carbon atoms in the hydrocarbon radicals of the solute and solvent, respectively, A_{12} is a constant depending on the nature of the functional groups of both the solute and the solvent, B_2 is a coefficient depending on the solvent functional group only, C_1 is a coefficient depending on the solute functional group only, D is a coefficient independent of the functional group of either the solvent or the solute, and F_2 is a coefficient which essentially depends only on the solvent functional group.

The constants for several solute-solvent systems are listed in Table 5.8. The values are given at discrete temperatures and may be interpolated for use at intermediate temperatures. For compounds in which the functional group is not at the end of the hydrocarbon chain, for branched hydrocarbon chains, and for compounds that have more than one hydrocarbon radical, specific counting rules must be used to determine the values of n_1 and n_2. In such cases n_1 and n_2 are split into $n_1', n_1'', \ldots$ and $n_2', n_2'', \ldots$. For example, for a nonterminal functional group, n_1' and n_1'' are the numbers of carbon atoms in each alkyl group originating at the functional group. Thus for t-butyl alcohol, the central carbon atom is counted in each branch, so that $n_1' = n_1'' = n_1''' = 2$. A similar counting rule is used for the alkyl groups in acetal structures, R''' (OR'') OR'.

Pierotti, Deal, and Derr also give correlations for families of hydrocarbon solutes in some specific solvents. In order to account for different hydrocarbon structures, such as paraffinic, olefinic, aromatic, cyclic, and multiringed molecules, the form of the correlations is somewhat different from that shown in (5.54). A detailed summary of Pierotti's correlation is given in the text by Treybal [75].

Group Contribution Methods

Almost 50 years ago, Langmuir [37] suggested that activity coefficient data may be correlated by considering a solution to be composed of functional groups rather than of chemical compounds. This suggestion has been developed by Wilson and Deal [79], and by Deal and Derr [15, 16].

The activity coefficient is split into contributions from entropy effects (due to size differences) and from group interactions (due to intermolecular forces) so that, for solute j in a solution,

$$\log \gamma_j = \log \gamma_j^S + \log \gamma_j^G \tag{5.55}$$

The entropy contribution is calculated from a Flory-Huggins expression:

$$\log \gamma_j^S = \log \frac{n_j}{\Sigma_i x_i n_i} + 0.4343 \left[1 - \frac{n_j}{\Sigma_i x_i n_i} \right] \tag{5.56}$$

where n_i is number of atoms (other than hydrogen) in component i, and x_i is mole fraction of component i.

The interaction contribution is expressed as a function of the differences of the activity coefficients of the functional groups in the given solution and those at a standard state. If Γ_k is the activity coefficient of group k in the solution, and Γ_k^* that in the standard state, the interaction contribution is given by:

$$\log \gamma_j^G = \Sigma_k \nu_{kj} (\log \Gamma_k - \log \Gamma_k^*) \tag{5.57}$$

Here, ν_{kj} is number of groups of type k in component j. The group activity coefficient Γ_k is assumed to be a single function of the group number fractions of the various groups for both the solution and the standard state:

$$\Gamma_k = f(X_1, X_2, \ldots) \tag{5.58}$$

where $X_k = \Sigma_j \nu_{kj} x_j / \Sigma_j (\Sigma_k \nu_{kj}) x_j$. The functions Γ_k are evaluated from experimental data; they may be normalized to a group standard state such that $\Gamma_k^* = 1$ for pure group k.

The solution-of-groups concept provides a promising method which appears to work well for solutions containing only a few functional groups. Extensions to multigroup solutions, however, have been developed for only a few cases [16].

Extensions of Regular Solution Theory

It is sometimes possible to predict activity coefficients from minimal data by using the solution theory of Scatchard and Hildebrand. For regular solutions

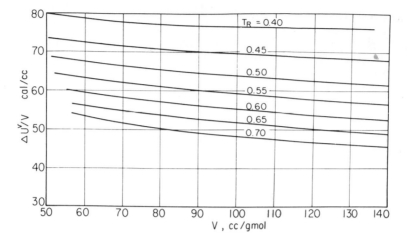

Fig. 5.21. Energy density of n-paraffins (homomorph plot).

this theory expresses the infinite dilution activity coefficient in terms of the solute molar volume V_1 and the solute and solvent solubility parameters δ_1 and δ_2 by

$$\ln \gamma_1^\infty = (\delta_1 - \delta_2)^2 V_1 / RT \tag{5.59}$$

However, in addition to many other assumptions, the regular-solution theory does not take into consideration intermolecular forces arising from the polar nature of the solute or solvent; in its original form, therefore, it is valid only for nonpolar systems. For hydrocarbons in polar solvents, Weimer and Prausnitz [77] extended the regular-solution theory by introducing nonpolar and polar solubility parameters (as defined earlier by (5.37)). They obtained the following relation for γ_1^∞:

$$\ln \gamma_1^\infty = [(\lambda_1 - \lambda_2)^2 + \tau_2^2 - 2\psi_{12}] V_1/RT + \left(\ln \frac{V_1}{V_2} + 1 - \frac{V_1}{V_2} \right) \tag{5.60}$$

where λ_1 and λ_2 are the nonpolar solubility parameters of the solute and solvent, respectively, τ_2 is the polar solubility parameter of the solvent, and ψ_{12} is an induction parameter characteristic of the nature of the solvent and the type of hydrocarbon. [The last three terms in (5.13) come from the Flory-Huggins theory for the entropy of mixing of liquids of widely different molar volumes.]

Weimer analyzed experimental data for a large number of hydrocarbon-polar solvent systems. The nonpolar solubility parameter of a polar solvent was obtained from its homomorph, the polar solubility parameter then being obtained by difference:

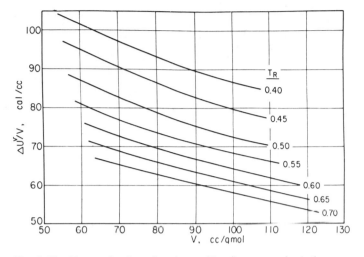

Fig. 5.22. Energy density of cycloparaffins (homomorph plot).

$$\tau^2 = \delta^2 - \lambda^2 \tag{5.61}$$

As originally proposed by Bondi and Simkin [7], the homomorph of a polar molecule is that hydrocarbon whose structure is the same as that of the polar molecule; the properties of the homomorph are evaluated at the same reduced temperature as that of the polar fluid. Anderson [1] suggested that, in addition, the homomorph should have the same molar volume as that of the polar fluid. Since the physical properties of a family of similar liquids vary in a smooth manner with their molar volumes, Weimer prepared "homomorph plots" of the cohesive energy density versus reduced temperature for n-paraffins, cycloparaffins, and aromatic hydrocarbons; these are shown in Fig. 5.21 to 5.23. The curves obtained have been expressed analytically by Thompson [73]:

For n-paraffins,

$$\Delta U^v/V = \frac{77 \; v^{-0.127} e^{-0.00161 \, V}}{T_r^{(0.2566 \; \ln \, (V) \, - \, 0.3553)}} \tag{5.62}$$

For cycloparaffins,

$$\Delta U^v/V = \frac{225.3 \, V^{-0.35}}{T_r^{0.7}} \tag{5.63}$$

For aromatic hydrocarbons,

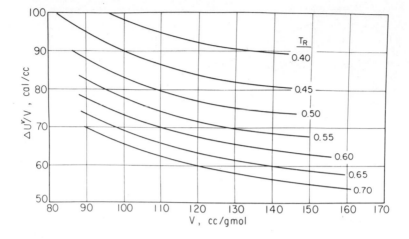

Fig. 5.23.   Energy density of aromatic hydrocarbons (homomorph plot).

$$\Delta U^v/V = \frac{1616 \; V^{(0.0017 \, V - 0.73)} e^{\;(-0.00872 \, V)}}{T_r^{\,(0.003 \; V + 0.4)}}$$ (5.64)

where ΔU^v is the energy of vaporization, V is the molar volume, and T_r is the reduced temperature.

To illustrate the use of the homomorph plots, let us calculate the polar and nonpolar solubility parameters for acetone at 25°C. From the vapor pressure data of Dreisbach and Martin [17], the enthalpy of vaporization is 7720 cal mol^{-1}. The molar volume at 25°C is given by the same investigators as 74.0 ml mol^{-1}. The total cohesive energy density is then

$$\Delta U^v/V = \frac{\Delta H - RT}{V} = 96.3 \; \text{cal ml}^{-1}$$

The critical temperature of acetone is 508.2 K [25]; therefore at 25°C the reduced temperature is 0.586. From Fig. 5.21 the nonpolar cohesive energy density is approximately 59 cal ml^{-1}. Then

$$\lambda = \sqrt{59} = 7.68 \; (\text{cal ml}^{-1})^{\frac{1}{2}}$$

and

$$\tau = \sqrt{96.3 - 59} = 6.10 \; (\text{cal ml}^{-1})^{\frac{1}{2}}$$

A similar procedure may be used for other polar fluids.

Weimer found that activity-coefficient data for linear and cyclic saturated hydrocarbons in polar solvents could be correlated by expressing the induction parameter ψ_{12} as a function of the polar solubility parameter of the solvent:

Saturated hydrocarbons: $\psi_{12} = 0.396 \ \tau_2{}^2$

Similarly, from the data of Gerster, Gordon, and Eklund [22] on the activity coefficients of 1-pentene in polar solvents:

Olefins: $\psi_{12} = 0.415 \ \tau_2{}^2$

and from the benzene-polar solvent data of Deal and Derr [14]:

Aromatics: $\psi_{12} = 0.450 \ \tau_2{}^2$

With these correlations for the induction parameter, experimental data were fit to within ±10% of ln $\gamma_1{}^\infty$ for almost all saturated hydrocarbon-polar solvent systems.

Table 5.9 lists the molar volumes and nonpolar and polar solubility parameters of some hydrocarbons and solvents at 25°C, taken from the work of Weimer and Prausnitz. These investigators also tabulated some values at 0, 45, 60, and 100°C.

Helpinstill and van Winkle [30] suggested an extension of Weimer's correlation to systems in which both the solute and solvent are polar. Both components were assigned polar and nonpolar solubility parameters. The infinite dilution activity coefficient was then expressed as

$$\ln \ \gamma_1{}^\infty = [(\lambda_1 - \lambda_2)^2 + (\tau_1 - \tau_2)^2 - 2\psi_{12}] \ V_1/RT + \left[\ln \left(\frac{V_1}{V_2} \right) + 1 - \frac{V_1}{V_2} \right] (5.65)$$

The nonpolar part of the solubility parameter was obtained as before from the homomorph concept. The induction parameter ψ_{12} was correlated as a function of $(\tau_1 - \tau_2)^2$. The average error obtained by these workers was 5.8% in ln $\gamma_1{}^\infty$ or 11.6% in $\gamma_1{}^\infty$ for saturated hydrocarbons in polar solvents; for olefin-solvent systems it was 8.5% in $\gamma_1{}^\infty$; for aromatic-solvent systems the error was 13.5% in $\gamma_1{}^\infty$.

Null and Palmer [49] extended the concept of polar and nonpolar contributions to the solubility parameter to include the effect of component association in the cohesive energy density:

$$\delta^2 = \frac{\Delta U^v}{V} = \lambda^2 + \tau^2 + \zeta^2 \tag{5.66}$$

Table 5.9 Molar Volumes, Polar and Nonpolar Solubility Parameters of Some Hydrocarbons and Solvents at 25°C [77]

	V (ml mol^{-1})	λ (cal ml^{-1})$^{1/2}$	τ (cal ml^{-1})$^{1/2}$
Hydrocarbon			
Propane	89.47	6.56	—
n-Butane	101.43	6.94	—
n-Pentane	116.16	7.17	—
n-Hexane	131.56	7.33	—
n-Heptane	147.44	7.47	—
n-Decane	195.92	7.85	—
n-Hexadecane	294.08	8.75	—
Cyclopentane	94.71	8.20	
Methylcyclopentane	113.11	7.91	—
Cyclohexane	108.74	8.23	—
Methylcyclohexane	128.33	7.85	—
Ethylcyclohexane	143.13	8.01	—
1-Butene	95.28	7.08	—
1-Pentene	110.37	7.23	—
1,3-Butadiene	87.97	7.42	—
Benzene	89.40	9.20	—
Toluene	106.84	8.95	—
Ethylbenzene	123.06	9.01	—
p-Xylene	123.92	8.83	—
Mesitylene	139.58	8.88	—
Solvent			
Acetophenone	117.4	9.44	3.69
Tetrahydrofuran	81.7	8.32	3.71
Pyridine	80.9	9.88	3.71
Cyclohexanone	104.2	8.84	4.04
Chloroethane	74.1	7.38	4.32
Diethyl ketone	106.4	7.75	4.44
Diethyl carbonate	121.9	7.89	4.49
Bromoethane	75.1	7.63	4.83
Nitrobenzene	102.7	9.70	4.89
Di-(2-chloroethyl) ether	117.8	8.34	5.22
Trimethyl phosphate	116.2	8.46	5.22
Iodoethane	81.1	7.66	5.24
Methyl ethyl ketone	90.1	7.64	5.33
Cyclopentanone	89.5	8.70	5.37

Table 5.9 Molar Volumes, Polar and Nonpolar Solubility Parameters of Some Hydrocarbons and Solvents at 25°C [77] (Continued)

	V (ml mol^{-1})	λ (cal ml^{-1})$^{1/2}$	τ (cal ml^{-1})$^{1/2}$
2,4-Pentanedione	103.0	8.06	5.69
2,5-Hexanedione	117.7	8.45	5.88
Diethyl oxalate	136.2	8.37	5.94
2-Nitropropane	90.7	7.95	6.02
Methoxyacetone	93.2	7.91	6.11
Acetone	74.0	7.66	6.14
Dimethyl carbonate	85.0	7.77	6.20
Butyronitrile	87.9	7.96	6.28
2,3-Butanedione	87.8	7.73	6.35
Aniline	91.5	9.85	6.37
1-Nitropropane	89.5	8.06	6.40
N-Methylpyrrolidone	96.6	9.15	6.55
Acetic anhydride	95.0	7.85	7.11
Propionitrile	70.9	7.97	7.17
Citraconic anhydride	89.7	9.42	7.22
Methoxyacetonitrile	75.2	8.06	7.33
Furfural	83.2	9.04	7.62
Nitroethane	72.1	8.04	7.66
Dimethylacetamide	93.2	8.29	7.69
γ-Butyrolacetone	77.1	9.50	8.01
Dimethylformamide	77.4	8.29	8.07
3-Chloropropionitrile	77.7	8.44	8.73
Acetonitrile	52.6	8.03	8.98
Ethylenediamine	67.3	8.10	9.40
Nitromethane	54.3	8.08	9.44
Dimethyl sulfoxide	71.3	8.56	9.47

where λ and τ are defined as before, and ζ represents the contribution due to association. Null and Palmer used the Antoine vapor pressure equation for calculating δ. The parameter λ was back-calculated from a regression analysis to fit an equation of the Antoine form. The parameter ζ was obtained from the entropy and enthalpy of association as defined by Wiehe and Bagley [78]; these investigators expressed the activity coefficients in binary alcohol-hydrocarbon systems as a function of the mixture composition, the ratio of the molar volumes of the two components, and a hydrogen-bonding equilibrium constant. This equilibrium constant was in turn written in terms of the enthalpy and entropy of hydrogen-bond formation. Setting the enthalpy of the hydrogen bond equal to 5.9 kcal mol^{-1}, the entropy of association was obtained by trial-and-error fitting of experimental activity coefficient data. Null and Palmer then calculated τ from (5.66), and the infinite dilution activity coefficient was calculated by means of an equation similar to (5.60). The average error in the prediction of γ_1^∞ for about 300 binary systems was 25%. However, the average error for all saturated hydrocarbons in all solvents was less than 9%. The error was found to be somewhat greater when the polar or associating component is the solute rather than the solvent.

8 SOLVENT SELECTION FOR SEPARATION PROCESSES

For extraction, and azeotropic and extractive distillation, the choice of solvent depends on a variety of factors. While good selectivity is essential, other solvent properties must also be considered, as briefly discussed in the following paragraphs.

Selectivity

The primary requirement of an effective solvent is that it be selective with respect to one or more components of the feed mixture; selectivity is generally the first property to be considered when selecting a solvent. The greater the selectivity of a solvent, the greater the separation factor. A greater selectivity therefore leads to the requirement of a fewer number of stages to effect the desired separation. Although the calculation of solvent selectivity requires activity coefficients at the conditions of operation of the process, a relative ordering of the selectivities can generally be obtained from the infinite-dilution activity coefficients of the key components. Tassios [71] describes a fairly rapid chromatographic technique for estimation of a for given key components at infinite dilution in several different solvents.

For separating a nonpolar substance, such as a paraffin, from a close-boiling solution with a polar substance, such as an alcohol, or a highly polarizable substance such as an aromatic, a polar solvent usually gives good selectivity. The more polar the solvent, the greater the interaction between it and the polar or highly polarizable component, and therefore the greater the resultant selectivity. Tassios [70] plotted the selectivity data of Gerster, Gorton, and Eklund [22] for n-pentane and 1-pentene in various polar solvents versus the polarity of the solvent. The ratio of the infinite dilution activity coefficients of the two components was found to be an approximately linear function of the polar part of the cohesive energy density of the solvent, as shown in Fig. 5.24.

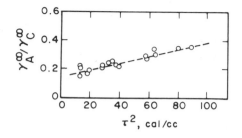

Fig. 5.24. Variation of selectivity with polar cohesive energy density of solvent for the system *n*-pentane-1-pentene at 25°C.

Other measures of the polarity of a solvent have been suggested. Organic chemists interested in the effect of solvents on chemical reactions assign empirical polarity values to various solvents according to different scales. These scales, and the procedures used for establishing them, have been reviewed by Reichardt [61], and by Dack in Part II of this volume. Perhaps the most extensive of these are the Z scale and the E_T scale. The energy of the peak absorption band of the intermolecular charge-transfer band of 1-ethyl-4-carbomethoxypyridinium iodide when dissolved in a polar solvent is called its Z value. The energy of the absorption band of the intramolecular charge transfer in pyridinium N-phenolbetaine in a solvent is defined as the E_T value of the solvent. However, there appears to be no useful connection between these polarity scales and solvent selectivity in separation processes.

Capacity

The capacity of the solvent to dissolve some or all of the components present in the feed mixture determines the amount of solvent required per unit feed. A solvent with a high capacity is preferred, as it requires smaller equipment size as well as a smaller solvent inventory. However, in most cases solvents with a higher capacity tend to have a lower selectivity..For example, as shown in Fig. 5.20, the solvent selectivity rises with γ_A^∞, that is, with increasing nonideality between the solvent and the solutes. Mixtures with large nonidealities tend to have limited miscibilities. In general therefore a compromise must be made between high selectivity and high capacity.

Recoverability

The solvent used in a separation process must be recovered for reuse, and also to meet product purity requirements. Solvent recovery is frequently a major cost item in the total separation process. Often the solvent is recovered by distillation; in this case the solvent volatility and its enthalpy of vaporization become important. High volatility and low heat of vaporization are desirable. However, if the solvent is highly volatile, relatively expensive operation at high pressure may be required. The heat capacity of the solvent is also important, since the recovered solvent must in general be cooled before it can be returned to the separator. The temperature swing between the hottest and the coolest points in the solvent circuit (determined chiefly by the volatilities of the solvent and the solutes), the solvent circulation rate, and the solvent heat capacity determine the amount of heat that must be removed in the process. Because of the conservation of enthalpy in the process, these terms also determine to a large extent the amount of heat that must be supplied to the process. Both addition and removal of heat involve equipment and processing costs; a low heat capacity minimizes both.

An illustration of the importance of the above-mentioned factors is provided by Kumar, King, and Prausnitz [36], who considered the separation of propane-propylene mixtures. Consideration was given to the significance of different variables in the design of an extractive distillation process; the basis for the design of the process is shown in Table 5.10. The Lewis-Matheson stage-to-stage method was used for distillation column computations. Equilibrium data for potential solvents were assumed to be similar to those shown in Fig. 5.20. Two of the variables studied were solvent concentration in the extractive distillation column and solvent selectivity. For a given solvent, higher solvent concentrations

Table 5.10   Design Basis for Propane-Propylene Separation

Feed		
	Rate	415 lb mol/hr
	Composition	50 mol % C_3H_8, 50 mol % C_3H_6
	Condition	Saturated vapor at the tower pressure
Propylene		
	Purity	99.0%
	Recovery	95.0%
	Capacity	75 million lb/yr
Solvent		
	Amount in each hydrocarbon product	$\leqslant 0.01$ mol %
	Normal boiling point	$140°F$
	Other physical properties	Those of acetone
	Concentration at solvent feed plate	0.85-0.90 mol fraction
	S^∞	1.7-3.1
	$\gamma^\infty_{C_3H_8}$	6.0-16.0

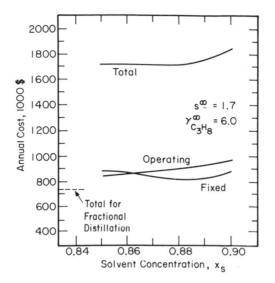

Fig. 5.25. Effect of solvent concentration on the costs of propane-propylene separation by extractive distillation.

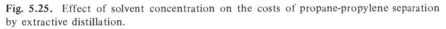

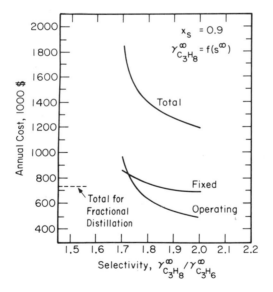

Fig. 5.26. Effect of solvent selectivity on annual costs of propane-propylene separation by extractive distillation.

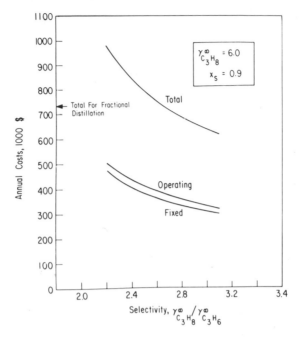

Fig. 5.27. Effect of solvent selectivity on annual costs of propane-propylene separation by extractive distillation. Constant propane activity coefficient.

result in the requirement of fewer equilibrium stages but a greater solvent/feed ratio. The two effects tend to compensate with respect to process cost; net process cost is not sensitive to the solvent concentration, as shown in Fig. 5.25. Rising solvent selectivity decreases the solvent/feed ratio, as well as the equilibrium-stage requirements in the fractionating section of the main column. The net effect on process cost is shown in Fig. 5.26. A greater selectivity gives a higher solvent activity coefficient at low solvent concentration in the hydrocarbon. This makes it more difficult to remove the solvent from the overhead of the extractive distillation column, leading to greater process costs. If the solvent selectivity could be increased without a corresponding rise in the activity coefficients, the total process cost would be as shown in Fig. 5.27. Figures 5.25 to 5.27 also show the cost for the same separation by fractional distillation.

The above study showed that the costs of separating paraffin-olefin mixtures by extractive distillation are strongly affected by solvent selectivity, the paraffin activity coefficient, and solvent volatility; they are thus mostly dependent on solvent characteristics. For a solvent to be effective for the separation of key components A and C, the solvent must have a high value of S^∞, while at the same time it must have low γ_A^∞. Further, the solvent vapor pressure should be lower (but not several orders of magnitude lower) than that of the less volatile hydrocarbon.

For the propane-propylene-solvent system, for the solvents exhibiting the relation between S^∞ and γ_A^∞ shown in Fig. 5.20, it does not seem likely that extractive distillation costs can be reduced below those of binary fractional distillation. Figure 5.26 shows decreasing costs with increasing selectivity, but the slope of the curve of cost versus S^∞ seems to be flattening out at a value of S^∞ of about 2, where extractive distillation costs are still about 65% greater than the costs for binary distillation. However, for solvents that give a high selectivity at relatively lower γ_A^∞, Fig. 5.27 shows a cost-versus-S^∞ curve which indicates a lower cost for the extractive distillation process at S^∞ higher than about 2.6.

Miscellaneous Considerations

For long-term use, the solvent must be chemically stable and it must be chemically inert toward the feed mixture. Many organic solvents are susceptible to polymerization and/or oxidation, particularly at the high temperatures encountered in the recovery columns. In many cases these reactions only represent loss of active solvent in the process, but sometimes explosive or corrosive products are formed. Such possibilities must be carefully considered before a solvent is finally chosen.

The solvent should not be corrosive toward the material of construction used in the separation equipment. Fortunately, corrosion is not as severe a problem as it once was, because of the (relatively recent) availability of corrosion-resistant materials at competitive cost. For instance, several commercial extraction plants have been built entirely of plastic materials or of plastic-lined metals [75].

Solvents with low viscosity are desirable, because they have low power requirements for pumping and agitation and because they usually permit high heat and mass transfer rates. A low freezing point facilitates outdoor storage and handling of the solvent. The solvent should not be flammable and, if combustible, should have a high flash point and close concentration limits for explosive mixtures with air. Toxic materials should be avoided. Solvent costs should be low, because solvent makeup in the process contributes to operational costs and because solvent inventory represents investment.

The choice of a solvent in a specific application may also depend on some other considerations. For example, in the case of liquid-liquid extraction, the density difference between the two phases must be as large as possible for ease of phase separation. A satisfactory density difference must exist between the two phases throughout the complete range of operating conditions. In the system methyl ethyl ketone-water-trichloroethylene, there is a reversal of sign in the density difference between equilibrium phases with increasing fraction of ketone [47]. A continuous contacting operation therefore could not be used for this system. Similarly, if the interfacial tension between the two phases is very low, stable emulsions may form, and therefore extraction with a particular solvent may not be practically feasible even though all other criteria are satisfied.

In liquid-liquid extraction the distribution coefficient of the component extracted may also be important aside from the solvent selectivity. For a process in which component C is separated from component A, using solvent B, the separation factor may be written as

$$a_{CA} = \frac{x_{CB}}{x_{CA}} \frac{x_{AA}}{x_{AB}} \qquad (5.67)$$

where x_{ij} is the mole fraction of component i in the j-rich phase. The ratio x_{CB}/x_{CA} is defined as the distribution coefficient of component C between the B-rich phase and the A-rich phase. Denoting the distribution coefficient by m, the separation factor may be written as

$$a_{CA} = m \frac{x_{AA}}{x_{AB}} \qquad (5.68)$$

Since x_{AA} is necessarily greater than x_{AB}, whether or not the separation factor is greater than unity depends upon the value of m. This is of particular interest in the extraction of ionic components, since the value of m can be changed by altering the pH of the solution (cf. the separation of tantalum from niobium cited earlier).

Finally, the use of specified solvents in specific separations is sometimes covered by current patents, and possible patent infringement must therefore be taken into consideration.

9   GENERAL CONSIDERATIONS

Air Pollution Regulations

Organic solvents make only a relatively small contribution to overall air pollution and smog formation in metropolitan areas. Therefore many air pollution control agencies do not have specific regulations concerning industrial solvents. As of 1972, the most restrictive regulation in the United States was Rule 66 of the Los Angeles County Air Pollution Control District; similar regulations have been adopted by the San Francisco Bay Area Air Pollution Control District and the Orange County (California) Air Pollution Control District.

Los Angeles County Air Pollution Control District Rule 66 restricts the emission of photochemically reactive organic solvents. A photochemically reactive solvent is one whose vapor produces measurable eye irritation in the presence of nitrogen oxides and ultraviolet light; alternatively, solvents that support ozone formation may be classified as photochemically reactive. From studies of the reactivity of individual hydrocarbons with nitric oxide and air under controlled conditions, olefins were found to be most reactive. Aromatics are much less reactive; paraffins have very low, if any, reactivity.

To take into account reactivities of different substances, Section *k* of Rule 66 defines a photochemically reactive solvent as one which has any of the following:

"1. More than 5 vol% of a combination of hydrocarbons, alcohols, aldehydes, esters, ethers or ketones having an olefinic or cyclo-olefinic type of unsaturation.

2. More than 8 vol% of a combination of aromatic substances with eight or more carbon atoms (except ethyl benzene).

3. More than 20 vol% of a combination of ethyl benzene, ketones with branched hydrocarbon chains, trichloroethylene or toluene.

4. A combination of any of the above mentioned substances in excess of 20 vol%."

San Francisco Bay Area Air Pollution Control District Regulation 3 is somewhat less restrictive in that the maximum permissible concentration of group-1 substances is 8 vol % rather than 5 vol %. Further, no limit is placed on the trichloroethylene content of a solvent.

Rule 66 requires that no more than 15 lb per day of any solvent be emitted from a baking oven in which the solvent comes in contact with the flame or is baked, heat-cured, or heat-polymerized in the presence of oxygen, unless this effluent is less than 15% of the effluent prior to the adoption of this rule. When a coating is baked only to speed up the drying time, the limit for reactive solvents is 40 lb per day, unless reduced by 85%.

Attempts to comply with the regulations have been of two kinds. The more common method has been to reformulate solvent blends to restrict the reactive components to their maximum permissible limits. Larson and Sipple [38] and Nelson et al. [45] discuss the reformulation of blends to obtain nonreactive solvent mixtures. The other method has been to control baking oven emissions by means of condensers, adsorbers, absorbers, and incinerators. These regulations have also provided an added impetus to the use of water-based latex paints.

Physical Properties

A comprehensive listing of the physical properties of organic solvents is given in Volume II of this series [63]. The listed properties include the normal boiling point, vapor pressure, density, coefficient of thermal expansion, refractive index, viscosity, surface tension, enthalpy of vaporization, critical constants, heat capacity, optical activity, cryoscopic and ebullioscopic constants, electrical properties, dipole moment, flash point, and references to spectral data. In addition, data are given for the variation in certain properties with temperature, enthalpies of sublimation, fusion, transition, formation, combustion and polymerization, and acidity and basicity. However, not all these properties are tabulated for each solvent.

Volume II also discusses purification and health and safety guidelines for handling of solvents.

Solvent Prices

Current solvent prices are available in the weekly issues of the *Chemical Marketing Reporter* (formerly *Oil, Paint and Drug Reporter*). Table 5.11 lists approximate prices of some solvents at the end of January 1972. The prices are based on large shipments; the tabulated values are dollars per pound unless specifically marked otherwise. The prices shown are for the East Coast; West Coast prices are generally higher by about 1 cent/lb, whereas Gulf Coast prices are somewhat lower. Solvent prices are subject to change, and the values shown should be used only as a guide. Prices for proprietary solvents are not included in Table 5.11.

Table 5.11   Prices of Some Solvents, January 1972, from *Chemical Marketing Reporter*

Solvent	Price[a]	Solvent	Price[a]
Acetaldehyde	0.09	2-Ethyl butylalcohol	0.31
Acetic acid	0.13	Ethyl butyl ketone	0.40
Acetic anhydride	0.14	Ethyl ether	0.47/gal
Acetone	0.06	Ethylbenzene	0.06
Acetonitrile	0.24	Ethylenediamine	0.26
Acetophenone	0.29	Ethylene dichloride	0.09
Acrylonitrile	0.15	Ethylene glycol	0.09
Allyl alcohol	0.28	Formaldehyde	0.04
Amyl acetate	0.16	Formamide	0.12
Amyl alcohol	0.18	Formic acid	0.13
Aniline	0.15	Furfural	0.17
Benzene	0.20/gal	Furfural alcohol	0.19
Benzyl alcohol	0.33	Glycerine	0.22
1-Butene	0.06	Heptane	0.20/gal
2-Butene	0.06	Hexane	0.20/gal
n-Butylacetate	0.14	1-Hexanol	0.15

Table 5.11 Prices of Some Solvents, January 1972, from *Chemical Marketing Reporter* (Continued)

Solvent	Price[a]	Solvent	Price[a]
n-Butyl alcohol	0.12	Isobutyl alcohol	0.08
s-Butyl alcohol	0.14	Isopropyl acetate	0.12
t-Butyl alcohol	0.15	Isopropyl alcohol	0.49/gal
n-Butyl ether	0.33	Lacquer diluent (petroleum)	0.18/gal
Butyraldehyde	0.20	Methanol	0.11/gal
Butyric acid	0.33	Methyl ethyl ketone	0.10
Butyrolactone	0.41	Mineral oil (50-65 SSU)	0.64/gal
Butyronitrile	0.54	Mineral spirits (odorless)	0.31/gal
Carbon disulfide	0.05	Mineral spirits (regular)	0.17/gal
Carbon tetrachloride	0.11	Naphtha, VM & P	0.20/gal
Cctyl alcohol	0.35	Nitroethane	0.35
Chloroform	0.18	Nitromethane	0.31
Cleaner's naphtha	0.23/gal	1-Nitropropane	0.16
Cresol	0.18	1-Octanol	0.23
Cyclohexane	0.24/gal	n-Octane	0.27/gal
Cyclohexanol	0.28	Paraffin	0.06
Cyclohexanone	0.18	Phenol	0.08
Diethanolamine	0.12	n-Propyl acetate	0.14
Diethylamine	0.28	n-Propyl alcohol	0.13
Diethylbenzene	0.15	Propylene glycol	0.11
Diethylene glycol	0.09	Rubber solvent (petroleum)	0.19/gal
Diisobutyl ketone	0.15	Solvent naphtha	0.28/gal
Dimethyl sulfide	0.16	Tetrachloroethylene	0.22
Dimethyl sulfoxide	0.33	Tetraethylene glycol	0.17
Dimethylacetamide	0.40	Tetrahydrofuran	0.37
Dimethylamine	0.12	Toluene	0.23/gal
Dimethylformamide	0.22	Trichloroethylene	0.09
Ethyl acetate	0.12	Triethanolamine	0.16
Ethyl acetoacetate	0.56	Triethylene glycol	0.14
Ethyl alcohol (95%)	0.54/gal	Xylene (isomer mixture)	0.23/gal
Ethyl alcohol (100%)	0.61/gal		

[a] Prices are given in dollars per pound unless otherwise marked.

Acknowledgment

The authors are grateful to the National Science Foundation, The Paint Research Institute, and the donors of the Petroleum Research Fund for financial support.

References

1. R. Anderson, Ph.D. Thesis, Chemical Engineering, University of California, Berkeley, 1961.
2. E. B. Bagley, T. P. Nelson, and J. W. Barlow, S. -A. Chen, *Ind. Eng. Chem. Chem. Fundam.,* **9**, 93 (1970).
3. R. Battino and H. L. Clever, *Chem. Rev.,* **66**, 395 (1966).
4. M. T. Beck, *Chemistry of Complex Equilibria,* van Nostrand Reinhold, New York, 1970.
5. L. Berg, S. V. Buckland, W. B. Robinson, R. L. Nelson, T. W. Wilkinson, and J. W. Petrin, *Hydrocarbon Process.,* **45**(12), 103 (1966).
6. R. F. Blanks and J. M. Prausnitz, *Ind. Eng. Chem., Fundam.,* **3**, 1 (1964).
7. A. Bondi and D. J. Simkin, *J. Chem. Phys.,* **25**, 1073 (1956);*Am. Inst. Chem. Eng. J.,* **3**, 473 (1957).
8. P. L. T. Brian, *Ind. Eng. Chem., Fundam.,* **4**, 101 (1965).
9. H. C. Brown and J. O. Brady, *J. Am. Chem. Soc.,* **74**, 3570 (1952).
10. H. Burrell, *Off. Dig.,* **27**, 748 (1955).
11. H. Burrell, *J. Paint Technol.,* **40**, 197 (1968).
12. H. Cheung and D. Wang, *Ind. Eng. Chem., Fundam.,* **3**, 355 (1964).
13. J. D. Crowley, G. S. Teague, Jr., and J. W. Lowe, Jr., *J. Paint Technol.,* **38**, 269 (1966); **39**, 19 (1967).
14. C. H. Deal and E. L. Derr, *Ind. Eng. Chem., Process Des. Dev.,* **3**, 394 (1964).
15. C. H. Deal and E. L. Derr, *Ind. Eng. Chem.,* **60**(4), 28 (1968).
16. C. H. Deal Jr., and E. L. Derr, *Proc. Int. Symp. Distill.,* **3**, 40 (1969).
17. R. R. Dreisbach and R. A. Martin, *Ind. Eng. Chem.,* **41**, 2875 (1949).
18. R. H. Ewell, J. M. Harrison, and L. Berg, *Ind. Eng. Chem.,* **36**, 871 (1944).
19. P. J. Flory, *J. Chem. Phys.,* **10**, 51 (1942).
20. A. W. Francis, *Liquid-Liquid Equilibriums,* Interscience, New York, 1963.
21. J. L. Gardon, *J. Paint Technol.,* **38**, 43 (1966).
22. J. A. Gerster, J. A. Gorton, and R. -B. Eklund, *J. Chem. Eng. Data,* **5**, 423 (1960).
23. W. Gordy, *J. Chem. Phys.,* **7**, 93 (1939); **8**, 170 (1940); **9**, 204 (1941).
24. E. Hala, J. Pick, V. Fried, and O. Vilim, *Vapour-Liquid Equilibrium,* 2nd Engl. ed., Pergamon, New York,1967. See also I. Wichterle, J. Linek, and E. Hala, *Vapor-Liquid Equilibrium Data Bibliography,* Elsevier, Amsterdam, 1973.
25. *Handbook of Chemistry and Physics,* 39th ed., Chemical Rubber, Cleveland, 1957.
26. C. Hanson, Ed., *Recent Advances in Liquid-Liquid Extraction,* Pergamon, Oxford, 1971.
27. C. Hansen and A. Beerbower, "Solubility Parameters," in *Encyclopedia of Chemical Technology,* 2nd ed., Suppl. Vol., H. F. Mark, J. J. McKetta, Jr., and D. F. Othmer, Eds., Interscience, New York, 1971.
28. C. M. Hansen, *J. Paint Technol.,* **39**, 104, 505 (1967).
29. C. M. Hansen, and K. Skaarup, *J. Paint Technol.,* **39**, 511 (1967).

30. J. G. Helpinstill and M. van Winkle, *Ind. Eng. Chem., Process Des. Dev.,* 7(2), 213 (1968).

31. J. H. Hildebrand and R. L. Scott, *Solubility of Nonelectrolytes,* 3rd ed., Reinhold, New York, 1950; reprinted with corrections by Dover, New York, 1964.

32. J. H. Hildebrand, J. M. Prausnitz, and R. L. Scott, *Regular and Related Solutions,* Van Nostrand Reinhold, 1970.

33. L. H. Horsley, *Advan. Chem. Ser.,* 6 (1952); 35 (1962).

33.a. *Hydrocarbon Process.,* 45(12), 103 (1966).

34. C. J. King, *Separation Processes,* McGraw-Hill, New York, 1971.

35. A. H. Konstam and W. R. Feairheller, Jr., *Am. Inst. Chem. Eng. J.,* 16, 837 (1970).

36. R. Kumar, C. J. King, and J. M. Prausnitz, *Advan. Chem. Ser.,* 115 16 (1972).

37. I. Langmuir, *Third Colloid Symposium Monograph,* Chemical Catalog, New York, 1925.

38. E. C. Larson and H. E. Sipple, *J. Paint Technol.,* 39, 258 (1967).

39. G. N. Lewis, M. Randall., K. S. Pitzer, and L. Brewer, *Thermodynamics,* 2nd ed., McGraw-Hill, New York, 1961.

40. E. P. Leibermann, *Off. Dig.,* 34, 30 (1962).

41. E. McLaughlin and H. A. Zainal, *J. Chem. Soc.,* 1959, 863.

42. B. J. Mair, A. R. Glasgow, Jr., and F. D. Rossini, *J. Res. Nat. Bur. Stand.,* 27, 39 (July 1941).

43. I. Mertl, *Collect. Czech. Chem. Commun.,* 37, 375 (1972).

44. E. F. Meyer and R. E. Wagner, *J. Phys. Chem.,* 70, 3162 (1966); 75, 642 (1971).

45. R. C. Nelson, V. F. Figurelli, J. G. Walsham, G. D. Edwards, *J. Paint Technol.,* 42, 644 (1970).

46. R. C. Nelson, R. W. Hemwall, and G. D. Edwards, *J. Paint Technol.,* 42, 636 (1970).

47. M. Newman, C. B. Hayworth, and R. E. Treybal, *Ind. Eng. Chem.,* 41, 2039 (1949).

48. H. R. Null, *Phase Equilibrium in Process Design,* Wiley-Interscience, New York, 1970.

49. H. R. Null and D. A. Palmer, *Chem. Eng. Prog.,* 65(9), 47 (1969).

50. D. Patterson, *Macromolecules,* 2(6), 672 (1969).

51. R. H. Perry, C. H. Chilton, and S. D. Kirkpatrick, Eds., *Chemical Engineers' Handbook,* 4th ed., McGraw-Hill, New York, 1963.

52. G. J. Pierotti, C. H. Deal, and E. L. Derr, *Ind. Eng. Chem.,* 51, 95 (1959).

53. G. C. Pimentel and A. L. McClellan, *The Hydrogen Bond,* W. H. Freeman, San Francisco, 1960.

54. J. M. Prausnitz, *Molecular Thermodynamics of Fluid-Phase Equilibria,* Prentice-Hall, Englewood Cliffs, N.J., 1969.

55. J. M. Prausnitz and P. L. Chueh, *Computer Calculations for High-Pressure Vapor-Liquid Equilibria,* Prentice-Hall, Englewood Cliffs, N.J., 1968.

56. J. M. Prausnitz, C. A. Eckert, R. V. Orye, and J. P. O'Connell, *Computer*

Calculations for Multicomponent Vapor-Liquid Equilibria, Prentice-Hall, Englewood Cliffs, N.J., 1967.

57. G. T. Preston and J. M. Prausnitz, *Ind. Eng. Chem., Process Des. Dev.,* **9**, 264 (1970).

58. G. T. Preston, E. W. Funk, and J. M. Prausnitz, *J. Phys. Chem.,* **75**, 2345 (1971).

59. I. Prigogine, *The Molecular Theory of Solutions,* North Holland, Amsterdam, 1957.

60. I. Prigogine and R. Defay, *Chemical Thermodynamics,* Longmans, Green, New York, 1954.

61. C. Reichardt, *Angew. Chem., Int. Ed.,* **4**, 29 (1965).

62. H. Renon, L. Asselineau, G. Cohen, and C. Raimbault, *Calcul sur Ordinateur des Equilibres Liquide-Vapeur et Liquide-Liquide,* Technip, Paris, 1971.

63. J. A. Riddick and W. B. Bunger, *Organic Solvents,* 3rd ed., *Techniques of Chemistry,* Vol. II, A. Weissberger, Ed., Wiley-Interscience, New York, 1971.

64. G. M. Schneider, *Advan. Chem. Phys.,* **18**, 1 (1970).

65. F. H. Shair and J. M. Prausnitz, *Am. Inst. Chem. Eng. J.,* **7**, 682 (1961).

66. C. J. Sheehan and A. L. Bisio, *Rubber Chem. Technol.,* **39**(1), 149 (1966).

67. P. A. Small, *J. Appl. Chem.,* **3**, 71 (1953).

68. M. Souders, *Chem. Eng. Prog.,* **60**(2), 75 (1964).

69. M. Steinberg and B. Manowitz, *Ind. Eng. Chem.,* **51**, 49 (1959).

70. D. Tassios, *Chem. Eng.,* **76**(3), 118 (1969).

71. D. Tassios, *Ind. Eng. Chem. Process, Des. Dev.,* **11**(1), 43 (1972).

72. J. P. Teas, *J. Paint Technol.,* **40**, 19 (1968).

73. R. W. Thompson, Ph.D. Thesis, Chemical Engineering, University of California, Berkeley, 1972.

74. J. Timmermans, *The Physico-Chemical Constants of Binary Systems in Concentrated Solutions,* Interscience, New York, 1959-1960.

75. R. E. Treybal, *Liquid Extraction,* 2nd ed., McGraw-Hill, New York, 1963.

76. A. E. van Arkel, *Trans. Faraday Soc.,* **42B**, 81 (1946).

77. R. F. Weimer and J. M. Prausnitz, *Hydrocarbon Process.,* **44**(9), 237 (1965).

78. I. A. Wiehe and E. B. Bagley, *Ind. Eng. Chem., Fundam.,* **6**, 209 (1967).

79. G. M. Wilson and C. H. Deal, *Ind. Eng. Chem., Fundam.,* **1**, 20 (1962).

80. J. W. Wyart and M. Dante, "Solvents, Industrial," in *Encyclopedia of Chemical Technology,* R. E. Kirk, and D. F. Othmer, Eds., 2nd ed., Interscience, New York, 1968.

81. L. Yen and J. J. McKetta, *Am. Inst. Chem. Eng. J.,* **8**, 501 (1962).

Chapter **VI**

SOLVENT EFFECTS ON MOLECULAR COMPLEX EQUILIBRIA

Sherril D. Christian and Edwin H. Lane

1 INTRODUCTION

In some of the earliest studies of molecular association, it was recognized that the solvent is not merely a passive agent which provides a place for adduct formation to occur. A century of research on complexes has provided ample evidence that the truly inert solvent is only a fiction, and that all the properties of a dissolved complex are affected to some extent by physical or chemical interactions with the solvent.

This article deals mainly with the thermodynamics of complex formation reactions and with the modification of thermodynamic properties caused by interactions of a solvent with complex molecules and with the separated, monomeric species that undergo association. The constants of interest are the equilibrium constants and energies of formation of adducts in various media, including the gas phase.

The association reactions of most concern here are those producing adducts stabilized in large measure by hydrogen-bonding or charge-transfer interactions. Weakly bonded complexes, having formation constants considerably less than $1 \, \ell \, mol^{-1}$, are generally avoided because of difficulties inherent in attempts to obtain reliable information about them [50, 155]. Solvents are restricted primarily to nonpolar and slightly polar organic liquids which are not appreciably self-associated. Aqueous solutions and other solutions in which solvent molecules interact strongly with each other through hydrogen bonds are not considered. However, solvents that possess either proton-donating or proton-accepting groups, but not both, are included.

This chapter consists of a brief historical review of important studies of solvent effects, a summary of the basic thermodynamics of association equilibria in vapor and condensed phases, tables of data illustrating the influence of various media on complex-formation equilibria, and a discussion of several existing methods for treating specific and nonspecific interactions between complexing solutes and the solvent.

Several recent reviews [4, 20, 62a, 69, 70, 124, 147, 156, 191, 195a] provide information on theoretical and experimental aspects of molecular complex equilibria. Solvent effects on spectra of molecular complexes are intimately related to the subject of this chapter; they are discussed in several of the above reviews, but are not considered here.

2   HISTORICAL REVIEW

General Considerations

One of the earliest detailed discussions of association equilibria is that given by Gibbs [78]. He considered information about the anomalous vapor densities of compounds like acetic acid and nitrogen dioxide, and concluded that these compounds are strongly associated. By attributing deviations from the expected vapor densities solely to the formation of associated molecules, and by assuming that the individual monomeric and associated species obey the ideal gas equation, he deduced equilibrium constants and formation heats for reactions such as

$$2NO_2 \rightleftharpoons N_2O_4$$

$$2CH_3COOH \rightleftharpoons (CH_3COOH)_2$$

and

$$PCl_5 \rightleftharpoons PCl_3 + Cl_2$$

The basic concept that information about chemical association can be inferred by observing deviations from an ideal relation, such as the perfect gas law, has persisted almost without modification to the present. In fact, it is fair to say that virtually all the reported thermodynamic constants pertaining to complex-formation reactions have been obtained by comparing experimental results on associating systems with results predicted by some ideal law or combination of laws. In condensed-phase systems, either Raoult's law or Henry's law is almost universally assumed to apply to the individual species involved in a complex-formation reaction; and when spectral methods are used to determine formation constants, it is generally assumed that Beer's law, as well as the laws of dilute solution, are obeyed by the individual species.

It is not difficult to identify fairly distinct schools of thought regarding the propriety of various methods for handling physical data to obtain association constants. At one extreme are those who, with Dolezalek, ascribe all deviations from ideality to the formation of chemical compounds between interacting molecules. Negative deviations from Raoult's law are rationalized in terms of the assumed existence of heteromolecular complexes; positive deviations are attributed to the formation of homomolecular complexes between molecules of the individual components. However, there have been frequent and continuing challenges to the assumption that all deviations from the ideal solution laws are chemical in origin. (For a discussion of this controversy, see Refs. 185 and 102.) Quite early, van Laar forcefully advocated the concept that physical forces that are not capable of stabilizing discrete molecular aggregates lead to nonideality in mixtures of nonelectrolytes. Current molecular theories of pure nonpolar liquids and their mixtures [100, 169] can account for sizable deviations from Raoult's law in systems where chemical effects are presumably absent. Consequently, it is often argued that complex formation should be considered to occur only to the extent that solution nonideality exceeds that predicted by physical interaction theories. However, there is still considerable ground for disagreement concerning the quantitative contributions of physical effects in associating systems or, stated differently, the degree to which activity coefficient effects complicate the interpretation of data. These ambiguities clearly make it difficult to develop meaningful descriptions and explanations of solvent effects; we shall see that there continue to be lively arguments about the relative importance of nonspecific (physical) and specific (chemical) effects of the solvent on complex-formation equilibria.

A related controversy deals with *solvent shifts* (the change in the position of an absorption band or a nuclear magnetic resonance (NMR) signal on a change of medium) as evidence for or against specific interactions between the solvent and a solute species. Our concern with this is limited to the implications it has for interactions between the solvent and the species undergoing association. References to experimental and theoretical work in the area of solvent shifts of absorption bands are given elsewhere [43, 69, 105, 166, 194]. Laszlo, Engler, and Soong [61, 135, 136] give numerous references to explanations of analogous shifts in NMR studies.

Fortunately, in the case of moderately strong complexes (for which the formation constant K_c is appreciably greater than 1 ℓ mol^{-1}), there is relatively little fundamental conflict among the various schools. In solution studies of strong complexes, total solute concentrations can be limited to the less than 1 M range, and Henry's law, rather than Raoult's law, can serve as the basis for reckoning the departures from ideality that are attributable to complex formation. For sufficiently strong complexes, it is generally agreed that physical effects or activity coefficient correction terms of nonspecific origin are small compared to the chemical association effects of interest. In such cases reliable thermodynamic constants can be inferred for reactions of the type D(S) + A(S) $\rightleftharpoons$ DA(S) where D(S), A(S), and DA(S) denote donor, acceptor, and complex molecules, respectively, in the ideal dilute solution standard states in the solvent S.

It is shown later that the problem of predicting solvent effects is essentially equivalent to that of developing satisfactory methods for calculating thermodynamic results for the transfer reactions

$$D(V) \rightleftharpoons D(S) \quad A(V) \rightleftharpoons A(S) \quad DA(V) \rightleftharpoons DA(S)$$

where the state (V) refers to the ideal gaseous state for each species. Therefore our ability to predict the effects of a solvent on molecular complex-formation equilibria is in general no better than our capacity to derive thermodynamic properties of solute molecules in ideal dilute solution standard states in solvents of interest. Considerable attention is given to the general problem of predicting energies and free energies of the important transfer reactions.

Early Research on Solvent Effects

Early workers tended to believe that the effect of a solvent on an equilibrium is analogous to that of an added inert gas on a gas-phase equilibrium. For example, if a reaction involves no change in the number of moles between reactants and products, the solvent should have very little influence on the equilibrium; however, investigation proved that the solvent often has a considerable effect.

van't Hoff [190] showed that one could allow for these solvent effects by taking into consideration changes in *active mass* (activity) of the substances involved in the reaction. In his treatment the activity of a solute in a given solvent is taken as the ratio of the solute's concentration to its solubility in that solvent. Moelwyn-Hughes [143] and Glasstone [79] discuss van't Hoff's approach; several investigators, notably Dimroth [52], although not studying associating systems, showed the utility of this procedure. Kortüm [125] used a similar method to relate the dissociation constants for the I_2-dioxane complex in heptane and cyclohexane.

Most of the early information about associating systems was obtained from vapor density, partition, and colligative property measurements. In the final decades of the nineteenth century, molecular weights were determined for several carboxylic acids, amides, alcohols, and phenols in various solvents [134].

Solvent effects were recognized to be important almost from the beginning of these studies. For example, from partition experiments it was inferred that benzoic acid exists largely as the dimer in benzene and as the monomer in water. The behavior in water was attributed to reaction of the solute, benzoic acid, with water, which in dilute solutions does not change the molal concentration of the solute. However, in benzene the equilibrium $2C_6H_5COOH \rightleftharpoons (C_6H_5COOH)_2$ predominates, presumably because the tendency of the solute molecules to associate is not opposed by a competitive reaction involving the solvent. Apparent molecular weights or degrees of polymerization of associating solutes were found to vary considerably from solvent to solvent. Table 6.1 contains data giving polymerization numbers (apparent molecular weight/formula weight) for benzoic acid and naphthalene in several solvents.

Table 6.1 Polymerization Numbers of Benzoic Acid and Naphthalene in Various Solvents [12][a]

	Polymerization Number	
Solvent	*Benzoic Acid*	*Naphthalene*
Acetone	0.99	1.00
Diethyl ether	1.04	0.98
Ethyl acetate	1.09	1.04
Chloroform	1.82	1.00
Benzene	1.96	1.10
Carbon disulfide	2.14	1.03

[a]From measurements of ebullioscopic constants; values extrapolated to 0.5 m solutions [134].

Oxygenated or nitrogenated solvents and solvents of high dielectric constant tend to dissociate dimers of benzoic acid and similar solutes. In mixed solvents, at fixed total concentration of the polymerizing solute, it is possible to observe the progressive depolymerization of a solute as the mole fraction of the active solvent increases [11]. Lassettre's review [134] summarizes early investigations of these phenomena for solutes now classified as hydrogen-bonded.

Much of the early work on what are now referred to as charge-transfer complexes was done with one of the reactants as the solvent. This work generally centered on colors of solutions and what variations in color implied about the existence of complexes; the early research on I_2 solutions is typical [121]. Lachmann [128] initiated much of this work with his explanation that the brown color of some I_2 solutions, as opposed to the violet color in such solvents as CCl_4 and aliphatic hydrocarbons, was due to complex formation between the solvent and I_2. In 1909, Hildebrand and Glascock [99] used a colorimetric technique to obtain equilibrium constants for I_2 complexes with ethyl alcohol and ethyl acetate in "inactive" solvents (CS_2, CCl_4, and $CHCl_3$); they found no evidence of solvent effects on the equilibria.

Little was done in the area of solvent effects on charge-transfer complex formation until much later. Summarizing the work done through 1960, Briegleb [21] noted the scarcity of data on complexes in different solvents. In addition, he observed that values measured by different investigators, even using the same methods, often differed significantly. These facts help to explain why theories of solvent effects on molecular complex equilibria evolved rather slowly.

3 GENERAL THERMODYNAMIC RELATIONS AND CONSIDERATIONS

Reactions in the Gas or Vapor Phase

Tamres [181] recently summarized most of the available thermodynamic data for gas-phase complexes which are thought to be stabilized (at least in part) by charge-transfer forces. References to reported gas-phase association constants and energies of association in hydrogen-bonded systems are more scattered, although several reviews [40, 148, 156] cite most of the reliable literature. In developing methods for predicting solvent effects, the gas-phase results are of primary importance, because they provide information about association reactions occurring in the absence of the complicating effects of solvent molecules. Unfortunately, because of experimental difficulties in studying gas-phase association, there are relatively few reliable data for systems that have also been extensively studied in condensed phases.

Consider the formation of a 1:1 donor-acceptor complex in the vapor phase by the reaction $D(V) + A(V) \rightleftharpoons DA(V)$. The equilibrium constant for the reaction may be expressed in terms of activities as $K_a = a_{DA}/a_D a_A$, and several alternative choices of standard state may be made. One may, for example, use pressure-based, ideal gaseous standard states at 1 atm or 1 torr. Or, concentration-based states may be employed, such as the unit-molarity ideal gaseous state. However, mole fraction states are not suitable, since there is no direct relation between the fugacity of a species in a gas-phase mixture and its mole fraction. Similarly, molality states are of no value in gas-phase studies.

Thermodynamic results reported in this review pertain to ideal gas molar concentration states, rather than pressure-based states. With this choice of states, there is no need to change the basis for activity in considering condensed phases, for which pressure states are not convenient. Moreover, there is evidence that volume-based states are preferable in treating the equilibria of associating solutes in condensed phases; K_c values for a given reaction are ordinarily more nearly constant throughout a range of similar solvents than are K_x values (based on mole fraction states) [127]. The latter constants may vary considerably with molar volume of the solvent, even within a series of quite similar aliphatic hydrocarbon solvents. This is illustrated by the results of Buchowski et al. [26] in Table 6.2. Foster [66, 69] has made a somewhat different argument in expressing a preference for concentration units of moles per liter or moles per kilogram of solution.

Table 6.2 Association Constants ($25°C$) for the Reaction 1-Heptyne + Acetone $\rightleftharpoons$ 1-Heptyne-Acetone[a]

Solvent	$K_c(\ell\ mol^{-1})$	K_x
Hexane	0.45	3.0
Decane	0.43	1.6
Tetradecane	0.43	0.9
Cyclohexane	0.44	3.7

[a] K_x values have been calculated from experimental K_c values [26] by using equation (6.23b) and the molar volumes of solvent and acetone. See footnote on p. 357.

In most studies of complex formation in the gas phase, at total pressures less than 1 atm, it has been assumed that each species individually—donor, acceptor, and complex—obeys the ideal gas equation. This assumption is quite analogous to the Dolezalek assumption (see Section 2) in condensed phases, although it is perhaps more defensible theoretically in the case of gaseous systems. The problem of dividing virial coefficients into separate terms, attributable to physical and to chemical interactions, is nearly as formidable as that of assigning part of the deviation from Raoult's law in solutions to physical and part to chemical effects. Moreover, the formalism of statistical mechanics as applied to nonideal gases does not provide a way to distinguish between the chemical approach, in which all deviations from ideality are ascribed to the formation of discrete complexes, and the physical approach, which relates the virial coefficients to an intermolecular potential function [139, 162].

Some workers [129, 171] have estimated the physical contribution to the virial coefficient (from Berthelot's equation of state) and subtracted this term to obtain a residue attributable solely to specific effects. Others, for example, Rice and his colleagues [8, 162, 163], have corrected only for the generally small contribution of molecular volume or excluded volume to the virial coefficient in calculating the association constant. K values determined in this way include both physical and chemical effects. The K_c values used here to characterize gas-phase association reactions are constants obtained by ascribing all the deviation from ideality to formation of complexes; ordinarily it is not important to correct for excluded volume contributions to the virial coefficient.

Energies of adduct formation in gaseous systems are generally obtained by using some form of the van't Hoff relation. If a pressure constant K_p is employed, the enthalpy change of the reaction may be calculated from $\Delta H^\circ = -R\,[d \ln K_p/d(1/T)]$, whereas if K_c is measured, the internal energy change is deduced from the analogous relation $\Delta E^\circ = -R[d \ln K_c/d(1/T)]$. ΔH° and ΔE° differ in the gas phase by ΔnRT, where Δn is the change in number of moles in the chemical equation representing the association. This difference is equal to $-RT$ in the case of the 1:1 complex formation reaction.

Regardless of the choice of standard state, the relation $\Delta G^\circ = \Delta H^\circ - T\Delta S^\circ$ must be satisfied, although it is important to observe that the different standard states lead to differing values of ΔG° and ΔS°. However, neither the enthalpy nor the energy of formation of a complex differs from one to another of the ideal gaseous states. This should be obvious if one recalls that the energy and the enthalpy of an ideal gas are constants independent of pressure or concentration at constant temperature. We consider ΔE° for a complex formation reaction to be more fundamental than ΔH°, since the latter constant includes a PV (or RT) term not related to the binding energy of the complex. In solution studies, ΔE° and ΔH° can be used virtually interchangeably, since $\Delta(PV)$ terms are small in magnitude. However, it is necessary to recognize that, in condensed-phase studies of simple 1:1 association reactions, ΔH° is not given by $-R[d \ln K_c/d(1/T)]$, but by $-R[d \ln K_c/d(1/T)] - RT^2a$, where a is the coefficient of thermal expansion of the solution [195]. The RT^2a term typically contributes 0.1 to 0.3 kcal mol^{-1} to the calculated value of ΔH°.

Condensed-Phase Reactions

In considering a given molecular association reaction that occurs in both the vapor phase (V) and in a solvent (S), we find it convenient to examine the thermodynamic cycle

$$
\begin{array}{ccccc}
D(V) & + & A(V) & \rightleftharpoons & DA(V) \\
\updownarrow & & \updownarrow & & \updownarrow \\
D(S) & + & A(S) & \rightleftharpoons & DA(S)
\end{array}
$$

where the activity of each species is referred to the 1 M, ideal dilute solution standard state (whether in the vapor or condensed phase). In addition to the thermodynamic equilibrium constants for association in the two phases ($K_c^{(S)}$ and $K_c^{(V)}$), it is necessary to consider the equilibrium constants for the transfer reactions ($K_{d,D}, K_{d,A}$, and $K_{d,DA}$). The $K_{d,i}$ values are dimensionless partition or distribution coefficients for a solute i distributed between the ideal gaseous state and the infinitely dilute state in the solvent. Thus we have the five equilibrium relations that must be satisfied simultaneously in all treatments of solvent effects in the ideal dilute solution region:

$$K_c^{(S)} = \frac{C_{DA}^{(S)}}{C_D^{(S)} C_A^{(S)}} \tag{6.1}$$

$$K_c^{(V)} = \frac{C_{DA}^{(V)}}{C_D^{(V)} C_A^{(V)}} \tag{6.2}$$

$$K_{d,D} = \frac{C_D^{(S)}}{C_D^{(V)}} \tag{6.3}$$

$$K_{d,A} = \frac{C_A^{(S)}}{C_A^{(V)}} \tag{6.4}$$

$$K_{d,DA} = \frac{C_{DA}^{(S)}}{C_{DA}^{(V)}} \tag{6.5}$$

Equations (6.1) to (6.5) are of course not all independent; they may be combined to give the useful relation

$$\frac{K_c^{(S)}}{K_c^{(V)}} = \frac{K_{d,DA}}{K_{d,D} K_{d,A}} \tag{6.6}$$

The thermodynamic identities

$$\Delta E_S^\circ = \Delta E_V^\circ + \underset{V \to S}{\Delta E_{DA}^\circ} - \underset{V \to S}{\Delta E_D^\circ} - \underset{V \to S}{\Delta E_A^\circ} \tag{6.7}$$

and

$$\Delta G_S^\circ = \Delta G_V^\circ + \underset{V \to S}{\Delta G_{DA}^\circ} - \underset{V \to S}{\Delta G_D^\circ} - \underset{V \to S}{\Delta G_A^\circ} \tag{6.8}$$

must also be satisfied, where ΔE_S°, ΔE_V°, ΔG_S°, and ΔG_V° refer to the standard energy and free-energy changes for the formation reactions in the two phases, and $\underset{V \to S}{\Delta E_i^\circ}$ and $\underset{V \to S}{\Delta G_i^\circ}$ denote standard energies and free energies of transfer of individual species (where i = DA, D, or A) from the vapor phase into the solvent phase. From a purely thermodynamic viewpoint, therefore, one may assert that $K_c^{(S)}$ and $K_c^{(V)}$ will be equal only if $K_{d,DA} = K_{d,D}K_{d,A}$, and that the standard energy (or free energy) of the formation reaction will be the same in vapor as in solution only if the standard transfer energy (or free energy) of the complex is equal to the sum of the standard transfer energies (or free energies) of the donor and acceptor. In fact, either explicitly or implicitly, numerous investigators have assumed the validity of these equalities in situations in which supposedly inert solvents are employed.

There seems to have been some confusion in the literature regarding the meaning of K_c and ΔH° or ΔE° values for complex-formation reactions in condensed phases, particularly in the case of solvents in which relatively strong solute-solvent interactions are known to exist. Some workers have tended to view the derived equilibrium constants for reactions in "inert" media (such as aliphatic hydrocarbons) as true, or thermodynamic constants, and to regard K_c values in more reactive solvents as nonthermodynamic constants. However, it is worthwhile noting here that the reaction $D(S) + A(S) \rightleftharpoons DA(S)$ involves well-defined thermodynamic states, regardless of the nature or strength of solute-solvent interactions. If valid experimental and computational methods have been used to determine sets of thermodynamic constants for a given association reaction involving the ideal dilute states in several widely differing media, there is no reason to treat any one of these sets as more fundamental or better thermodynamically than another. The constants are of course different, but so are the reference states, which in the condensed phases at infinite dilution are solutes completely surrounded by solvent molecules. From a theoretical point of view, thermodynamic constants for gas-phase association reactions are inherently more valuable than the corresponding condensed-phase constants, because theoretical predictions of binding energies, vibration frequencies, and stabilities of adducts are limited almost exclusively to gas-phase systems.

There is much recent evidence that significant changes can be expected in K_c and ΔE° for complex-formation reactions when the medium is changed from the gas phase to solvents of whatever complexity. Examples can be found in which

$$\left|\Delta E_{DA}^\circ\right| < \left|\Delta E_D^\circ + \Delta E_A^\circ\right|$$
$$V \to S V \to S V \to S$$

In fact, this situation seems to prevail for relatively weak complexes. This effect has been attributed [38, 40] to a reduction in the extent of direct contact between DA and the solvent, as compared to contact of the isolated molecules D and A with the solvent. In other words, the formation of DA in solution requires that some of the solvent be "squeezed out" or excluded from contact with the reacting parts of the separated D and A molecules. Consequently, $|\Delta E_S^o|$ is less than $|\Delta E_V^o|$ and, correspondingly, $|\Delta G_S^o|$ and $K_c^{(S)}$ are usually less than $|\Delta G_V^o|$ and $K_c^{(V)}$, respectively. Kroll [126] has related similar arguments in discussing properties of complexes in gas and solution phases.

However, there exist complexes for which

$$\left|\Delta E_{DA}^o\right| > \left|\Delta E_D^o + \Delta E_A^o\right|$$
$$\quad V \to S \qquad V \to S \quad V \to S$$

and for which $K_c^{(S)} > K_c^{(V)}$, even in aliphatic hydrocarbon solvents. In the case of the moderately strong 1:1 complex between trimethylamine (TMA) and SO_2, this effect has been attributed [88, 89] to the existence of a strong dipole-induced dipole interaction between the very polar TMA-SO_2 and the polarizable solvent; solute dipole-solvent forces are not so important in contributing to the energy of interaction between the weakly polar monomers, TMA and SO_2, and the solvent. Intermediate examples can be found in which

$$\left|\Delta E_{DA}\right| \simeq \left|\Delta E_D + \Delta E_A\right|$$
$$\quad V \to S \qquad V \to S \quad V \to S$$

and in which $K_c^{(S)} \simeq K_c^{(V)}$. Such cases seem to occur when the complex is somewhat more polar than the monomers, and one may speculate that the effect of squeezing out solvent molecules is nearly counterbalanced by an excess dipole-solvent interaction involving the complex.

Christian et al. [33, 38, 40] have stressed the utility of the dimensionless constants a and a', defined by the relations

$$a = \frac{\underset{V \to S}{\Delta E_{DA}^o}}{\underset{V \to S}{\Delta E_D^o} + \underset{V \to S}{\Delta E_A^o}} \tag{6.9}$$

$$a' = \frac{\underset{V \to S}{\Delta G_{DA}^o}}{\underset{V \to S}{\Delta G_D^o} + \underset{V \to S}{\Delta G_A^o}} \tag{6.10}$$

in correlating and predicting solvent effects on K_c and ΔE^o for complex-formation reactions (see Section 6, p. 366). Using these constants, one can rewrite (6.6) to (6.8) as

$$\frac{K_c^{(S)}}{K_c^{(V)}} = (K_{d,D}K_{d,A})^{a'-1} \tag{6.11}$$

$$\Delta E_S^\circ = \Delta E_V^\circ + (a-1)(\underset{V \to S}{\Delta E_D^\circ} + \underset{V \to S}{\Delta E_A^\circ}) \tag{6.12}$$

$$\Delta G_S^\circ = \Delta G_V^\circ + (a'-1)(\underset{V \to S}{\Delta G_D^\circ} + \underset{V \to S}{\Delta G_A^\circ}) \tag{6.13}$$

These relations are merely rearranged forms of (6.6) to (6.8) and contain no new information; however, they have proved to be useful in characterizing solvent effects. It is shown later that the constants a and a' can be predicted with some confidence from thermodynamic and statistical mechanical theories of liquids: the magnitudes of these constants are shown to provide qualitative information about the geometry and polarity of adducts.

4 TYPICAL RESULTS ILLUSTRATING THE EFFECTS OF SOLVENTS ON COMPLEX EQUILIBRIA

The results in Tables 6.3 and 6.4 illustrate the influence of the solvent on both K_c and ΔE° for complex-formation reactions. This compilation is not intended to represent data selected critically, or to be exhaustive. The data included here are chosen from those cases in which a particular research group has reported results for a complex in more than one medium.

All values of association constants have been converted to molar units using the following relations (see footnote on p. 357):

$$K_c = K_x \overline{V}_S + (\overline{V}_S - \overline{V}_D)$$

$$K_c = \frac{K_r + M_D/1000}{\rho_S} - \overline{V}_D$$

K_c, K_x, and K_r are association constants with concentrations expressed in moles per liter, mole fractions, and moles per kilogram of solution, respectively; ρ_S is the density of the solvent (at the appropriate temperature) in grams per milliliter, $\overline{V}_S$ and $\overline{V}_D$ are molar volumes of (pure) solvent and donor in liters per mole (calculated from the molecular weight and the density); M_D is the molecular weight of the donor.

The ΔE° values in Tables 6.3 and 6.4 have been taken without change from literature values of ΔH° in the case of condensed-phase studies and converted, where necessary, via an RT term from ΔH° in the case of gas-phase studies.

Table 6.3 contains thermodynamic data for charge-transfer complexes; Table 6.4 contains similar data for hydrogen-bonded complexes. The first entry is

Table 6.3 K_c (ℓ mol⁻¹) and $\Delta E°$ (kcal mol⁻¹) Values for Charge-Transfer Complexes

Complex	Temp. (°C.)	Method[b]	Ref.	Solvent[a]							
				$c\text{-}C_6H_{12}$	$n\text{-}C_7H_{16}$	CCl_4	Dioxane	Benzene	1,2-DCE	CH_2Cl_2	CH_3CN
I_2-triphenylarsine	20	UV-VIS	9	—	—	1400 (−9.4)	—	—	—	21,600 (−9.9)	—
I_2-tetramethylurea	20	UV-VIS	132	—	19 (−5.0)	—	—	—	—	4.6 (−3.3)	—
I_2-tetramethyl-thiourea	20	UV-VIS	132	—	13,100 (−9.9)	—	—	—	—	49,000 (−9.0)	—
I_2-benzene	22	UV-VIS	15	—	0.23	0.17	—	—	—	—	—
I_2-benzene	24	UV-VIS	17	0.260 (−1.72)	0.246 (−1.62)	0.184 (−1.41)	—	—	0.116 (−0.99)	—	—
I_2-mesitylene	22	UV-VIS	15	—	0.79	0.65	—	—	—	—	—
$I_2\text{-}Et_2S$	20	UV-VIS	80	—	317 (−8.0)	233 (−8.0)	—	—	—	—	—
$I_2\text{-}Et_2S_2$	20	UV-VIS	80	—	5.7 (−7.1)	4.6 (−5.8)	—	—	—	—	—
$I_2\text{-}HC(S)N(CH_3)_2$	20	UV-VIS	145	—	1850 (−10.3)	1270 (−7.8)	610 (−8.3)	—	—	6400 (−10.1)	27,700 (−10.1)
I_2-N-methyl-ε-thio-caprolactam	20	UV-VIS	164	—	5200 (−10.5)	3115 (−8.8)	—	—	—	19,400 (−14.3)	35,000
I_2-N,N-dimethylace-tamide	25	UV-VIS	54	—	—	—	—	2.6 (−3.30)	—	1.4 (−2.55)	—

(continued)

339

Table 6.3 K_c (ℓ mol⁻¹) and $\Delta E°$ (kcal mol⁻¹) Values for Charge-Transfer Complexes

				Solvent							
Complex	Temp. (°C)	Method	Ref.	Vapor	Squalene	n-C_7H_{16}	c-C_6H_{12}	CCl_4	1,2-DCE	$CHCl_3$	CH_2Cl_2
I_2-pyridine	25	UV-VIS	111	–	185	142	135	102	–	–	–
I_2-pyridine	25	UV-VIS	137	–	–	157 (−8.16)	–	105 (−7.47)	189 (−7.77)	77.7 (−7.82)	151 (−8.59)
I_2-pyridine	25	UV-VIS	16	–	–	140	126	111	–	63 (−8.4)	–
Br_2-pyridine	25	UV-VIS	51	–	–	–	–	96.8	273.0	–	–
Br_2-3-methylpyridine	25	UV-VIS	51	–	–	–	–	157.7	441.1	–	–
Br_2-2,4-dimethyl-pyridine	25	UV-VIS	51	–	–	–	–	295.5	1230.8	–	–
Br_2-2,4,6-trimethyl-pyridine	25	UV-VIS	51	–	–	–	–	149.9	530.5	–	–
Br_2-3-chloropyridine	25	UV-VIS	51	–	–	–	–	9.3	26.8	–	–
SO_2-trimethylamine	25	UV-VIS	88	340 (−9.1)	–	2550 (−11.0)	–	–	–	–	–
SO_2-trimethylamine	20	UV-VIS	89	–	–	–	–	–	–	36,100 (−14.2)	63,500 (−15.2)

Solvent

Complex	Temp. (°C)	Method	Ref.	Vapor	n-C_6H_{14}	CCl_4	$CHCl_3$	CH_2Cl_2	1,2-DCE	Et_2O
CO(CN)$_2$-diethyl-ether	Not given	UV-VIS	157	~45	3.6	—	—	—	—	—
CO(CN)$_2$-benzene	Not given	UV-VIS	157	~14	0.4	—	—	—	—	—
Tetracyanoethylene-o-xylene	22	UV-VIS	141	—	—	—	0.72	0.39	—	0.29
Tetracyanoethylene-triphenylene	25	UV-VIS	122	—	—	8.6	2.6	0.86	0.43	—
Tetracyanoethylene-mesitylene	25	UV-VIS	62	—	—	8.54	—	1.61	—	—
Tetracyanoethylene-hexamethylbenzene	25	UV-VIS	62	—	—	123	—	22.9	—	—
Fluoranil-hexamethyl-benzene	33.5	NMR	25, 67	—	—	9.76 (−5.4)	2.6 (−3.3)	2.4	2.9 (−3.8)	—
Fluoranil-durene	33.5	NMR	25, 67	—	—	3.1 (−3.9)	0.83 (−3.1)	0.60	0.64 (−3.2)	—
1,4-Dicyano-2,3,5,6-tetrafluorobenzene-hexamethylbenzene	33.5	NMR	25	—	—	3.3	0.59	0.52	0.56	—
1,3,5-Trinitrobenzene-aniline	33.5	NMR	75	—	—	1.48	0.35	—	—	—
1,3,5-Trinitrobenzene-N-ethylaniline	33.5	NMR	75	—	—	1.98	0.48	—	—	—

(continued)

Table 6.3 K_c (ℓ mol^{-1}) and $\Delta E°$ (kcal mol^{-1}) Values for Charge-Transfer Complexes

Complex	Temp. (°C)	Method	Ref.	Solvent							
				$c\text{-}C_6H_{12}$	$n\text{-}C_7H_{16}$	Decalin	CCl_4	CS_2	$CHCl_3$	CH_2Cl_2	1,2-DCE
1,3,5-Trinitrobenzene-N,N-dimethylaniline	33.5	NMR	75	—	—	—	1.89	—	0.49	—	—
1,3,5-Trinitrobenzene-N,N-dimethylaniline	33.5	NMR	71	—	—	—	2.02	—	0.45	0.27	—
1,3,5-Trinitrobenzene-N,N-dimethylaniline	~20	UV-VIS	73	9.6	8.2	7.2	3.4	—	1.3	—	—
1,3,5-Trinitrobenzene-N,N-diethylaniline	33.5	NMR	75	—	—	—	1.61	—	0.45	—	—
1,3,5-Trinitrobenzene-naphthalene	20	UV-VIS	183	9.15 (-4.16)	9.58 (-3.31)	—	5.16 (-3.02)	3.25 (-2.12)	1.82 (-2.38)	—	—
1,3,5-Trinitrobenzene-quinoline	33.5	NMR	74	—	—	—	1.7	—	0.2	—	—
1,3,5-Trinitrobenzene-mesitylene	20	UV-VIS	184	3.51 (-2.30)	3.02 (-2.90)	—	1.36 (-2.81)	—	0.17	—	—
1,3,5-Trinitrobenzene-pentamethylbenzene	20	UV-VIS	184	10.45 (-3.55)	8.69 (-3.05)	—	3.09 (-2.44)	—	1.02	—	—
1,3,5-Trinitrobenzene-pentamethylbenzene	33.5	NMR	25	—	—	—	1.9	—	0.40	—	0.31
1,3,5-Trinitrobenzene-hexamethylbenzene	33.5	NMR	25, 67	—	—	—	3.2 (-3.7)	—	0.55 (-1.9)	—	0.46 (-3.6)
1,3,5-Trinitrobenzene-hexamethylbenzene	20	UV-VIS	68	13.5	—	—	5.7	—	0.76	—	—

Solvent

Complex	Temp. (°C)	Method	Ref.	$c\text{-}C_6H_{12}$	$n\text{-}C_6H_{14}$	$n\text{-}C_7H_{16}$	CCl_4	$CHCl_3$	1,2-DCE	Benzene	CH_2Cl_2
1,3,5-Trinitrobenzene-hexamethylbenzene	20	UV-VIS	184	17.5 (−4.37)	—	14.7 (−3.77)	4.86 (−4.18)	0.90	—	—	—
1,3,5-Trinitrobenzene-hexamethylbenzene	33.5	NMR	66	9.6	8.9	7.9	—	—	—	—	—
1,2,3,5-Tetranitrobenzene-hexamethylbenzene	20	UV-VIS	68	24.6	—	—	9.4	—	—	—	—
Tetrachlorophthalic anhydride-hexamethylbenzene	Not given	UV-VIS	46	—	34	—	14.0	—	—	2.3	—
Tetrachlorophthalic anhydride-acenaphthene	20	UV-VIS	149	—	—	—	5.6	—	2.1	1.2	—
Tetrachlorophthalic anhydride-triethylamine	20	UV-VIS (KIN)	150	—	—	—	—	—	8.0	5.3	15.0
2,5-Dichloro-p-benzoquinone-hexamethylbenzene	33.5	NMR	72	—	—	—	1.17	—	0.48	—	—
Pyromellitic dianhydride-fluoroanthrene	25	UV-VIS	115	—	—	—	—	23.8	—	9.8	7.9
Pyromellitic dianhydride-anthracene	25	UV-VIS	115	—	—	—	—	5.5	—	3.9	3.7
3-Nitro-1,8-naphthalic anhydride-fluoranthrene	25	UV-VIS	114	—	—	—	—	4.1	—	—	6.2
3-Nitro-1,8-naphthalic anhydride-anthracene	25	UV-VIS	114	—	—	—	—	7.7	—	—	2.1

[a] Abbreviations used for solvents are: 1,2-DCE, 1,2-dichloroethane; Et$_2$O, diethyl ether; o-DCB, o-dichlorobenzene; $\emptyset$Cl, chlorobenzene; CH$_2\emptyset_2$, diphenylmethane.

[b] Abbreviations used for methods are: UV-VIS, Ultraviolet and/or visible spectroscopy; NMR, nuclear magnetic resonance spectroscopy; PART, partition studies; IR, infrared spectroscopy; VP, vapor pressure measurements; CAL, calorimetric measurements; DIELEC, dielectric constant measurements; KIN, kinetic measurements; SOL, solubility measurements.

Table 6.4 K_c (ℓ mol^{-1}) and ΔE^0 (kcal mol^{-1}) Values for Hydrogen-Bonded Complexes[a]

Complex	Temp. (°C)	Method	Ref.	Solvent						
				c-C_6H_{12}	n-C_7H_{16}	CS_2	CCl_4	Benzene	$CHCl_3$	CH_2Cl_2
Phenol-pyridine	20	IR	82	–	–	71.5 (-7.5)	59.8 (-7.0)	–	–	–
Phenol-pyridine	25	DIELEC	18	–	80 (-6)	–	45 (-7)	36 (-9)	–	–
Phenol-pyridine	20	IR	170	–	–	73	59	–	17	17
Phenol-diethyl sulfide	25	CAL	192	2.0 (-4.6)	–	–	1.1 (-3.6)	–	–	–
Phenol-tetrahydro-thiophene	25	CAL	192	2.1 (-4.9)	–	–	1.4 (-3.7)	–	–	–
Phenol-isophorone	25	NMR	151	69.8	–	39.8	29.6	–	–	39.7
Phenol-triphenylphosphine oxide	20	IR	81	–	–	1372 (-7.9)	1055 (-6.7)	–	–	–
Phenol-N,N-dimethylformamide	20	IR	84	–	–	116.7 (-6.0)	78.6 (-5.4)	–	–	–
Phenol-N,N-diethylacetamide	20	IR	84	–	–	259.1 (-6.8)	157.2 (-5.1)	–	–	–
Phenol-tributylamine	20	IR	82	–	–	34.3 (-7.9)	29.2 (-6.9)	–	–	–
(Phenol)$_3$	20	NMR	47	17.0 (-14.2)	–	–	4.4 (-9.6)	–	–	–

Solvent

Complex	Temp. (°C)	Method	Ref.	c-C_6H_{12}	CS_2	CCl_4-c-C_6H_{12} (80:20 w/w)	CCl_4	CCl_4-CBr_4 (80:20 w/w)	o-DCB
Phenol-p-NO_2-C_6H_4-C(O)N(CH_2)$_5$	20	IR	83	—	—	59.2	45.3	39.2	—
Phenol-triethylphosphine oxide	20	IR	83	—	—	2191	2522	1052	—
Phenol-pyridine	20	IR	83	—	—	63.0	59.8	42.1	—
Pentachlorophenol-p-NO_2-C_6H_4-C(O)-N-(CH_2)$_5$	20	IR	83	—	—	32.2	30.3	21.7	—
Pentachlorophenol-pyridine	20	IR	82, 83	—	108.1 (-7.3)	109.3	111.4 (-5.8)	65.4	—
Pentachlorophenol-triphenylphosphine oxide	20	IR	81	—	1072 (-7.8)	—	674 (-5.7)	—	—
Pentachlorophenol-N,N-dimethylformamide	20	IR	84	—	74.3 (-5.8)	—	54.1 (-4.6)	—	—
Pentachlorophenol-N,N-diethylacetamide	20	IR	84	—	172.6 (-5.7)	—	126.5 (-5.6)	—	—
p-Chlorophenol-pyridine	Not given	CAL	56	211 (-8.1)	—	—	116 (-7.0)	—	—
m-CF_3-phenol-N,N-dimethylacetamide	Not given	CAL	154	971 (-9.8)	—	—	768 (-7.3)	—	347 (-6.9)
m-CF_3-phenol-tetrahydrothiophene	~22	CAL	192	6.0 (-5.7)	—	—	2.9 (-4.1)	—	—

(continued)

345

Table 6.4 K_c (ℓ mol⁻¹) and ΔE^0 (kcal mol⁻¹) Values for Hydrogen-Bonded Complexes[a]

Complex	Temp. (°C)	Method	Ref.	Solvent							
				c-C_6H_{12}	n-C_6H_{14}	CCl_4	$\varnothing Cl$	o-DCB	Benzene	1,2-DCE	CH_2Cl_2
p-Fluorophenol-pyridine	25	NMR	119	107	—	—	40	43	—	19.5	18
p-Fluorophenol-N,N-dimethylformamide	25	NMR	119	200	—	—	55	50	—	18.6	15.2
p-Fluorophenol-dimethyl sulfoxide	25	NMR	119	360	—	—	160	150	—	44.7	27.6
p-Fluorophenol-hexamethylphosphoramide	25	NMR	119	6300	—	—	1150	1150	—	355	234
m-Fluorophenol-ethyl acetate	24	CAL	152	34 (−6.7)	—	19 (−5.2)	—	10.3 (−4.7)	5.0 (−4.0)	2.5 (−3.7)	—
m-Fluorophenol-dimethyl sulfoxide	24	CAL	152	—	—	470 (−7.2)	—	321 (−6.7)	254 (−6.1)	73 (−5.4)	—
m-Fluorophenol-pyridine	24	CAL	152	262 (−8.4)	—	106 (−7.5)	—	116 (−6.9)	53 (−6.3)	32 (−6.4)	—
m-Fluorophenol-di-n-butyl ether	24	CAL	152	17.0 (−6.5)	—	11.1 (−6.0)	—	7.4 (−5.7)	—	3.3 (−4.5)	—
2,6-Di-t-butylphenol-hexamethylphosphoramide	25	NMR	86	44	—	3	—	—	—	—	—
1,1,3,3,3-Hexafluoroisopropanol-pyridine	Not given	CAL	159	—	(−9.8)	(−8.4)	—	—	—	—	—
2,2,2-Trifluoroethanol-pyridine	Not given	CAL	175	—	(−7.8)	(−6.7)	—	—	—	—	—

Solvent

Complex	Temp. (°C)	Method	Ref.	Vapor	$n\text{-}C_{14}H_{30}$	$n\text{-}C_6H_{14}$	$c\text{-}C_6H_{12}$	CCl_4	Benzene	$CHCl_3$	$CH_2\phi_2$
3-Bromo-1-propyne-N,N-diethylacetamide	28	IR	161	—	—	—	1.17	0.60	—	—	—
(ε-Caprolactam)$_2$	25	IR	77	—	—	—	—	78.8 (−6.8)	—	1.55	—
(Benzoic acid)$_2$	60	IR	1	(−15.5)	—	—	1210 (−12.8)	710 (−11.0)	150 (−7.6)	—	—
(Benzoic acid)$_2$	~24	IR	113	—	—	—	—	2330	—	396	—
(Acetic acid)$_2$	35	PART	49	—	—	58 (−9.0)	—	322 (−7.6)	80.6 (−8.2)	—	—
(Acetic acid)$_2$	29.5	NMR	116	—	—	—	7515 (−12.5)	2467 (−11.4)	—	—	—
(Propanoic acid)$_2$	29.5	NMR	117	—	—	—	8196 (−12.2)	2853 (−11.1)	—	—	—
(Propanoic acid)$_2$	35	PART	49	—	—	621	—	444 (−7.4)	87.7 (−7.8)	19.4 (−6.7)	—
(Trifluoroacetic acid)$_2$	25	IR	39	5660	—	—	—	128 (−9.0)	—	—	—
(Trifluoroacetic acid)$_2$	25	IR-PART	39	—	—	—	192 (−11.7)	149 (−9.2)	2.6	—	—
(Trifluoroacetic acid)$_2$	40	VP	97	—	377.0 (−18.1)	—	—	—	—	—	2.84 (−7.45)

(continued)

347

Table 6.4 K_c (ℓ mol^{-1}) and ΔE^0 (kcal mol^{-1}) Values for Hydrogen-Bonded Complexes[a]

Complex	Temp. (°C)	Method	Ref.	Solvent							
				Vapor	n-C$_{16}$H$_{34}$	c-C$_6$H$_{12}$	n-C$_6$H$_{14}$	CS$_2$	CCl$_4$	Benzene	CH$_2\phi_2$
Perfluoro-t-butanol-pyridine	Not given	CAL	176	—	—	—	(−12.5)	—	(−10.3)	—	—
n-Butanol-triethyl-amine	~20	CAL, IR	131, 196	—	—	5.07	—	—	3.17	—	—
Methanol-triethylamine	31	IR	103	(−7.6)	—	—	—	—	6.8 (−6.0)	—	—
Methanol-diethylamine	25	VP	40, 187	13.6 (−6.7)	8.7; 300[b] (−4.5; −13.8)	—	—	—	—	—	4.8; 22.7[b] (−3.6; −10.2)
(n-Butanol)$_2$	25	IR, NMR	112	—	—	4.74	—	3.47	2.32	0.83	—
H$_2$O-diethylamine	25	VP	40, 187	7.9 (−6.0)	11.0 (−6.1)	—	—	—	—	—	8.5 (−4.5)
H$_2$O-pyridine	25	PART	118	—	—	5.3; 6.8[c]	—	—	2.8; 2.7[c]	1.5; 1.5[c]	—
H$_2$O-triethylamine	25	SOL, PART	87	—	—	7.0	—	—	—	3.5	—
CHCl$_3$-[(CH$_3$)$_2$N]$_3$PO	20	NMR	85	—	—	13.4 (−4.1)	—	—	3.0 (−2.7)	—	—
CHCl$_3$-(C$_2$H$_5$O)$_2$P(O)-i-C$_3$H$_7$	20	NMR	85	—	—	6.13 (−4.0)	—	—	1.41 (−2.3)	—	—

[a] For abbreviations see footnote to Table 6.3.
[b] Second values pertain to 2:1 complex.
[c] Second values pertain to 1:2 complex.

348

K_c (in liters per mole), and the entry in parentheses is ΔE° (in kilocalories per mole).

The results in Tables 6.3 and 6.4 exemplify the types of solvent effects commonly observed for systems of molecular complexes. Of the charge-transfer systems, weak aromatic hydrocarbon donor-halogen complexes almost invariably exhibit decreases in K and $-\Delta E^\circ$ as more reactive solvents are employed. Complexes of halogens with aliphatic sulfides, aromatic amines, and aliphatic amines have formation constants and energies that either remain nearly constant or increase (in magnitude) with increasing solvent reactivity. The highly polar complexes of aliphatic amines with both SO_2 and iodine are important examples of adducts that are stabilized by more interactive media. In general, organic π-donor-π-acceptor complexes have considerably smaller K and $-\Delta E^\circ$ values in more polar and more polarizable media.

Hydrogen-bonded adducts almost without exception become less stable in more strongly interactive solvents. Particularly large decreases in K and $-\Delta E^\circ$ are observed for the dimers of carboxylic acids, which are thought to be cyclic species. The sections that follow are devoted to theoretical and correlational methods for rationalizing solvent effects exhibited in the preceding tables.

5 SPECIFIC SOLVATION METHODS

Large solvent effects on equilibrium constants and other thermodynamic functions were noted early in the work on hydrogen-bonded complexes. In fact, even when the association constants were not determined with great accuracy, the influence of the solvent was often evident [156]. Influenced by the knowledge that many of the solvents that were exerting an influence on the association of interest were capable of hydrogen-bonding, Barger [11] postulated that, in these cases, the solvent was involved in an equilibrium which was competing with the equilibrium of interest. Despite the apparent acceptance of this point of view, evidently no one attempted corrections to the observed constants until much later. Perhaps Lassettre [134] expressed the attitude of many early researchers when, speaking about hydrogen-bonding data in different solvents, he said, "Presumably these data can be analyzed by applying the law of mass action, but the solutions are so concentrated and the equilibria are so complex that this does not seem worthwhile." Hobbs and Bates [104] noted the expected effect of a competing equilibrium involving the solvent, although they did not attempt to obtain corrected constants.

Similarly, solvent effects on thermodynamic and spectral functions were found from the beginning of studies on charge-transfer complexes [10, 15, 45]. Experimental techniques had improved enough by this time that significant differences were found in values of association constants, even in supposedly inert solvents; however, investigators were not able to find a simple explanation for these solvent effects on which most of them could agree. Many of these researchers became dissatisfied with their inability to isolate, chemically or mathematically, a complexation reaction from the effects of the interfering solvent. They felt that they could not study nonspecific solvent effects (i.e., correlate association constants or thermodynamic functions with such solvent properties as polarizability or dielectric constant) unless corrections were made for specific interactions involving the solvent [69, 137]. Two of the proposed approaches, which purport to correct for "specific solvation," are examined in detail here; they are (1) a single equilibrium involving solvated species, and (2) competing equilibria, some of which involve complex formation with the solvent.

Single Equilibrium between Solvated Species

In a discussion of solvent effects that could complicate the study of weak molecular complexes, Ross, Kelley, and Labes [167] wrote the complex-formation reaction with specifically solvated species. A detailed treatment of this idea was presented by Carter, Murrell, and Rosch (CMR) [28]. Their intention was to provide a theory that was applicable to weak complexes (e.g., contact charge-transfer complexes [153]), as well as to strong complexes. They assume that donor, acceptor, and complex occur in solution with well-defined solvent shells; thus the complexation reaction is written as

$$(A \cdot S_n) + (D \cdot S_m) \rightleftharpoons (DA \cdot S_p) + qS$$

where $q = n + m - p$. (Trotter and Hanna [186] simplify the above equation by assuming that the donor is not specifically solvated.) The equilibrium constant is defined as

$$K_{CMR} = \frac{[DA \cdot S_p](X_S)^q}{[A \cdot S_n][D \cdot S_m]} \qquad (6.14)$$

where brackets denote molar concentrations and X_S is the mole fraction of free solvent relative to the total number of moles of D, A, and S. K_{CMR} is defined in this way so that its units will agree with the units of $K_{B-H}*$.

*K_{B-H} designates the equilibrium constant calculated with the Benesi-Hildebrand (B–H) equation [15] or with one of its many modifications. ϵ_{B-H} represents the molar absorptivity of the charge-transfer complex calculated by the same method. Under the assumption of $[D]_0 \gg [A]_0$, the B–H equation is written as

$$\frac{[A]_0 \ell}{A} = \frac{1}{K_{B-H}\epsilon_{B-H}} \frac{1}{[D]_0} + \frac{1}{\epsilon_{B-H}}$$

where A = absorbance, ℓ = optical path length, and $[D]_0$ and $[A]_0$ represent analytical (total) concentrations of donor and acceptor. ϵ_{B-H} is then calculated from the intercept of a plot of $[A]_0 \ell/A$ versus $1/[D]_0$; K_{B-H} is equal to the intercept divided by the slope.

Assuming $[D]_0 \gg [A]_0$ and $[S]_0 > [D]_0$ one can show that

$$K_{CMR} = K_{B-H} + \frac{q(m+1)}{[S]_0} \qquad (6.15)$$

and

$$\epsilon_{CMR} = \epsilon_{B-H} \left(\frac{K_{B-H}}{K_{CMR}}\right) \qquad (6.16)$$

where $[S]_0$ is the total solvent concentration when $[D]_0 = 0$. For strong complexes the equilibrium constant is much greater than

$$\frac{q(m+1)}{[S]_0} \quad ,$$

and K_{B-H} is virtually the same as K_{CMR}. Zero or negative K_{B-H} values are possible for weak complexes if

$$K_{CMR} \leqslant \frac{q(m+1)}{[S]_0} \quad .$$

By making plots of ϵ_{CMR} versus K_{CMR} for a series of similar complexes [using different values of $q(m+1)$], the term $q(m+1)$ can be optimized so that the ϵ_{CMR}-versus-K_{CMR} plot passes through the origin (as required by the Mulliken theory [146, 147]). Carter, Murrell, and Rosch give examples in which the association constant is apparently underestimated, using the B–H approach, by up to ~3 ℓ mol⁻¹.

Tamres and Duerksen [59, 181] observe that the CMR treatment requires the $K\epsilon$ product to be the same before and after correction for solvent displacement. They are critical of the CMR approach because they feel it does not adequately account for the difference generally found between the gas-phase value of $K_{B-H} \, \epsilon_{B-H}$ and its value in solution. In contrast, Carter, Murrell, and Rosch [28] claim that, while ϵ_{CMR} should be nearly constant and approximately the same as ϵ_{B-H} measured in a gas-phase experiment, K_{CMR} should be solvent-dependent.

Another aspect of the CMR treatment which has been interpreted in different ways is the degree of solvation required of the solute species. Carter et al. assume a "well-defined solvent shell" involving solvent molecules "bound in such a way that they cease to behave like free solvent." Foster [69, 72] calls this specific solvation. In contrast, Scott [173, 174] maintains that the CMR theory "has not received the attention which it deserves because it seems to require a specific interaction .... Actually the solvent molecule occupies space (e.g., a lattice site); it must be included in the bookkeeping because it is there." Scott's application of the Guggenheim quasi-lattice model in the treatment of this problem is considered later.

Foster et al. [60, 69] do not feel that the CMR treatment can account for the differences between optically and nonoptically determined values of association constants; they say that NMR methods yield the correct K value for a 1:1 complex, while optical methods are likely to yield this value only under the condition $[D]_0 = [A]_0$. In reply to these suggestions, Carter [27] claims that the CMR technique applies to NMR methods as well as to optical methods and offers the explanation that Foster's findings were complicated by higher-order complexes.

Emslie et al. [60] quote values of K_{B-H} and ϵ_{B-H} for complexes of tetra-chlorophthalic anhydride with hexamethylbenzene [46], and of N,N-dimethyl-aniline with 1,3,5-trinitrobenzene [73] in different solvents; for a given complex ϵ_{B-H} is relatively constant in several solvents. They suggest that, according to the CMR treatment, these ϵ_{B-H} values should change with the solvent. However, Carter [27] claims that, since these complexes are "strong" (K_{B-H} ranges from 1.3 to 14.0 ℓ mol^{-1}), a B—H treatment should be valid (i.e., the CMR correction is not needed, and the constant ϵ_{B-H} values are in accord with the CMR approach).

Other researchers have used their results to question the utility of the CMR approach in correcting for solvent effects on complex formation. Bhowmik [17] attempted to use the CMR approach on his results for the interaction of benzene and iodine in different solvents. He comments that "the relation between the stability and the intensity of the complex is not improved from the B—H treatment." Hunter and Norfolk [109] used the CMR method on their data for complexes of chloranil with methylbenzenes. They found that, whereas both their original data (from the B—H equation) and their data corrected by the use of a competing-equilibrium technique yielded a "reasonable relationship" between K and ϵ, the CMR approach, especially at higher temperatures, yielded a relationship that was "inadequate."

In view of the disagreement concerning the assumptions and results of the CMR treatment, perhaps it is well to proceed to a related approach in which the solvent molecules are assumed to be involved in competitive complex formation with the solute molecules.

Competing Equilibria

During the early studies of charge-transfer complexes, some researchers came to agree with investigators of hydrogen-bonded complexes that one of the most important causes of variation in formation constants with solvent was the competitive formation of solvent-solute complexes. Merrifield and Phillips [141] attempted corrections for complexes of solvent with the (electron) acceptor. They claimed that, for their study of complexes of tetracyanoethylene (TCNE) with aromatic molecules, the formation of a TCNE-solvent complex accounted for the major part of the apparent variation with the solvent of equilibrium constants for the donor-acceptor interaction. The equations that Merrifield and Phillips used were essentially the same as those employed by Corkill, Foster, and Hammick [44] for studies of two donors interacting with a single acceptor. The procedure is limited to the situation in which both donors (or donor and solvent) are in large excess.

Tamres [180] has considered in detail the basic equations for dealing with a competing equilibrium involving the solvent. He demonstrates that, assuming $[S]_0 \gg [SA]$, $[A][D] \gg [DA]^2$, and $[D] \gg [A]$, one can arrive at the following equation, which relates K_c^{DA} (the formation constant for $D + A \rightleftharpoons DA$) and K_c^{SA} (the formation constant for $S + A \rightleftharpoons SA$):*

$$\frac{1}{K_c^{DA}} = \frac{1}{(1 + K_c^{SA}[S]_0)} \left[\frac{[D]_0 [A]_0 \epsilon \ell}{A} - [D]_0 \right] \tag{6.17}$$

where ℓ is the optical path length, A is the absorbance of DA, and ϵ is the molar absorptivity of DA. If written in the form of the B—H equation [15], (6.17) becomes

$$\frac{[A]_0 \ell}{A} = \frac{1 + K_c^{SA}[S]_0}{K_c^{DA} \epsilon} \frac{1}{[D]_0} + \frac{1}{\epsilon} \tag{6.18}$$

Thus, according to Tamres' analysis, ϵ_{B-H} will be correct, and $K_{c,B-H}$ will be related to K_c^{DA} by

$$K_{c,B-H} = \frac{K_c^{DA}}{1 + K_c^{SA}[S]_0} \tag{6.19}$$

If mole fractions are used and one assumes that $x_S + x_D = 1$, the analogous equation [141] is:

$$K_{x,B-H} = \frac{K_x^{DA} - K_x^{SA}}{1 + K_x^{SA}} \tag{6.20}$$

In general, it can be shown [18] that

*The same considerations are applicable if the solvent S is an acceptor rather than donor; K_c^{SD} (the formation constant for $S + D \rightleftharpoons SD$) would replace K_c^{SA}.

$$K_c^{DA} = K_{c,B-H} \ (1 + K_c^{SA}[S]_0) \ (1 + K_c^{SD}[S]_0) \qquad (6.21)$$

where $K_{c,B-H}$ is obtained from a B–H-type analysis, and the other symbols are as above. Most investigators use the above equations or related expressions when attempting corrections for the effects of competitive equilibria.

It should be noted that an equilibrium involving the complex and the solvent may also be included. La Planche, Thompson, and Rogers [133] have included this idea in their model of self-associating systems. Others have postulated an interaction between complex and solvent (e.g., Refs. 13, 14, 115) to account for variations in spectral properties with solvent. Moreover, this specific interaction is included explicitly in other types of treatments [28].

It can easily be seen that the major obstacle to using this procedure lies in finding reliable values for the equilibrium constants involving the solvent. Various techniques for deriving and using these values have been developed; several of them are mentioned here.

Merrifield and Phillips [141] assumed that, when $CHCl_3$ was used as solvent, there was no interaction between TCNE and $CHCl_3$ (i.e., $K_x^{SA} = 0$); thus, when (6.20) was used, the observed association constant $K_{x,B-H}$ was equivalent to K_x^{DA}. If the donor in this first case can be used as a solvent in other studies of TCNE complexation, K_c^{SA} will be known (assuming these corrected values are all independent of the identity and amount of other solution components), and $K_{x,B-H}$ can be used to give values for K_x^{DA}.

Ewall and Sonnessa [62] sought to improve on the assumption that $CHCl_3$ is inert. They assumed that CCl_4 did not interact with TCNE. They worked with TCNE and aromatics in mixtures of CCl_4 and CH_2Cl_2. For each pair of $K_{c,B-H}$ and $[CH_2Cl_2]_0$ values, values of K_c^{DA} were calculated using (6.19) and trial values of K_c^{SA}; then a K_c^{DA}-versus-K_c^{SA} line was plotted. "Best values" for K_c^{DA} and K_c^{SA} were taken from the intersection of several of the K_c^{DA}-versus-K_c^{SA} lines. K_c^{DA} values obtained in this manner were shown to agree with values calculated from solutions using only CCl_4 as solvent.

Higuchi et al. [98] used (6.19) as well as observed K_c values in "inert" solvents and an interacting solvent (CCl_4) to obtain a value for K_c^{SD}. Other investigators (e.g., Refs. 3, 53, 193) have spectrally determined equilibrium constants for aromatic donors with CCl_4 in a "noninteracting" solvent. However, because these constants are rather small, their significance may be questioned [50, 96, 155, 158, 168a]. Nevertheless, Drago et al. [54, 152] believe that using a value of K_c^{SA} determined for S and A in CCl_4 solution is justified when the absorbance data for A in pure S passes through the common intersection on a K-versus-ϵ plot [165] for the data obtained in CCl_4. However, they note that good intersections often do not occur.

Takahashi et al. [179] used a rearranged form of (6.19) for complexes in solvent mixtures (in which only one of the solvents was believed to interact specifically with one of the solutes). They employed the following equation:

$$K_{c,B-H} = K_c^{DA} - K_c^{SD} K_{c,B-H} [S]_0$$

where $[S]_0$ is the total concentration of *interacting* solvent. Thus a plot of $K_{c,B-H}$ versus $K_{c,B-H}[S]_0$ yields values of K_c^{DA} and $-K_c^{SD}$ as the intercept and limiting slope, respectively.

Tamres [182] employed the competing equilibria approach to make a correction for 2:1 solvent-acceptor complexes rather than for 1:1 complexes. Using Childs' data [29] for I_2-aromatic complexes, he found that a correction for 2:1 SA complexes in the case of pyridine and I_2 in aromatic solvents brings the corrected equilibrium constant into better agreement with values in heptane than did a correction for 1:1 SA complexes.

Other investigators (e.g., Refs. 109, 123, 137) used a published value for K_c^{SD} or K_c^{SA} to correct their observed results. However, the unreliability and scarcity of the necessary values make this calculation possible in only a few cases. Even when corrected equilibrium constants are obtained, there is disagreement as to their significance. While some investigators [108, 123, 137] have noted nonspecific solvent effects on K's corrected for specific solvation, others [17, 17a, 141, 149] apparently believe the correction for solvent competition produces K_c^{DA} values free from all other solvent effects.

Huong and Lascombe [110] introduced a procedure which is similar to the competing-equilibria approach, although it also resembles in some respects the method of Carter, Murrell, and Rosch [28]. They write the complexation reaction as

$$DS + A \rightleftharpoons DA + S$$

in which the (proton) donor is always complexed with either solvent or acceptor. They define $K_R = [DA] X_S^0/[DS] X_A^0$, where $[S]_0$, $[A]_0 \gg [D]_0$. They propose procedures for determining K_R by following the decrease in the "DS" band or by following the increase in the DA band. K_R, K_x^{DA}, and K_x^{SD} are related by the equation

$$K_R = \frac{K_x^{DA}}{K_x^{SD}} \tag{6.22}$$

Although treatment of the observed data in terms of competing equilibria often seems to account for a large portion of the variation in equilibrium constants with solvent, many researchers agree with Drago [54] that "it is by no means the only effect to consider in order to describe the variation in K with solvent." In fact, some investigators [41, 145] believe that considering the solvent a reactant and introducing mass action expressions explicitly involving the solvent will not prove worthwhile. They feel that an unequivocal separation of the effects of specific interactions from nonspecific effects cannot be achieved with the techniques available at present.

6 NONSPECIFIC INTERACTION MODELS

The energies and free energies of transfer of donor, acceptor, and complex molecules from the gas phase into even relatively inert solvents, such as the aliphatic hydrocarbons, are frequently as large in magnitude as the energies and free energies of the complex formation reactions themselves. Therefore it is not surprising that, even in the absence of strong specific interactions, minor changes in solvent properties can induce relatively large variations in adduct stabilities and energies.

Clearly, an adequate theory of the effects of solvents on the thermodynamic properties of individual solutes would permit one to predict the effects of solvents on complex-formation equilibria. Attempts have been made to treat nonspecific solvent effects on molecular complex equilibria by employing current theories of nonelectrolyte solutions, by considering variations in activity coefficients, and with several other procedures. Included in this section are discussions of (1) treatments in which the concentration dependence of solute activity coefficients is considered explicitly, (2) methods for correlating effects of media on complex stability with bulk properties of solvents, (3) methods for treating solvent effects which are based on current theories of nonelectrolyte solutions, (4) attempts to compensate for the effects of nonspecific interactions by comparing data for actual complexing systems with results for analogous or homomorphic systems in which presumably only nonspecific solvent-solute interactions are involved, and (5) methods that attempt to relate the transfer energies and free energies of a complex in a range of solvents to those of the donor and acceptor molecules.

Activity Coefficient Effects

In their classic article [15] on aromatic hydrocarbon-I_2 complexes, Benesi and Hildebrand noted that the equations derived for inferring K and ϵ from spectral absorbance-concentration data should apply only if the activity coefficient ratio $(\gamma_{DA}/\gamma_D\gamma_A)$ remains constant throughout the range of concentrations employed. Using solubility parameter theory, they estimated that the solubility parameter of the complex δ_{DA} is approximately the arithmetic mean of δ_D and δ_A and noted that γ_{DA} is approximately canceled by $\gamma_D\gamma_A$.

However, they noted that K for a given donor-acceptor system should change with a change in solvents. Alley and Scott [2] have argued that if δ_{DA} is equal to the volume-weighted average of the solubility parameters of donor and acceptor,

$$\delta_{DA} = \frac{(\overline{V}_D \delta_D + \overline{V}_A \delta_A)}{(\overline{V}_D + \overline{V}_A)} ,$$

K_c should be nearly independent of total solute concentration or, stated alternatively, γ_{DA} should nearly equal $\gamma_D \gamma_A$. In the case of complexes that are weak and nonpolar or slightly polar, such a relation between δ_{DA} and the solubility parameters of the donor and acceptor seems reasonable. Despite numerous claims (e.g., Refs. 55, 120) that activity coefficient effects cancel, there have been many challenges to the assumption that the activity coefficient ratio will always fortuitously equal unity, even if measurements are restricted to the very dilute solution range. Mulliken and Person [147], for example, have concluded that uncertainties in estimating activity coefficient corrections can lead to errors as large as a factor of 2 in the K value inferred for the benzene-I_2 complex in heptane.

Scott [172], Hanna, Rose, and Trotter [95, 186], and others [4, 127] have observed that the choice of standard state (or concentration scale) may strongly influence the values of K and ϵ derived from relations of the B–H type. Thus values of K based on different concentration scales are frequently not equivalent when a simple conversion of units is made from one scale to another; the ϵ's also are often not equivalent. In the case of weak complexes, discrepancies of this type are particularly important.* Some investigators have attempted to associate these discrepancies with activity coefficient effects. For example in an

---

*It is possible to derive definite mathematical relations between K and ϵ values obtained from B–H-type plots utilizing different concentration scales with the same data [5, 127, 131a]. It is also possible to demonstrate that if a B–H plot is linear on one concentration scale (mole fraction, molarity, or molality), and if the volume change on mixing is zero, the plot will be linear on the other two scales *regardless* of the extent of deviations from solution ideality (cf. Ref. 186). The K's and ϵ's derived from the slopes and intercepts of the B–H plots can be related by [131a]

$$K_m = (K_x + 1) \frac{M_S}{1000} \quad \text{and} \quad \epsilon_m = (\frac{K_x}{K_x + 1})\epsilon_x \qquad (6.23a)$$

$$K_c = K_x \overline{V}_S + (\overline{V}_S - \overline{V}_D) \quad \text{and} \quad \epsilon_c = \left[\frac{K_x \overline{V}_S}{\overline{V}_S(K_x + 1) - \overline{V}_D} \right] \epsilon_x \qquad (6.23b)$$

$$K_c = \frac{K_m}{\rho_S} - \overline{V}_D \quad \text{and} \quad \epsilon_c = (\frac{K_m}{K_m - \rho_S \overline{V}_D})\epsilon_m \qquad (6.23c)$$

attempt to relate K_x and K_m values calculated from NMR data for the benzene-caffeine complex in CCl_4, Hanna and Rose [95] corrected for the nonideality of the donor. (For comments on the procedure of Hanna and Rose, see refs. 131a and 176a.)

Christian, Lane, and Childs [131a] have modified B—H equations to include activity coefficients. They develop relations between thermodynamic K's (i.e., equilibrium constants expressed in terms of activities) and K's obtained from spectral and solubility studies. They demonstrate that activity coefficient corrections enter in different ways when different physical methods are employed to study the same complex equilibrium. In considering how these differences can arise, it is instructive to compare results obtained with the solubility method with those inferred from the more commonly used spectral methods. The solubility method has been employed to study diverse systems, including complexes of water [40] and of HCl [23, 24] with polar solutes in nonaqueous media, and complexes of iodine with donor molecules in organic solvents [30, 125] and in the vapor phase [31]. A recent extension of the solubility method has been to use gas-chromatographic techniques to study complexing systems [140, 160]. (In fact, Purnell [160a] suggests that gas-chromatographic methods may be particularly useful in isolating and studying solvent effects.)

Ordinarily, at least in dilute solution studies, it is assumed that the increase in solubility of a component at constant activity (e.g., solid I_2 or HCl at 1 atm) that occurs on addition of a second solute (e.g., an aromatic donor) may be attributed to formation of a complex. Although it has been recognized that nonspecific interactions may also contribute to the solubility increases, it is

$$K_m = K_r + \frac{M_D}{1000} \quad \text{and} \quad \epsilon_m = \left[\frac{K_r}{K_r + (M_D/1000)} \right] \epsilon_r \qquad (6.23d)$$

where the subscripts m, c, x, and r indicate units expressed in molality, molarity, mole fraction, and moles per kilogram of solution, respectively. M_S and M_D are the molecular weights of solvent (S) and donor (D); ρ_S is the density of solvent (in grams per milliliter); $\overline{V}_S$ and $\overline{V}_D$ are the molar volumes of pure solvent and pure donor in liters.

Therefore, if a set of data gives K values that are not in the (infinite dilution) ratio required for consistency of the concentration scales, or ϵ values that are not equal, the "discrepancies" are not related to solution nonideality, but are dictated by the definite relations among the various concentration units. One should note that in some instances nonpositive K or ϵ values can result from the above relations. Obviously, if the K values are large ($K \gg 1$), (6.23a) to (6.23d) reduce to the expected relations among the concentration scales (i.e., $K_m = K_x M_S/1000$, $K_c = K_x \overline{V}_S$, $K_c = K_m/\rho_S$, $K_m = K_r$, and $\epsilon_m = \epsilon_c = \epsilon_x = \epsilon_r$).

It may be argued that molality standard states should be avoided in treating data for concentrated nonelectrolyte solutions, since it can be shown that, even in the case of binary systems which rigorously obey Raoult's law at all concentrations, activity coefficients based on the molality scale are strongly concentration-dependent. Thus K_m values do not seem to be generally useful in providing information about weak molecular complexes.

commonly assumed that the concentration of uncomplexed solute (at constant activity) does not change. Consequently, the concentration of the complex is equated to the total difference between the augmented solubility and the original solubility in the absence of the added second solute. Recently, Childs and co-workers [29, 30, 31, 34] utilized mixtures of solid polyiodides (e.g., tetramethylammonium triiodide and pentaiodide) to maintain a constant activity of iodine in equilibria established in heptane solution. The increase in iodine concentration in the region of very dilute solutions of donor in heptane approaches linearity with donor concentration C_D. From the limiting slope and intercept of a plot of total I_2 concentration against C_D, an overall K value is obtained, based on the assumption that γ_{I_2} (the activity coefficient of *uncomplexed* iodine) remains constant at high dilution. When variations in activity coefficients are explicitly considered, it can be shown [29, 131a] that the total K determined from solubility data is the sum of a thermodynamic K and a term that is a function of γ_{I_2} only. (That is, the activity coefficients of the donor and of the donor-I_2 complex are eliminated as unknowns in this treatment.)

Table 6.5 Equilibrium Constants for Several Donor-Iodine Complexes in Heptane at $25°C^a$

Donor	$K_c^{(s)}$ (ℓ mol^{-1})^b	$K_c^{(corr)}$ (ℓ mol^{-1})^c	$K_c^{(B-H)}$ (ℓ mol^{-1})^d
Benzene	0.37	0.09	0.20
Toluene	0.50	0.24	0.32
o-Xylene	0.64	0.34	0.42
p-Xylene	0.64	0.38	0.41
m-Xylene	0.70	0.44	0.54
Mesitylene	0.98	0.70	0.74

^{a}From refs. 29 and 131a.
$^b K_c^{(s)}$ values are obtained directly from polyiodide solubility results, assuming $\gamma_{I_2} = 1$.
$^c K_c^{(corr)}$ values have been corrected for γ_{I_2}, estimated from solubility parameter theory [100].
$^d K_c^{(B-H)}$ values have been inferred using the B–H analysis [15], assuming $\gamma_{AD} = \gamma_D \gamma_A$.

Solubility parameter theory [100] has been employed to estimate the effect of added donor on the activity coefficient of I_2 in donor-solvent mixtures. Table 6.5 shows some of the results obtained by the solubility method, in comparison with constants for the same systems obtained by using the conventional B–H spectral method. For the stronger complexes the discrepancy between the two types of equilibrium constants is relatively less important.

Christian, Lane, and Childs [131a] have recently shown that unambiguous values of activity coefficients of weak molecular complexes in dilute solution can be inferred by combining the results of spectral and solubility experiments with activity coefficient data for the donor. The same systems that are shown in Table 6.5 were studied. It was found that solvation of the aromatic donor-I_2 complexes is considerably greater than the solvation of the uncomplexed donor molecules and that, even in the submolar concentration range, activity coefficient effects can have a significant influence. However, a word of caution is still in order; any of the methods proposed to correct for activity coefficient effects are subject to considerable error and, in the case of systems in which K_c is much less than 1 ℓ mol^{-1}, these errors may be relatively large compared to K_c itself.

Effects of Bulk Properties of Solvent

Attempts have been made to relate changes in K and ΔH for association reactions to bulk properties of the solvent, notably the dielectric constant D. Davies et al. [48, 49] noted the decrease in K and $-\Delta H$ for the dimerization of acetic and propionic acids in the series of media: vapor, hexane, benzene, CCl_4, CS_2, nitrobenzene, chlorobenzene. The $-\Delta H$ values were shown to vary linearly with

$$\left\{ 1 + \left[\frac{\partial \ln D}{\partial \ln T} \right]_p \right\} /D$$

for the solvent, as is predicted from electrostatic theories and the Gibbs-Helmholtz equation for an association in which the attractive forces are purely electrostatic in origin.

Franzen, Franzen, and Stephens [76, 77] found a linear relation between ΔG of dimerization and D^{-1} of the solvent for N-methylacetamide and ϵ-caprolactam in mixtures of *cis*- and *trans*-dichloroethylene and also for ϵ-caprolactam in mixtures of CCl_4 and 1,1,1-trichloroethane. A linear relation between ΔH of dimerization and D^{-1} was also found for both amides in the *cis*- and *trans*-dichloroethylene mixtures. Enthalpy values were corrected for changes in D with T, so that this correlation is equivalent to the linear relation of Davies et al. Hirano and Kozima [103] studied the association of methanol and triethylamine in CCl_4, chlorobenzene, and dichloromethane, as well as in the gas phase. They noted that values of ΔE for hydrogen-bond formation in solution varied linearly with D^{-1} of the solvent; however, the gas-phase value of ΔE did not fall on the line extrapolated from the solution points.

Such relations do not generally hold for other hydrogen-bonded and charge-transfer complex equilibria. Ordinarily, there is a general decrease in stability of hydrogen-bonded adducts and of the majority of charge-transfer complexes in solvents of greater polarity and polarizability. However, it has often been stated that simple correlations of K or ΔH with D do not obtain because of the complicating effects of specific solvent-solute interactions.

It is worth noting here the influence of the polarity of complexes on their stability in more reactive media. Most weak charge-transfer complexes and virtually all hydrogen-bonded complexes that have been studied in diverse media have dipole moments that are not significantly greater than the vector sums of the moments of the monomers that associate to form them. For such systems, the effect of squeezing out solvent molecules (as the complex is formed through union of the reacting sites of the monomers) usually predominates, leading to positive changes in ΔH and in ΔG as more reactive media are employed. However, strong complexes of the charge-transfer type are often considerably more polar than the donors and acceptors that form them, and for such systems the use of more reactive solvents frequently leads to increases in K and/or $-\Delta H$ for the complex formation reactions. Examples of complexes that behave in this way are tetramethylthiourea-I_2 [132], triphenylarsine-I_2 [9], trimethylamine-SO_2 [88, 89, 90], and several substituted pyridine-Br_2 complexes [51]. Undoubtedly, in many of these systems specific solute-solvent interactions act to complicate the interpretation of solvent effects in terms of bulk dielectric properties. However, in the case of the trimethylamine-SO_2 complex (which has a dipole moment of 4.5 D), there are significant differences between values of K and $-\Delta H$ in the gas phase and those obtained at high dilutions in the solvent heptane [88]. The complex formation constant (at $25°C$) increases from 340 ℓ mol^{-1} in the gas phase to 2550 ℓ mol^{-1} in heptane; the corresponding ΔE values for complex formation in the two phases are -9.1 kcal mol^{-1} and -11.0 kcal mol^{-1}, respectively. These changes in thermodynamic properties, which presumably occur in the absence of any specific (orientation) effects between solvent and solute, apparently depend on the stabilizing influence of inductive interactions between the complex and the solvent.

Applications of Nonelectrolyte Solution Theories

In place of deriving expressions for γ_D, γ_A and γ_{DA}, it is possible to utilize equations for the chemical potential of D, A, and DA derived from various theories of nonelectrolyte solutions. Moelwyn-Hughes [144] has considered the effects of total solute concentration on reactions of the type A $\rightleftharpoons$ B occurring in a dilute solution in solvent S, where the solution is assumed to follow the van Laar theory [188, 189] of mixtures. He derives the expression

$$K_x = K_x{}^0 \exp\left[\left(\frac{\Delta\mu_{AS}-\Delta\mu_{BS}}{kT}\right)(x_A+x_B)\right]$$

where K_x and $K_x{}^0$ are the mole fraction equilibrium constants at finite concentrations of solute and at infinite dilution, respectively; x_A and x_B are the mole fractions of A and B; and $\Delta\mu_{AS}$ and $\Delta\mu_{BS}$ are interchange energies defined by

$$\Delta\mu_{AS} = Z\epsilon_{AS} - \frac{Z}{2}(\epsilon_{AA} + \epsilon_{SS})$$

and

$$\Delta\mu_{BS} = Z\epsilon_{BS} - \frac{Z}{2}(\epsilon_{BB} + \epsilon_{SS})$$

Z is the number of nearest neighbors of A or of B; $\epsilon_{AS}, \epsilon_{BS}, \epsilon_{AA}, \epsilon_{BB}$, and ϵ_{SS} are potential energies of pair interactions of AS, BS, AA, BB, and SS contacts, respectively. In dilute solution, application of the van Laar theory predicts that K_x will vary linearly with $x_A + x_B$.

A similar treatment of the effects of solvents on complex formation equilibria has been presented by Buchowski et al. [26]. These workers assume that solubility parameter theory [100, 101], modified by Flory-Huggins corrections [64, 65, 106, 107] to account for unequal sizes of solute and solvent molecules, applies to D, A, and DA individually in solution. The solubility parameters of D, A, and DA are assumed to be the same in a series of different solvents, and the molecular volume of DA is taken to equal the sum of the volumes of D and A. Buchowski et al. derive the relation $\log K_c = a + b\delta_S$, where K_c is the formation constant expressed in molar concentration units, δ_S is the solubility parameter of the solvent, and a and b are constants depending only on properties of D, A, and DA. The linearity of plots of $\log K_c$ against δ_S supports this model for the complexes 1-heptyne-acetone [26] and pyridine-I_2 [111] in several nonpolar media and for the complex phenothiazine-chloranil in solvents of varying polarity [10a]. (Results for the butyronitrile-I_2 complex do not give a linear plot [137a].) The derivation of the linear relation between $\log K_c$ and δ_S requires inclusion of the Flory-Huggins entropy corrections for differences in molar volume between solutes and solvent. An analogous linear relation between $\log K_x$ and δ_S (where K_x is the mole fraction equilibrium constant) is predicted from the simpler solubility parameter theory (which does not account for size differences between solvent and solute molecules) [26]; such a relation does not hold for the systems to which this model has been applied. Using solubility parameter theory to calculate solute activity coefficients, Christian [32] showed the similarity between the above approach and the a method (Section 6, p. 366). In contrast, Fletcher [63] claims that the results of Buchowski et al. [26] and Huong, Platzer, and Josien [111] can be interpreted in terms of specific solvent-solute interactions.

Recently, Scott [173] applied the Guggenheim quasi-lattice model [92] for nonelectrolyte mixtures to systems in which the reaction $D + A \rightleftharpoons DA$ occurs in a solvent. All interactions by pairs of the individual "faces" of D, A, DA, and S molecules are assumed to have the same interaction energies, except for the specific interfacial contact between the reactive faces of the D and A molecules. If it is assumed that this particular interaction leads to an observable spectral effect, one can derive relations that indicate that the observed values of K_x obtained with the B–H equation (or various modifications of it) are equal to the excess of interactions over the number expected if mixing were random, or what Guggenheim [93] calls "sociation" constants. Scott notes that his equations are substantially equivalent to those of Carter, Murrell, and Rosch [28] (Section 5), who assumed that donor, acceptor, and complex exist in solution with well-defined solvent shells. In the newer treatment, Scott points out that it is not necessary to assume that specific interactions occur between the solutes and the solvent. The solvent molecule is included in the bookkeeping system simply because it occupies space (lattice sites) around the interacting solutes.

Use of Homomorphs or Analogs

Numerous investigators have proposed that the nonspecific contributions to solute-solute or solute-solvent interactions can be inferred by comparing results for associating systems with those for systems in which the molecules interacting specifically are replaced by nonpolar molecules having nearly equivalent dimensions. The word homomorph has been introduced by Brown and co-workers [22] to designate a molecule similar to another molecule in size and shape. Bondi and Simkin [19] used aliphatic hydrocarbons as homomorphs of liquid aliphatic alcohols in estimating the specific contribution of hydrogen bonding to the heats of vaporization of the alcohols.

Martire and Riedl [138] measured gas-chromatographic retention volumes of volatile, low-molecular-weight alcohols in the electron donor solvents di-n-octyl ether and di-n-octyl ketone; these results were compared with retention volumes of the same alcohols in n-heptadecane. The latter solvent is molecularly quite similar to the donor solvents except for the polar group. Differences in retention volumes of the alcohols in the polar and nonpolar solvents, corrected for small differences in activity coefficients of the uncomplexed solute, were used directly to infer association constants for the alcohol solvent interactions. Sheridan, Martire, and Tewari [174a] used this technique to study interactions of haloalkanes with di-n-octyl ether and di-n-octyl thioether.

The pure base method introduced by Arnett and co-workers [6, 6a, 7] has been used to infer enthalpies of hydrogen-bond formation of alcohols and phenols with proton-acceptor solvents, such as aromatic hydrocarbons, alkyl halides, aromatic and aliphatic ethers, amines, amides, and ketones. Calorimetric enthalpies of each proton-donor solute (A) in a given solvent are compared with those for a model compound (M) that cannot hydrogen-bond with the solvent. For example, p-fluoroanisole is used as a model compound for p-fluorophenol, anisole is used for phenol, and n-butyl chloride is used for n-butyl alcohol.

Calorimetric data lead to enthalpy values for the dilute solution transfer reactions M(reference solvent) $\rightleftharpoons$ M(basic solvent) and A(reference solvent) $\rightleftharpoons$ A(basic solvent). In the pure base method, it is assumed that neither M nor A is hydrogen-bonded to the reference solvent (CCl_4 or other nonpolar solvent), and that A is hydrogen-bonded completely in the basic solvent. The transfer enthalpy of A minus that of M is equated to the specific enthalpy of formation of the hydrogen bond between A and the basic solvent. Values of hydrogen-bond enthalpies obtained in this way agree quite well with those inferred from calorimetric or spectrophotometric studies of the same proton donor-acceptor reactions in dilute solution in CCl_4 [6]. Moreover, in the systems studied, the initial results seemed to indicate that there is little variation in hydrogen-bond enthalpies when different model compounds or reference solvents (including cyclohexane, methylcyclohexane, hexane, carbon tetrachloride, chloroform, dichloromethane, and isooctane) are used. This is surprising, in view of the large variations in ΔH for hydrogen-bond formation enthalpies that are generally observed in the range of solvents employed as references.

Duer and Bertrand [58] checked the reliability of the pure base method for phenol-base interactions by using three different reference solvents (cyclohexane, heptane, and carbon tetrachloride) and by using different model compounds. They found that enthalpies of formation of hydrogen bonds determined by the method are strongly dependent on the choice of reference solvent and model compound. An alternative method is proposed to compensate for solvent effects through use of model compounds. It is suggested that ΔH for the reaction $AB'(B') + B(B) + M(B) \rightarrow AB(B) + B'(B') + M(B')$ can be equated to the difference in hydrogen-bond formation enthalpies of AB and AB', where B and B' are two bases, and (B) and (B') denote the solvent phases B and B', respectively. Relative heats determined in this way (for the proton donors phenol, acetic acid, and benzoic acid, with the bases pyridine, p-dioxane, and acetone) are not significantly different when different model compounds are employed. An advantage of the procedure is that it facilitates the study of acceptors, such as acetic and benzoic acids, that are highly associated even in dilute solution in nonpolar solvents. Dilute solution values of ΔH for several phenol-base adduct formation reactions have been shown [130] to be consistent with values of ΔH obtained using the pure base method as modified by Duer and Bertrand.

A similar method, but one that does not require the use of model compounds, has been proposed by Drago, Nozari, and Vogel [57, 152] to eliminate the effect of solvents on hydrogen-bond enthalpies. It is suggested that, in the absence of specific interactions between the basic molecules and the solvent, ΔH for the displacement reaction

$$BA + B' \rightarrow B'A + B$$

is constant in nonpolar and weakly basic solvents. For complexes of m-fluoro-phenol with ethyl acetate, dimethyl sulfoxide, and triethylamine in several nonpolar solvents, the enthalpy change for the displacement reaction appears to be constant; however, when pyridine is employed as a base, the procedure requires modification. Discrepancies are attributed to the specific interaction of pyridine with CCl_4, o-dichlorobenzene, and benzene. Results for the solvent 1,2-dichloroethane indicate that heats for the displacement reaction in that solvent do not agree with those in cyclohexane, CCl_4, o-dichlorobenzene, and benzene. Drago et al. have extended their procedure to include the polar solvent nitrobenzene [93a] and have shown that reactions in which the acid is displaced can be studied in a weakly acidic solvent [57a, 152a].

Recently, a model was proposed [35, 42] for relating thermodynamic properties of polar solutes involved in complex formation equilibria to those of analogous nonpolar solutes in the same media. In the nonpolar analog (NPA) method, a polar solute (P) is replaced by a hypothetical nonpolar molecule which has the same molecular volume and the same total energy of interaction with a nonpolar solvent as does P. The NPA molecule must have a molecular polarizability greater than that of P in order to interact as strongly with S as does P through a combination of dipole-induced dipole and dispersion forces. Since orientation forces are presumably absent in dilute solution of P in S, it may be argued that the molar entropy and free energy (as well as the energy) of the transfer reactions P(ideal gas) $\rightleftharpoons$ P(S) and NPA(ideal gas) $\rightleftharpoons$ NPA(S) are equal. When the NPA model is applicable, it makes possible the prediction of ΔE_i° (the transfer energy of any solute i from the ideal gaseous state into $V{\rightarrow}S$ ideal dilute solution in solvent S) from ΔG_i°. The NPA model is therefore $V{\rightarrow}S$ consistent with the earlier empirical observation [33, 36] that energies of transfer for both polar and nonpolar solutes from the gas phase into a given solvent vary linearly with the free energies of transfer, and that the slope of a plot of ΔG_i° versus ΔE_i° is a function of solvent properties only. The model may be $V{\rightarrow}S$ $V{\rightarrow}S$ used to predict changes in ΔG° (or K) for a formation reaction in a series of solvents from knowledge of changes in ΔE° for the reaction and the individual internal energies of transfer of donor and acceptor from the gas phase into the solvents [35].

Transfer Energy and Free-Energy Relationships

Linear free-energy relationships have been widely used to correlate kinetic data with physical properties of reactants and media (e.g., see Ref. 94). In addition, several types of linear free-energy relations have been proposed recently for predicting equilibrium constants for complex-formation reactions.

Taft and co-workers [119, 142, 178] related K_c for the 1:1 complexes of a given proton donor with a series of bases in CCl_4 by the equation

$$\log K_c = m(pK_{HB}) + C \tag{6.24}$$

where pK_{HB} is the logarithm of the formation constant (in CCl_4) for the reference proton donor (p-F–C_6H_4OH) with each base; m and C are empirical constants. For several hydroxylic proton donors with a wide variety of bases, an excellent correlation is obtained. Recently, a test of (6.24) was made for complexes of p-F–C_6H_4OH in the solvent cyclohexane and in several polar aprotic solvents [119]. Results for cyclohexane followed the expected linear relation, but formation constants for amine complexes in chlorinated solvents (chlorobenzene, o-$C_6H_4Cl_2$, $Cl(CH_2)_2Cl$, and CH_2Cl_2) were generally significantly greater than the values predicted from the linear relation inferred from data for other complexes of p-F–C_6H_4OH with electron-donor molecules. The enhancement of stability is attributed to an increased extent of proton transfer in the amine-phenol complexes, which is induced by polar solvents. Similar effects have been proposed previously to explain the unusual influence of solvent on stabilities, spectra, and dipole moments of amine-iodine [123] and trimethylamine-SO_2 [88, 89] complexes. These complexes presumably become more dative in character as more reactive solvents are employed.

Christian and co-workers [36, 38, 40] have emphasized the importance of obtaining thermodynamic data for the transfer reactions involved in the solvation cycle

$$D(V) + A(V) \rightleftharpoons DA(V)$$
$$\updownarrow \qquad \updownarrow \qquad \updownarrow$$
$$D(S) + A(S) \rightleftharpoons DA(S)$$

Equations (6.11) to (6.13) are thermodynamic identities expressed in terms of the constants

$$a = \frac{\Delta E^{\circ}_{V \to S}{}_{DA}}{\Delta E^{\circ}_{V \to S}{}_{D} + \Delta E^{\circ}_{V \to S}{}_{A}} \quad \text{and} \quad a' = \frac{\Delta G^{\circ}_{V \to S}{}_{DA}}{\Delta G^{\circ}_{V \to S}{}_{D} + \Delta G^{\circ}_{V \to S}{}_{A}}$$

These dimensionless constants represent the fraction of the energy (or free energy) of transfer of the donor plus acceptor that is retained by the complex. Initially, it was proposed that a and a' would be nearly equal for a given complex dissolved in a given solvent [36], and that a and a' would depend little on the choice of solvent or temperature [38]. It was supposed that as a general rule both a and a' would be less than unity, because interfacial contact between a complex and a solvent would be less than that between the isolated donor and acceptor molecules and the solvent. Later it was noted [88] that, in the case of highly polar complexes, a and a' can exceed unity.

Table 6.6 Experimental Values of a and a' for Typical Molecular Complexes

Complex	Solvent	a	a'	Ref.[a]
$(CF_3COOH)_2$	Cyclohexane	0.72	0.55	39, 177
$(CF_3COOH)_2$	CCl_4	0.59	0.55	39, 177
$(CF_3COOH)_2$	Benzene	0.43	0.39	39, 177
$(CF_3COOH)_2$	1,2-Dichloroethane	0.48	0.39	39, 177
Pyridine-H_2O	[b]	—	0.71	118
Diethylamine-H_2O	Hexadecane	1.01	1.05	40, 187
Diethylamine-H_2O	Diphenylmethane	0.85	1.01	40, 187
Diethylamine-H_2O	Benzyl ether	0.80	0.90	40, 187
Diethylamine-CH_3OH	Hexadecane	0.74	0.93	187
Diethylamine-CH_3OH	Diphenylmethane	0.73	0.89	187
Diethylamine-CH_3OH	Benzyl ether	0.70	0.85	187
m-F-phenol-dimethyl sulfoxide	[c]	0.58	—	152
m-F-phenol-ethyl acetate	[c]	0.45	—	152
m-F-phenol-pyridine	[c]	0.61	—	152
m-F-phenol-triethylamine	[c]	0.68	—	152
Benzene-I_2	CCl_4 or heptane	~1.0	0.8–0.9	8, 17, 29, 59
Diethyl ether-I_2	Heptane	1.0	0.9	29, 37, 91
Pyridine-I_2	[b]	—	0.93	32
Trimethylamine-SO_2	Heptane	1.30	1.21	88, 89
Trimethylamine-SO_2	Chloroform	1.43	1.44	89
Trimethylamine-SO_2	Dichloromethane	1.62	1.51	89
TCNE-triphenylene	[b]	—	0.65	122

[a] Reference to calculation of a' and a (if available) and to original data.
[b] Estimated from free-energy data for several nonpolar and slightly polar solvents; gas-phase data not available.
[c] Estimated from enthalpy data for several nonpolar and slightly polar solvents; gas-phase data not available.

The logarithmic form of (6.11),

$$\log K_c^{(S)} = \log K_c^{(V)} + (a' - 1) \log (K_{d,D}K_{d,A})$$

was used to correlate data for pyridine-I_2 [32] and pyridine-H_2O [38]. Plots of $\log K_c^{(S)}$ against $\log (K_{d,D}K_{d,A})$ in various hydrocarbon and chlorinated hydrocarbon solvents appear to be linear for these complexes; from the slopes of these plots the values $a' = 0.93$ and 0.71 have been inferred for pyridine-I_2 and pyridine-H_2O, respectively. When $K_c^{(V)}$ is known, as well as $K_c^{(S)}$ and the transfer free energies of the donor and acceptor from vapor to solvent, a' may be calculated directly from (6.11) or (6.13). Similarly, a may be calculated from (6.12), provided ΔE_S°, ΔE_V°, and the transfer energies of the donor and acceptor are known. Table 6.6 summarizes some of the values of a and a' that have been obtained for typical molecular complexes in a range of media. These results show that a and a' are not strictly equal, nor independent of solvent. In general, large discrepancies between a and a' are not observed. However, it is clear that both constants increase significantly in the case of the polar adduct trimethylamine-SO_2 as the solvent is varied from heptane to the chlorinated hydrocarbons.

It has been suggested that the constants a and a' can be used to infer something about the physical nature of complexes and their interaction with solvents [33]. Several tentative conclusions may be drawn from results such as those exhibited in Table 6.6:

1. Moderately weak complexes are ordinarily characterized by values of a and a' less than unity. For such systems the effect of squeezing out solvent molecules as the complex forms from the monomers is not compensated by a correspondingly large excess energy of interaction between the dipole of the complex and the medium. a and a' for weak complexes do not seem to change greatly from one solvent to another.

2. In the case of hydrogen-bonded systems, even moderately strong complexes rarely have values of a and a' greater than unity. In general, dipole enhancements* of hydrogen-bonded complexes are smaller than those for charge-transfer complexes of comparable strength; thus the squeezing-out effect almost always predominates over the opposing effect of complex dipole-solvent interactions.

3. Cyclic complexes should be characterized by abnormally small values of a and a'. Carboxylic acid dimers, for example, are generally much less polar than the monomers that form them and in addition, if they are cyclic in solution

---

*The dipole enhancement is defined as the magnitude of the dipole moment of the complex less that of the vector sum of the dipole moments of the donor and acceptor, oriented as in the complex.

as is commonly supposed, they must form with considerable loss of interfacial contact with the solvent. Both of these effects should act to diminish the magnitude of the dimer solvation energy compared to the solvation energies of the unreacted monomers. Where there is doubt regarding the probable geometry of a complex, examination of observed values of a and a' may be useful in determining whether the complex exists in cyclic or linear form.

4. Values of a and a' greatly in excess of unity are indicative of strong complexes in which the dipole moment greatly exceeds the vector sum of the dipoles of the separated donor and acceptor. Charge-transfer systems in which the polar dative structure contributes greatly to the stability of the ground state of the complex should fall into this category. It seems likely that a and a' for complexes of this sort generally tend to increase as more polar and more polarizable solvents are employed; however, specific complex formation between polar solvents and D, A, or DA could act to complicate this interpretation of solvent effects. The tendency of polar solvents to induce a further separation of charge [123] should contribute to the increase in a and a'.

5. For a given complex, a and a' are usually approximately equal. If a and a' are found to differ significantly for a given system, this fact could be interpreted as indicating the existence of unusual entropy effects, which is another way of saying that the energy and free-energy changes do not vary in direct proportion.

The near equality of a and a' for most complexes is predictable from the observed linearity of plots of transfer free energy against transfer energy. It has been shown [33, 36] that the relation

$$\Delta G_i^\circ \underset{V \to S^i}{} = \beta_S \, \Delta E_i^\circ \underset{V \to S^i}{} + 300 \, \text{cal mol}^{-1} \qquad (6.25)$$

adequately represents thermodynamic results for the transfer of solutes (i) into a given nonpolar or slightly polar solvent; β_S is a constant characteristic of the solvent and derivable from properties of the solvent alone. Nonpolar analog calculations (Section 6, p. 363) support the validity of the relation for polar as well as nonpolar solutes [35]. Given that $a' = \Delta G_{DA}^\circ \underset{V \to S}{} / (\Delta G_D^\circ \underset{V \to S}{} + \Delta G_A^\circ \underset{V \to S}{})$, and using (6.25) for the solutes D, A, and DA, it follows that

$$a' = \frac{\Delta E_{DA}^\circ \underset{V \to S}{} + \dfrac{300 \, \text{cal mol}^{-1}}{\beta_S}}{\Delta E_D^\circ \underset{V \to S}{} + \Delta E_A^\circ \underset{V \to S}{} + \dfrac{600 \, \text{cal mol}^{-1}}{\beta_S}}$$

Thus if the terms $300/\beta_S$ and $600/\beta_S$ can be neglected, a and a' will be nearly equal. (Since $\beta_S \cong 0.6$ for many common organic solvents, the additive terms in the numerator and denominator are about 0.5 and 1.0 kcal, respectively. In most cases the difference between a and a' produced by these terms does not exceed 5 or 10%.)

Lattice-model calculations have been used to estimate transfer energies for donor, acceptor, and complex molecules of interest [177]. A table of site interaction energies has been compiled [33, 177] for functional groups present in common molecular complexes and solvents. These parameters have been used with some success to predict values of $\Delta E^o_{i \atop V \rightarrow S^i}$ and a for complexes that are not highly polar. However, the lattice calculations alone do not account for the excess dipole-solvent interactions which must be considered in attempting to predict effects of solvents on complexes with large dipole moments.

CONCLUSION

This chapter has dealt with evidence that specific and nonspecific solvent effects can modify complex-formation equilibria. There are undoubtedly many systems for which both types of effects should be considered in a comprehensive examination of the influence of solvents; the methods reviewed here provide tests which may be helpful in deciding which effects are most important. We have emphasized that nonspecific effects should always be considered, even when solvents as nearly inert as aliphatic hydrocarbons are employed. In this regard we have undoubtedly betrayed our aversion to methods in which the formation of discrete solute-solvent complexes is proposed without adequate evidence.

Data on solvent effects are accumulating rapidly, and good experimental results in themselves will be useful in clarifying the mechanisms by which solvents act to alter complex equilibria. However, there is still a paucity of accurate results for associating systems in the gas phase. Inherently, information about gaseous complexes is important in providing reference data that apply in the absence of the complicating effects of solvent-solute interactions. Moreover, gas-phase data can be compared directly with thermodynamic constants derived from rapidly improving theoretical computations. We close this chapter with the plea that more effort be expended in obtaining reliable gas-phase results for complexing systems—systems for which both good theoretical calculations and condensed-phase data are available.

Acknowledgments

The authors are indebted to Dr. Milton Tamres for his helpful comments on this article and to numerous colleagues who provided reprints of their work.

References

1. G. Allen, J. G. Watkinson, and K. H. Webb, *Spectrochim. Acta,* **22**, 807 (1966).
2. S. K. Alley, Jr. and R. L. Scott, *J. Phys. Chem.,* **67**, 1182 (1963).
3. R. Anderson and J. M. Prausnitz, *J. Chem. Phys.,* **39**, 1225 (1963).
4. L. J. Andrews and R. M. Keefer, *Molecular Complexes in Organic Chemistry,* Holden-Day, San Francisco, 1964.
5. Ref. 4, p. 85.
6. E. M. Arnett, L. Joris, E. Mitchell, T. S. S. R. Murty, T. M. Gorrie, P. von R. Schleyer, *J. Am. Chem. Soc.,* **92**, 2365 (1970).
6.a. E. M. Arnett, E. J. Mitchell, and T. S. S. R. Murty, *J. Am. Chem. Soc.,* **96**, 3875 (1974).
7. E. M. Arnett, T. S. S. R. Murty, P. von R. Schleyer, and L. Joris, *J. Am. Chem. Soc.,* **89**, 5955 (1967).
8. D. Atack and O. K. Rice, *J. Phys. Chem.,* **58**, 1017 (1954).
9. E. Augdahl, J. Grundnes, and P. Klaeboe, *Inorg. Chem.,* **4**, 1475 (1965).
10. W. G. Barb, *Trans. Faraday Soc.,* **49**, 143 (1953).
10.a. M. Barigand, J. Orszagh, and J. J. Tondeur, *Bull. Soc. Chim. Fr.,* **1973**, 48.
11. G. Barger, *J. Chem. Soc.,* **1905**, 1042.
12. E. Beckmann, *Z. Phys. Chem.* (Leipzig), **6**, 437 (1890).
13. L. J. Bellamy, K. J. Morgan, and R. J. Pace, *Spectrochim. Acta,* **22**, 535 (1966).
14. L. J. Bellamy and R. J. Pace, *Spectrochim. Acta,* **A27**, 705 (1971).
15. H. A. Benesi and J. H. Hildebrand, *J. Am. Chem. Soc.,* **71**, 2703 (1949).
16. K. R. Bhaskar and S. Singh, *Spectrochim. Acta,* **A23**, 1155 (1967).
17. B. B. Bhowmik, *Spectrochim. Acta,* **A27**, 321 (1971).
17.a. B. B. Bhowmik and P. K. Srimani, *Spectrochim. Acta,* **A29**, 935 (1973).
18. R. J. Bishop and L. E. Sutton, *J. Chem. Soc.,* **1964**, 6100.
19. A. Bondi and D. J. Simkin, *J. Chem. Phys.,* **25**, 1073 (1956).
20. G. Briegleb, *Elektronen-Donator-Acceptor-Komplexe,* Springer-Verlag, Berlin, 1961.
21. Ref. 20, pp. 114-120.
22. H. C. Brown, G. K. Barbaras, H. L. Berneis, W. H. Bonner, R. B. Johannesen, M. Grayson, and K. L. Nelson, *J. Am. Chem. Soc.,* **75**, 1 (1953).
23. H. C. Brown and J. D. Brady, *J. Am. Chem. Soc.,* **74**, 3570 (1952).
24. H. C. Brown and J. J. Melchiore, *J. Am. Chem. Soc.,* **87**, 5269 (1965).
25. N. M. D. Brown, R. Foster, and C. A. Fyfe, *J. Chem. Soc.,* B, **1967**, 406.
26. H. Buchowski, J. Devaure, P. V. Huong, and J. Lascombe, *Bull. Soc. Chim. Fr.,* **1966**, 2532.
27. S. Carter, *J. Chem. Soc.,* A, **1968**, 404.
28. S. Carter, J. N. Murrell, and E. J. Rosch, *J. Chem. Soc.,* **1965**, 2048.
29. J. D. Childs, Ph.D. Dissertation, University of Oklahoma, 1971.

30. J. D. Childs, S. D. Christian, and J. Grundnes, *J. Am. Chem. Soc.*, **94**, 5657 (1972).
31. J. D. Childs, S. D. Christian, J. Grundnes, and S. R. Roach, *Acta Chem. Scand.*, **25**, 1679 (1971).
32. S. D. Christian, *J. Am. Chem. Soc.*, **91**, 6514 (1969).
33. S. D. Christian, Office of Saline Water, Research and Development Progress Report no. 706, July 1971.
34. S. D. Christian, J. D. Childs, and E. H. Lane, *J. Am. Chem. Soc.*, **94**, 6861 (1972).
35. S. D. Christian, R. Frech, and K. O. Yeo, *J. Phys. Chem.*, **77**, 813 (1973).
36. S. D. Christian and J. Grundnes, *Acta Chem. Scand.*, **22**, 1702 (1968).
37. S. D. Christian and J. Grundnes, *J. Am. Chem. Soc.*, **93**, 6363 (1971).
38. S. D. Christian, J. R. Johnson, H. E. Affsprung, and P. J. Kilpatrick, *J. Phys. Chem.*, **70**, 3376 (1966).
39. S. D. Christian and T. L. Stevens, *J. Phys. Chem.*, **76**, 2039 (1972).
40. S. D. Christian, A. A. Taha, and B. W. Gash, *Quart. Rev.* (London), **24**, 20 (1970).
41. S. D. Christian and E. E. Tucker, *J. Phys. Chem.*, **74**, 214 (1970).
42. S. D. Christian, K. O. Yeo, and E. E. Tucker, *J. Phys. Chem.*, **75**, 2413 (1971).
43. A. R. H. Cole and A. J. Michell, *Aust. J. Chem.*, **18**, 102 (1965).
44. J. M. Corkill, R. Foster, and D. L. Hammick, *J. Chem. Soc.*, **1955**, 1202.
45. T. M. Cromwell and R. L. Scott, *J. Am. Chem. Soc.*, **72**, 3825 (1950).
46. J. Czekalla and K. O. Meyer, *Z. Phys. Chem.* (Frankfurt am Main), **27**, 185 (1961).
47. A. J. Dale and T. Gramstad, *Spectrochim. Acta*, **A28**, 639 (1972).
48. M. Davies, in *Hydrogen Bonding*, D. Hadzi and H. W. Thompson, Eds., Pergamon, New York, 1959, p. 393.
49. M. Davies, P. Jones, D. Patnaik, and E. A. Moelwyn-Hughes, *J. Chem. Soc.*, **1951**, 1249.
50. D. A. Deranleau, *J. Am. Chem. Soc.*, **91**, 4044 (1969).
51. J. D'Hondt, C. Dorval, and T. Zeegers-Huyskens, *J. Chim. Phys.*, **69**, 516 (1972).
52. O. Dimroth, *Justus Liebigs Ann. Chem.*, **377**, 127 (1910).
53. F. Dörr and G. Buttgereit, *Ber. Bunsenges. Phys. Chem.*, **67**, 867 (1963).
54. R. S. Drago, T. F. Bolles, and R. J. Niedzielski, *J. Am. Chem. Soc.*, **88**, 2717 (1966).
55. R. S. Drago, R. L. Carlson, N. J. Rose, and D. A. Wenz, *J. Am. Chem. Soc.*, **83**, 3572 (1961).
56. R. S. Drago and T. D. Epley, *J. Am. Chem. Soc.*, **91**, 2883 (1969).
57. R. S. Drago, M. S. Nozari, and G. C. Vogel, *J. Am. Chem. Soc.*, **94**, 90 (1972).
57.a. R. S. Drago, J. A. Nusz, and R. C. Courtright, *J. Am. Chem. Soc.*, **96**, 2082 (1974).
58. W. C. Duer and G. L. Bertrand, *J. Am. Chem. Soc.*, **92**, 2587 (1970).
59. W. K. Duerksen and M. Tamres, *J. Am. Chem. Soc.*, **90**, 1379 (1968).

60. P. H. Emslie, R. Foster, C. A. Fyfe, and I. Horman, *Tetrahedron,* 21, 2843 (1965).

61. E. M. Engler and P. Laszlo, *J. Am. Chem. Soc.,* 93, 1317 (1971).

62. R. X. Ewall and A. J. Sonnessa, *J. Am. Chem. Soc.,* 92, 2845 (1970).

62.a. D. V. Fenby and L. G. Hepler, *Chem. Soc. Rev.,* 3, 193 (1974).

63. A. N. Fletcher, *J. Phys. Chem.,* 74, 216 (1970).

64. P. J. Flory, *J. Chem. Phys.,* 9, 660 (1941).

65. P. J. Flory, *J. Chem. Phys.,* 10, 51 (1942).

66. M. I. Foreman and R. Foster, *Rec. Trav. Chim. Pays-Bas,* 89, 1149 (1970).

67. M. I. Foreman, R. Foster, and C. A. Fyfe, *J. Chem. Soc., B,* 1970, 528.

68. R. Foster, *J. Chem. Soc.,* 1960, 1075.

69. R. Foster, *Organic Charge Transfer Complexes,* Academic, New York, 1969.

70. R. Foster, Ed., *Molecular Complexes,* Vol. 1, Crane, Russak, New York, 1973.

71. R. Foster and C. A. Fyfe, *Trans. Faraday Soc.,* 61, 1626 (1965).

72. R. Foster and C. A. Fyfe, *Trans. Faraday Soc.,* 62, 1400 (1966).

73. R. Foster and D. L. Hammick, *J. Chem. Soc.,* 1954, 2685.

74. R. Foster, and J. W. Morris, *J. Chem. Soc., B,* 1970, 703.

75. R. Foster and J. W. Morris, unpublished work cited in Ref. 69, p. 200.

76. J. S Franzen and B. C. Franzen, *J. Phys. Chem.,* 68, 3898 (1964).

77. J. S. Franzen and R. E. Stephens, *Biochemistry,* 2, 1321 (1963).

78. J. W. Gibbs, *Collected Works,* Vol. 1, Longmans, Green, New York, 1928.

79. S. Glasstone, *Textbook of Physical Chemistry*, 2nd ed., Van Nostrand, New York, 1946, pp. 844-845.

80. M. Good, A. Major, J. Nag-Chaudhuri, and S. P. McGlynn, *J. Am. Chem. Soc.,* 83, 4329 (1961).

81. T. Gramstad, *Acta Chem. Scand.,* 15, 1337 (1961).

82. T. Gramstad, *Acta Chem. Scand.,* 16, 807 (1962).

83. T. Gramstad, *Spectrochim. Acta,* 19, 1363 (1963).

84. T. Gramstad and W. J. Fuglevik, *Acta Chem. Scand.,* 16, 1369 (1962).

85. T. Gramstad and Ø. Mundheim, *Spectrochim. Acta,* A28, 1405 (1972).

86. T. Gramstad and O. R. Simonsen, unpublished work cited in Ref. 85.

87. M. D. Gregory, S. D. Christian, and H. E. Affsprung, *J. Phys. Chem.,* 71, 2283 (1967).

88. J. Grundnes and S. D. Christian, *J. Am. Chem. Soc.,* 90, 2239 (1968).

89. J. Grundnes and S. D. Christian, *Acta Chem. Scand.,* 23, 3583 (1969).

90. J. Grundnes, S. D. Christian, V. Cheam, and S. B. Farnham, *J. Am. Chem. Soc.,* 93, 20 (1971).

91. J. Grundnes, M. Tamres, and S. N. Bhat, *J. Phys. Chem.,* 75, 3682 (1971).

92. E. A. Guggenheim, *Proc. Roy. Soc.,* A148, 304 (1935).

93. E. A. Guggenheim, *Trans. Faraday Soc.,* 56, 1159 (1960).

93.a. R. M. Guidry and R. S. Drago, *J. Phys. Chem.,* 78, 454 (1974).

94. L. P. Hammett, *Physical Organic Chemistry,* 2nd ed., McGraw-Hill, New York, 1970.

95. M. W. Hanna and D. G. Rose, *J. Am. Chem. Soc.,* 94, 2601 (1972).

96. E. L. Heric, *J. Phys. Chem.,* **73**, 3496 (1969).
97. W. S. Higazy and A. A. Taha, *J. Phys. Chem.,* **74**, 1982 (1970).
98. T. Higuchi, J. H. Richards, S. S. Davis, A. Kamada, J. P. Hou, M. Nakano, N. I. Nakano, and I. H. Pitman, *J. Pharm. Sci.,* **58**, 661 (1969).
99. J. H. Hildebrand and B. L. Glascock, *J. Am. Chem. Soc.,* **31**, 26 (1909).
100. J. H. Hildebrand, J. M. Prausnitz, and R. L. Scott, *Regular and Related Solutions,* Van Nostrand Reinhold, New York, 1970.
101. J. H. Hildebrand and R. L. Scott, *The Solubility of Nonelectrolytes,* 3rd ed., Reinhold, New York, 1950.
102. Ref. 101, pp. 175-197.
103. E. Hirano and K. Kozima, *Bull. Chem. Soc. Jap.,* **39**, 1216 (1966).
104. M. E. Hobbs and W. W. Bates, *J. Am. Chem. Soc.,* **74**, 746 (1952).
105. M. Horak and J. Pliva, *Spectrochim. Acta,* **21**, 911 (1965).
106. M. L. Huggins, *J. Chem. Phys.,* **9**, 440 (1941).
107. M. L. Huggins, *Ann. N.Y. Acad. Sci.,* **43**, 1 (1942).
108. T. F. Hunter, N. C. Cutress, K. M. Mitchinson, G. N. Rowley, and M. Stillman, *Spectrochim. Acta,* **A27**, 1207 (1971).
109. T. F. Hunter and D. H. Norfolk, *Spectrochim. Acta,* **A25**, 193 (1969).
110. P. V. Huong and J. Lascombe, *J. Chim. Phys.,* **63**, 892 (1966).
111. P. V. Huong, N. Platzer, and M. L. Josien, *J. Am. Chem. Soc.,* **91**, 3669 (1969).
112. P. Huyskens, T. Zeegers-Huyskens, and A. M. Dierckx, *Ann. Soc. Sci. Bruxelles, Ser. I,* **78**, 175 (1964).
113. J. I'Haya and T. Shibuya, *Bull. Chem. Soc. Jap.,* **38**, 1144 (1965).
114. I. Ilmet and S. A. Berger, *J. Phys. Chem.,* **71**, 1534 (1967).
115. I. Ilmet and P. M. Rashba, *J. Phys. Chem.,* **71**, 1140 (1967).
116. U. Jentschura and E. Lippert, *Ber. Bunsenges. Phys. Chem.,* **75**, 556 (1971).
117. U. Jentschura and E. Lippert, *Ber. Bunsenges. Phys. Chem.,* **75**, 782 (1971).
118. J. R. Johnson, P. J. Kilpatrick, S. D. Christian, and H. E. Affsprung, *J. Phys. Chem.,* **72**, 3223 (1968).
119. L. Joris, J. Mitsky, and R. W. Taft, *J. Am. Chem. Soc.,* **94**, 3438 (1972).
120. R. M. Keefer and L. J. Andrews, *J. Am. Chem. Soc.,* **75**, 3561 (1953).
121. J. Kleinberg and A. W. Davidson, *Chem. Rev.,* **42**, 601 (1948).
122. D. R. Knudsen, Ph.D. Dissertation, North Dakota State University, 1970.
123. S. Kobinata and S. Nagakura, *J. Am. Chem. Soc.,* **88**, 3905 (1966).
124. P. A. Kollman and L. C. Allen, *Chem. Rev.,* **72**, 283 (1972).
125. G. Kortüm and W. M. Vogel, *Z. Elektrochem.,* **59**, 16 (1955).
126. M. Kroll, *J. Am. Chem. Soc.,* **90**, 1097 (1968).
127. I. D. Kuntz, Jr., F. P. Gasparro, M. D. Johnson, Jr., and R. P. Taylor, *J. Am. Chem. Soc.,* **90**, 4778 (1968).
128. A. Lachman, *J. Am. Chem. Soc.,* **25**, 50 (1903).
129. J. D. Lambert, G. A. H. Roberts, J. S. Rowlinson, and V. J. Wilkinson, *Proc. Roy. Soc.,* **A196**, 113 (1949).
130. L. Lamberts, *Z. Phys. Chem.* (Frankfurt am Main), **73**, 159 (1970).
131. L. Lamberts and T. Zeegers-Huyskens, *J. Chim. Phys.,* **60**, 435 (1963).

131.a E. H. Lane, S. D. Christian, and J. D. Childs, *J. Am. Chem. Soc.*, **96**, 38 (1974).

132. R. P. Lang, *J. Phys. Chem.*, **72**, 2129 (1968).

133. L. A. LaPlanche, H. B. Thompson, and M. T. Rogers, *J. Phys. Chem.*, **69**, 1482 (1965).

134. E. N. Lassettre, *Chem. Rev.*, **20**, 259 (1937).

135. P. Laszlo, *Prog. Nucl. Magn. Resonance Spectrosc.*, **3**, 231 (1967).

136. P. Laszlo and J. L. Soong, Jr., *J. Chem. Phys.*, **47**, 4472 (1967).

137. W. J. McKinney and A. I. Popov, *J. Am. Chem. Soc.*, **91**, 5215 (1969).

137.a.J. A. Maguire, J. J. Banewicz, R. C. T. Hung, and K. L. Wright III, *Inorg. Chem.*, **11**, 3059 (1972).

138. D. E. Martire and P. Riedl, *J. Phys. Chem.*, **72**, 3478 (1968).

139. E. A. Mason and T. H. Spurling, *The Virial Equation of State*, Pergamon, Oxford, 1969.

140. D. L. Meen, F. Morris, and J. H. Purnell, *J. Chromatogr. Sci.*, **9**, 281 (1971).

141. R. E. Merrifield and W. D. Phillips, *J. Am. Chem. Soc.*, **80**, 2778 (1958).

142. J. Mitsky, L. Joris, and R. W. Taft, *J. Am. Chem. Soc.*, **94**, 3442 (1972).

143. E. A. Moelwyn-Hughes, *Physical Chemistry*, 2nd rev. ed., Pergamon, Oxford, 1965, pp. 1023-1025.

144. Ref. 143. pp. 1031-1032.

145. H. Møllendal, J. Grundnes, and E. Augdahl, *Acta Chem. Scand.*, **23**, 3525 (1969).

146. R. S. Mulliken, *J. Am. Chem. Soc.*, **74**, 811 (1952).

147. R. S. Mulliken and W. B. Person, *Molecular Complexes: A Lecture and Reprint Volume*, Wiley-Interscience, New York, 1969.

148. A. S. N. Murthy and C. N. R. Rao, in *Applied Spectroscopy Reviews*, Vol. 2, E. G. Brame, Jr., Ed., Dekker, New York, 1969, p. 69.

149. O. B. Nagy, J. B. Nagy, and A. Bruylants, *Bull. Soc. Chim. Belg.*, **83**, 163 (1974).

150. J. B. Nagy, A. Bruylants, and O. B. Nagy, *Tetrahedron Lett.*, **1969**, 4825.

151. M. Nakano, N. I. Nakano, and T. Higuchi, *J. Phys. Chem.*, **71**, 3954 (1967).

152. M. S. Nozari and R. S. Drago, *J. Am. Chem. Soc.*, **94**, 6877 (1972).

152.a.M. S. Nozari, C. D. Jensen, and R. S. Drago, *J. Am. Chem. Soc.*, **95**, 3162 (1973).

153. L. E. Orgel and R. S. Mulliken, *J. Am. Chem. Soc.*, **79**, 4839 (1957).

154. W. Partenheimer, T. D. Epley, and R. S. Drago, *J. Am. Chem. Soc.*, **90**, 3886 (1968).

155. W. B. Person, *J. Am. Chem. Soc.*, **87**, 167 (1965).

156. G. C. Pimentel and A. L. McClellan, *The Hydrogen Bond*, W. H. Freeman, San Francisco, 1960.

157. J. Prochorow and A. Tramer, *J. Chem. Phys.*, **44**, 4545 (1966).

158. J. E. Prue, *J. Chem. Soc.*, **1965**, 7534.

159. K. F. Purcell, J. A. Stikeleather, and S. D. Brunk, *J. Am. Chem. Soc.*, **91**, 4019 (1969).

160. J. H. Purnell, in *Gas Chromatography 1966*, A. B. Littlewood, Ed.,

Institute of Petroleum, London, 1967, p. 3.

160.a. J. H. Purnell and O. P. Srivastava, *Anal. Chem.*, **45**, 1111 (1973).

161. R. Queignec and B. Wojtkowiak, *Bull. Soc. Chim. Fr.*, **1970**, 860.

162. O. K. Rice, *Statistical Mechanics, Thermodynamics, and Kinetics*, W. H. Freeman, San Francisco, 1967.

163. O. K. Rice, *Int. J. Quantum Chem., Suppl.*, **2**, 219 (1968).

164. A. Rogstad and E. Augdahl, *Acta Chem. Scand.*, **25**, 225 (1971).

165. N. J. Rose and R. S. Drago, *J. Am. Chem. Soc.*, **81**, 6138 (1959).

166. H. M. Rosenberg, E. Eimutis, and D. Hale, *Can. J. Chem.*, **44**, 2405 (1966).

167. S. D. Ross, D. J. Kelley, and M. M. Labes, *J. Am. Chem. Soc.*, **78**, 3625 (1956).

168. D. R. Rosseinsky and H. Kellawi, *J. Chem. Soc., A*, **1969**, 1207.

169. J. S. Rowlinson, *Liquids and Liquid Mixtures*, 2nd ed., Butterworths, London, 1969.

170. J. Rubin and G. S. Panson, *J. Phys. Chem.*, **69**, 3089 (1965).

171. K. Schäfer and O. R. Foz Gazulla, *Z. Phys. Chem.* (Leipzig), **B52**, 299 (1942).

172. R. L. Scott, *Rec. Trav. Chim. Pays-Bas*, **75**, 787 (1956).

173. R. L. Scott, *J. Phys. Chem.*, **75**, 3843 (1971).

174. R. L. Scott and D. V. Fenby, *Ann. Rev. Phys. Chem.*, **20**, 111 (1969).

174.a. J. P. Sheridan, D. E. Martire, and Y. B. Tewari, *J. Am. Chem. Soc.*, **94**, 3294 (1972).

175. A. D. Sherry and K. F. Purcell, *J. Am. Chem. Soc.*, **92**, 6386 (1970).

176. A. D. Sherry and K. F. Purcell, *J. Am. Chem. Soc.*, **94**, 1853 (1972).

176.a. F. L. Slejko and R. S. Drago, *J. Am. Chem. Soc.*, **94**, 6546 (1972).

177. T. L. Stevens, Ph.D. Dissertation, University of Oklahoma, 1968.

178. R. W. Taft, D. Gurka, L. Joris, P. von R. Schleyer, and J. W. Rakshys, *J. Am. Chem. Soc.*, **91**, 4801 (1969).

179. F. Takahashi, W. J. Karoly, J. B. Greenshields, and N. C. Li, *Can. J. Chem.*, **45**, 2033 (1967).

180. M. Tamres, *J. Phys. Chem.*, **65**, 654 (1961).

181. M. Tamres, in *Molecular Complexes*, Vol. 1, R. Foster, Ed., Crane, Russak, New York, 1973, p. 49.

182. M. Tamres and J. Yarwood, in *Spectroscopy and Structure of Molecular Complexes*, J. Yarwood, Ed., Plenum, New York, 1973, p. 217.

183. C. C. Thompson, Jr., and P. A. D. de Maine, *J. Am. Chem. Soc.*, **85**, 3096 (1963).

184. C. C. Thompson, Jr., and P. A. D. de Maine, *J. Phys. Chem.*, **69**, 2766 (1965).

185. J. Timmermans, *J. Chim. Phys.*, **19**, 169 (1921).

186. P. J. Trotter and M. W. Hanna, *J. Am. Chem. Soc.*, **88**, 3724 (1966).

187. E. E. Tucker, Ph.D. Dissertation, University of Oklahoma, 1969.

188. J. J. Van Laar, *Z. Phys. Chem.* (Leipzig), **72**, 723 (1910).

189. J. J. Van Laar, *Z. Phys. Chem.* (Leipzig), **83**, 599 (1913).

190. J. H. van't Hoff, *Vorlesungen über Theoretische und Physikalische Chemie*, Vol. I, F. Vieweg, Braunschweig, 1898.

191. S. N. Vinogradov and R. H. Linnell, *Hydrogen Bonding*, Van Nostrand Reinhold, New York, 1971.

192. G. C. Vogel and R. S. Drago, *J. Am. Chem. Soc.*, **92**, 5347 (1970).

193. R. F. Weimer and J. M. Prausnitz, *J. Chem. Phys.*, **42**, 3643 (1965).

194. R. L. Williams, *Ann. Rep. Chem. Soc.*, **58**, 34 (1961).

195. E. M. Woolley, J. G. Travers, B. P. Erno, and L. G. Hepler, *J. Phys. Chem.*, **75**, 3591 (1971).

195.a. J. Yarwood, Ed., *Spectroscopy and Structure of Molecular Complexes*, Plenum, New York, 1973.

196. T. Zeegers-Huyskens, *Bull. Soc. Chim. Belg.*, **69**, 282 (1960).

Chapter **VII**

THE SOLUBILITY OF GASES IN LIQUIDS

H. Lawrence Clever and Rubin Battino

---

1 INTRODUCTION

The solubility of gases in liquids is an area of active interest from both the theoretical and practical standpoints. Both the dilute solutions resulting from the low solubility of many gases in liquids, and the large variety of sizes, shapes, and polarities of gas molecules to act as "probes" have made the solubility of gases in liquids an excellent tool to investigate and test theories of liquid properties and liquid structure. A knowledge of the solubility of gases is of practical importance in various industrial processes, in the study of artificial atmospheres for divers and astronauts, in the interaction of gases with our environment (as in the biological oxygen demand in natural waters), in processes for saline water demineralization, and in the study of various biological fluids and tissues.

The literature on the solubility of gases in liquids has been reviewed several times in the past either directly [4, 79] or indirectly as it applies to salt effects [70b] and diffusion in liquids [56b].

Several recent books have valuable chapters on the solubility of gases in liquids [56a, 104]. There is a general chapter on experimental solubility techniques which contains some discussion of gas solubility techniques [76], and there is a chapter that reviews the diffusion and solubility of gases in polymers [86].

In addition to the gas solubility reviews cited above, there are useful general compilations of solubility data which contain gas solubility data [115, 121, 140].

This chapter is intended to be a guide for those interested in making experimental measurements of gas solubility in liquids. First, experimental methods useful in making reliable gas solubility measurements are reviewed;

second, the theoretical and empirical relations useful in either testing the reliability of experimental values or approximating gas solubility values in the absence of experimental data are presented.

The biochemist and clinical chemist have long used the constant-pressure manometry apparatus of Van Slyke and the constant-volume apparatus of Warburg to measure gas solubilities. The Van Slyke and Warburg apparatuses are satisfactory for many applications. In this chapter we plan to describe apparatuses especially designed to carry out the most reliable gas solubility measurements possible, as well as some apparatuses of lesser accuracy. Most of the apparatuses use constant-pressure manometry to measure the gas volume dissolved in a known volume of gas-free solvent. Other techniques discussed include a constant-volume (pressure-drop) apparatus; gas liquid chromatography, both as an analytical technique to measure the dissolved gas removed from a saturated solution and as a technique in which the chromatographic separation theory is used to obtain information on vapors dissolved in a solvent which is the supported liquid phase in the chromatograph; amperometric methods for oxygen and other gases; and chemical methods and mass spectrometry. It is not our intention to discuss gas analysis, although several of the methods are more appropriate for gas analysis than gas solubility determinations.

2 METHODS OF EXPRESSING GAS SOLUBILITY

The solubilities of gases in liquids have been expressed in terms of many different units, depending on the particular application. It would be useful if all investigators reported their data in one standard form, as well as in any other units that are specific to their application. To this end we recommend that the standard form be the mole fraction solubility at 1 atm partial pressure of gas at the temperature of measurement x_2. For those cases in which the solvent does not have an easily characterizable molecular weight (such as oils or biological fluids), the weight solubility C_w is to be recommended (see justification by Cook [32]). Since the latter unit is essentially a ratio of weights, it is easily determined and readily converted to other solubility units. To make the choice of units perfectly clear, each article should present a description of how the reported solubilities were calculated, along with a sample calculation. The principal methods of expressing gas solubility and their interconversions are described below.

The Bunsen Coefficient α

The Bunsen coefficient α is defined as the volume of gas reduced to STP (273.15 K and 760 torr) that is absorbed by a unit volume of solvent at the temperature of the measurement under a gas partial pressure of 760 torr. When the partial pressure of the gas in contact with the solvent differs from 760 torr, it is corrected to this pressure by Henry's law. (For most gases over the small pressure interval involved, Henry's law is a very good approximation.)

An illustration of how the Bunsen coefficient may be calculated is given by the equation

$$a = \left(V_g \; \frac{273.15}{T} \; \frac{P_g}{760} \right) \left(\frac{1}{V_s} \right) \left(\frac{760}{P_g} \right) \tag{7.1}$$

where P_g is the partial pressure (in torr) of the gas, T is the Kelvin temperature, V_g is the volume of gas absorbed (at T and at the total pressure of the measurement), and V_s is the volume of the absorbing solvent. Note that a is dimensionless. If the solvent has a nonnegligible vapor pressure, $P_g = P_T - P_s$, where P_T is the total pressure in the system, and P_s is the solvent vapor pressure. Equation (7.1) reduces to the simpler form

$$a = \frac{V_g}{V_s} \; \frac{273.15}{T} \tag{7.2}$$

The corrections to STP assume ideal-gas behavior. If other than the ideal-gas equation of state is used to correct to standard conditions, it is important to provide this information. For most gases near atmospheric pressure, corrections for nonideality are negligible, being less than 1%.

In the literature the Bunsen coefficient is sometimes called the absorption coefficient or the coefficient of absorption.

The Kuenen Coefficient S

The Kuenen coefficient S is closely related to the Bunsen coefficient and is defined as the volume of gas (in cubic centimeters) at a partial pressure of 760 torr reduced to 273.15 K and 760 torr dissolved by a quantity of solution containing 1 g of solvent. The Kuenen coefficient is thus proportional to gas molality.

The Ostwald Coefficient L

The Ostwald coefficient L is defined as the ratio of the volume of gas absorbed to the volume of the absorbing liquid, both measured at the same temperature, that is,

$$L = V_g/V_s \tag{7.3}$$

For the equilibrium:

$$\text{Gas(in liquid phase; } C_\ell) = \text{Gas(in gas phase; } C_g)$$

the Ostwald coefficient is $L = V_g/V_s = C_Q/C_g$, where C_Q and C_g are the concentrations of the gas in the liquid and gas phases, respectively. Thus the Ostwald coefficient is an equilibrium constant, and as such is independent of the partial pressure of the gas as long as ideality can be assumed. However, it is helpful to specify the temperature and total pressure of the measurements when reporting Ostwald coefficients.

The Absorption Coefficient β

If the total pressure is kept at 760 torr, the volume of gas absorbed (reduced to 273.15 K and 760 torr via the ideal-gas equation of state) per unit volume of liquid is the absorption coefficient β. Since a and β are very similar, it is important to specify the exact method of calculating the solubility.

The Mole Fraction x_2

The mole fraction solubility is given by

$$x_g = x_2 = n_g/(n_g + n_s) \qquad (7.4)$$

for two components, where n is the number of moles (an *amount* of substance). The difficulty with this equation is that gas solubility is directly proportional to the gas partial pressure via Henry's law. To be unambiguous then, both the partial pressure of the gas and the temperature of measurement must be specified to fix the mole fraction. The mole fraction solubility at any partial pressure of gas $x_2(P_g)$ is most conveniently calculated from the Ostwald coefficient as

$$x_2(P_g) = (RT/LP_g V_1^0 + 1)^{-1} \qquad (7.5)$$

where V_1^0 is the molar volume of the solvent at T. The recommended standard method for reporting gas solubilities is for a gas partial pressure of 760 torr. For some experimental techniques it may be convenient to calculate $x_2(P_g = 1$ atm) in ways other than indicated by (7.5).

The Henry's-Law Constant K_H, K_2, and K_c

The equation for a gas in equilibrium with a solution may be written as

$$\text{Gas(in liquid phase; } x_Q \text{ or } C_Q \text{ at } P_g) = \text{Gas(in gas phase; } P_g \text{ or } C_g) \quad (7.6)$$

The Henry's-law constants are dependent on the concentration variables used in the limiting law. Henry's law is strictly applicable only in the extrapolation to infinite dilution, that is,

$$\lim_{x_Q \to 0} \frac{P_g}{x_Q} \cong \lim_{x_Q \to 0} \frac{f_g}{x_Q} = K_H \qquad (7.7)$$

where f_g is the gas-phase fugacity. At about atmospheric pressure for most gases equating f_g with P_g involves negligible errors in gas solubility measurements. Over the concentration range (dilute solutions) where Henry's law is applicable, the following relations hold

$$K_H = P_g/x_\varrho(P_g) = P_g/x_2(P_g) = 1/x_2(P_g = 1) \tag{7.8}$$

Two other ways of representing Henry's law are

$$P_g = K_2 C_\varrho \tag{7.9}$$

and

$$C_g = K_c C_\varrho \tag{7.10}$$

From the last equation it is noted that $K_c = 1/L$. It is evident that the method of calculating Henry's-law constants and the gas partial pressure or concentration must be specified. The common practice of correcting experimentally determined gas solubilities to 760 torr involves negligible errors if the pressure range is reasonably small.

The Weight Solubility C_w

The weight solubility C_w is defined as the number of moles of gas, with $P_g = 1$ atm, per gram of solvent. This unit has the advantage of being a ratio of weights, thus making it easy to carry out conversions from one unit to another.

Other Gas Solubility Units

Depending on the application, units such as volume fraction, molarity, molality, and weight fraction have been used to express gas solubility. One special application of importance is the concentration of dissolved oxygen in natural waters. This concentration is normally expressed as milligrams of oxygen per cubic decimeter and is not usually a saturation solubility.

Some Additional Comments on Units

Table 7.1 is a presentation of the most commonly used interconversions. (See Ref. 4 for other interconversions.) Special symbols are defined in the footnotes.

For gas solubilities greater than about 0.01 mole fraction ($P_g = 1$ atm) there are several additional factors that must be taken into account to report gas solubilities unambiguously. Solution densities or the partial molar volumes of gases must be known. Corrections should be made for the possible nonideality of the gas phase or the nonapplicability of Henry's law. Some of these areas have been dealt with by Wilhelm and Battino [140] and Hayduk and Cheng [51]. The corrections and equations necessary for high-pressure gas solubility measurements have been covered by Prausnitz and co-workers [105, 116]. It may also be mentioned at this point that there is little distinction between high-pressure vapor-liquid equilibria and high-pressure gas solubilities.

Table 7.1 Formulas for the Interconversion of Gas Solubility Expressions[a]

$$L = aT/273.15$$
$$C_w = a/V_0\rho$$
$$K_H = \frac{17.033 \times 10^6 \rho_s}{aM_s} + 760$$
$$L - C_w V_t \rho$$
$$x_2(P_g = 1 \text{ atm}) = C_w M_s/(1+C_w M_s)$$
$$x_2(P_g) = [(RT/LP_g V_1^0) + 1]^{-1}$$

[a] V_0 = molal volume of the gas in cubic centimeters per mole at $0°C$; ρ = density of the solvent at the temperature of the measurement; M_s = molecular weight of solvent; ρ_s = density of solution; V_t = molal volume of the gas (in cubic centimeters per mole) at the temperature of the measurement.

3 SOME REPRESENTATIVE GAS SOLUBILITIES

The representative gas solubilities listed in Table 7.2 as mole fractions at $25°C$ and 1 atm partial pressure of gas were taken from the critical compilation of Wilhelm and Battino [140] for all solvents other than water, and from Miller and Hildebrand [87] for water.

4 GAS SOLUBILITY STANDARDS

The solubility of oxygen in water at $25°C$ has been more carefully measured than perhaps any other system. After a critical evaluation of the best measurements, Battino and Clever [4] recommended the following values as standard at $25°C$ and 1 atm:

$$a = 0.02847$$
$$L = 0.03108$$
$$x_2(P_g = 1 \text{ atm}) = 2.295 \times 10^{-5}$$
$$C_w = 1.274 \times 10^{-6}$$

The average deviation for seven recent measurements was 0.21% which was just about within the independent experimental error for each worker. Since

Table 7.2 Gas Solubilities as Mole Fraction at 25°C and 1 atm Partial Pressure of Gas[a]

Gas Solvent	He	Ne	Ar	H_2	N_2	O_2	CO_2	CH_4	$\overline{V}^0$ (cm³ mol⁻¹)
Water	0.068	0.082	0.254	0.142	0.119	0.231	—	0.248	18.07
n-Hexane	2.604	3.699	25.12	6.315	14.02	19.3	—	50.37	131.62
n-Octane	2.397	3.626	24.26	6.845	13.04	20.83	—	29.27	163.54
n-Tetradecane	2.249	3.340	26.50	—	—	—	—	—	261.31
Isooctane	3.083	4.593	29.21	7.832	15.39	28.14	138.7	29.66	166.10
Cyclohexane	1.217	1.792	14.80	4.142	7.61	12.48	76.0	32.75	108.75
Benzene	0.771	1.118	8.815	2.580	4.461	8.165	97.30	20.77	89.41
Toluene	0.974	1.402	10.86	3.171	5.74	9.09	101.3	24.14	106.86
m-Xylene	1.121	1.619	—	4.153	—	—	—	—	123.47
n-Perfluoroheptane	8.862	—	53.22	14.03	38.80	55.08	208.2	82.56	227.33
Hexafluorobenzene	2.137	3.455	23.98	—	17.95	24.18	220	38.42	115.79
Carbon tetrachloride	—	—	13.51	3.349	6.480	12.01	105.3	28.70	97.09
Chlorobenzene	0.691	0.979	8.609	2.609	4.377	7.910	98.06	20.47	102.27
Nitrobenzene	0.350	0.436	4.448	—	—	4.95	99.80	—	102.72
Methanol	0.595	0.814	4.491	—	2.747	4.147	55.78	8.695	40.73
Ethanol	0.769	1.081	6.231	2.067	3.593	5.841	63.66	12.80	58.68
Acetone	1.081	1.577	9.067	2.996	5.395	8.383	185.3	18.35	74.01
Dimethyl sulfoxide	0.284	0.368	1.54	0.761	0.833	1.57	90.8	3.86	71.37

[a] All mole fraction $\times 10^4$.

oxygen and water are readily purified, this system is easily checked. We recommend that workers in the field periodically check their apparatuses against this standard.

Battino and Clever [4] also tabulated the solubilities of nitrogen, oxygen, and argon in water over the range 0 to 50°C at 1 atm. The values for helium and neon are taken from Weiss [134]. Least-squares smoothing of these data gives the following temperature dependencies for the Bunsen coefficient and $\ln x_2 (P_g = 1$ atm):

$$a(N_2, 1.6\%) = 2.305 \times 10^{-2} - 4.5611 \times 10^{-4}t + 4.4210 \times 10^{-6}t^2$$
$$a(O_2, 0.8\%) = 4.8813 \times 10^{-2} - 1.2733 \times 10^{-3}t + 2.2238 \times 10^{-5}t^2$$
$$1.5957 \times 10^{-7}t^3$$
$$a(Ar, 4.4\%) = 5.3454 \times 10^{-2} - 1.1926 \times 10^{-3}t + 1.2695 \times 10^{-5}t^2$$
$$a(He, 0.3\%) = 9.4006 \times 10^{-3} - 4.9685 \times 10^{-5}t + 8.0182 \times 10^{-7}t^2$$
$$a(Ne, 0.4\%) = 1.2414 \times 10^{-2} - 1.2955 \times 10^{-4}t + 1.4457 \times 10^{-6}t^2$$
$$\ln x_2(N_2, 0.3\%) = 13.1500 - 4.2966 \ln T$$
$$\ln x_2(O_2, 0.3\%) = 17.6879 - 4.9742 \ln T$$
$$\ln x_2(Ar, 0.6\%) = 15.8548 - 4.6326 \ln T$$
$$\ln x_2(He, 0.1\%) = -8.7583 - 0.5443 \ln T$$
$$\ln x_2(Ne, 0.2\%) = -9.8730 - 1.8805 \ln T$$

The percentages indicate the average percentage deviations from the fitted curves. In each case the least number of coefficients necessary to fit the data adequately was chosen.

5 FACTORS AFFECTING THE PRECISION OF GAS SOLUBILITY MEASUREMENTS

The factors affecting the precision and accuracy of gas solubility measurements are: purity of materials, temperature measurement and control, pressure measurement and control, weighings, volume determinations, attainment of equilibrium (i.e., saturation), incomplete degassing of the solvent, ascertaining the true amount of gas dissolved, and devising procedures that avoid contamination of solvent or gas in any required transferring procedure. Cook [32] and Cook and Hanson [33] have provided one of the better analyses of these contributing factors (their analysis has been summarized [4]). The following discussion is directed toward the attainment of a precision of ±1% in gas solubility measurements. Although there have been a few notable examples of higher precision reported in the literature, a precision of 1% appears to be adequate for most practical and theoretical applications. Accuracy may be checked against the standards discussed in the previous sections.

The purity of the solvent is of relatively minor importance, 99 mol % purity being more than adequate, since impurities tend to be of the same molecular nature (size, shape, and polarity) as the solvent. This minimizes the effect of differences in solubility between the major component and the impurities. The same criteria essentially hold for gas purity, 99 mol % being generally adequate.

The measurement of pressure via manometers and barometers is one of the most accurate parts of gas solubility measurements, since the pressure can be readily determined to ±1 torr with rough equipment and ±0.1 torr with good but inexpensive equipment. Pressure control is another matter, and its importance depends on the apparatus and technique used. In some procedures pressure control is critical.

Cook [32] has provided a full analysis of the effect of temperature on gas solubility measurements. There are four factors to consider: (1) the temperature coefficient of the solvent vapor pressure; (2) the temperature coefficient of solubility, that is, the change in the equilibrium partial pressure of the dissolved gas with temperature at an approximately constant concentration; (3) the temperature level of the experiment; and (4) the pressure level of the experiment. The contributions these factors make obviously depend on the particular system under investigation and on the type of apparatus used. Cook found that for an overall precision of 0.05% for his apparatus for the system hydrogen-n-heptane in the range of -30 to 50°C that temperature control to ±0.1°C was quite adequate. For most purposes, it appears that temperature control to ±0.1°C (fairly easy to achieve) provides a more than adequate margin of error.

Weighings and volume determinations are readily done to a few parts per thousand and appear to contribute negligibly to the final error. However, in some designs the limiting error may be in the reading of small volumes of gas or liquid. If the apparatus is not clean, droplets and nonreproducible draining may lead to significant errors. Parallax, lighting, and calibration of the volume components (including meniscus corrections where applicable) are important factors.

The wide divergence of reported values of gas solubility for some systems are in all likelihood due to failure to attain equilibrium or failure to completely degas the solvent. Several methods of degassing the solvent and ways to check on this are discussed in Section 6. The attainment of equilibrium conditions is of prime importance, and it is well to remember that in any approach to saturation (be it solid, liquid, or gaseous solute) the saturation condition appears to be reached asymptotically with time. In flow systems the attainment of equilibrium can be checked by determining the solubility over a range of flow rates. For nonflow systems the vigor and duration of stirring or shaking can be varied, and adjustment of the pressure above and below the "saturation" pressure may be tried. The testing of an apparatus should include sufficient varying of the operating parameters to test for the attainment of equilibrium.

Determining the true amount of gas dissolved has presented difficulties. Where feasible, it is better to determine the initial quantity of gas in the dry state, that is, free of solvent vapors. This permits the use of more accurate equations of state where required. It also avoids the difficulties that surround the question whether or not the saturation vapor pressures of the solvent exist in the gas phase.

Finally, equipment should be so designed that transfers of gas and liquid components are made free of possible contamination. The newer all-Teflon vacuum stopcocks are helpful in this regard.

6   DEGASSING TECHNIQUES

Adequate degassing of the solvent is necessary for almost all gas solubility determinations. In some gas-chromatographic techniques saturation and removal of dissolved gases is accomplished by a purging process. Several methods of degassing and criteria for determination of the extent of degassing are described below.

A solvent has been degassed when sufficient gas has been removed such that the outgassing of any residual gas will have no effect on the measurement. Thus, a residual gas pressure of ca. 100 microns should be sufficient for most procedures.

Several criteria have been used for checking on the completeness of degassing. In a "reasoning-by-analogy" procedure many workers have used the reproduction of literature data as a criterion for adequate degassing. If one of the standard values (Section 3) is used, this criterion is reasonable although the possibility of the fortuitous canceling of several errors remains. Another check is to have a liquid-nitrogen trap and a vacuum thermocouple gauge positioned between the vacuum pump and the degassing apparatus. When the stopcock to the pump is closed first, and that to the apparatus second, then the pressure is due to the noncondensible gases in the degassing apparatus. These values can be monitored until a suitable level has been reached. The degassed solvent may be cushioned between mercury. Since the gases to be removed (usually air, hydrogen, and oxygen) are noncondensible and dissolve rather slowly, any visible gas bubbles must be proportional to gas in the solvent. Also, since bubbles as small as 0.001 cm^3 can be detected, the test is quite sensitive.

Perhaps the method most frequently used to degas a solvent is to boil away 10 to 20% of the solvent under vacuum. This method is effective, but is wasteful of solvent, frequently takes over an hour, does not work well with relatively nonvolatile solvents, and can lead to violent bumping (as in the

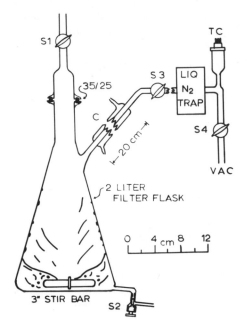

Fig. 7.1. Degassing apparatus made from a 3-ℓ filter flask [9]. (Reprinted by permission of the copyright owner, The American Chemical Society.)

particular case of water). Another method employs the procedure of pumping on the frozen solvent. Good results may be obtained from this procedure, but it is necessary to repeat the process at least three times (thawing then freezing), pumping on the frozen solvent for an hour or more in each cycle. This method is useful when it is necessary to minimize the loss of solvent. A somewhat similar technique [10] involves the vacuum condensation of a solvent onto a cold finger. This procedure is fairly slow (1 to 2 hr), and is readily adaptable only to relatively small quantities of solvent (40 cm^3).

The technique of spraying a solvent through a fine nozzle into an evacuated chamber has been used by several workers [27, 51]. Rapid and complete degassing is found by this technique, but it is accompanied by a considerable loss of solvent. Also, solvents like benzene frequently freeze in this process. A similar procedure was followed by Baldwin and Daniel [2], who let an oil sample drip slowly into an evacuated chamber, resulting in 97 to 98% degassing.

Battino and co-workers have described three degassing apparatuses [5, 8, 9]. The first [5] employs an all-glass pumping system to circulate the solvent by spraying it into an evacuated chamber. The second apparatus [8] consists of a horizontal cylinder about one-third full of solvent, which is agitated by Kel-F paddles. The large surface area leads to rapid degassing. The third apparatus [9] is shown in Fig. 7.1 and is the easiest one to fabricate and operate. In it, 500 cm^3 of olive oil, benzene, or water can be degassed down to the base pressure of the pump (5 to 10 microns) in 30 min or less with less than 4% loss of solvent. The apparatus is made from a 3-ℓ suction flask, has two all-Teflon stopcocks, one O-ring joint for introduction of the solvent and a 3-in stirring bar, and a condenser to minimize solvent loss. Vigorous stirring splashes solvent up on the walls of the flask, where it runs down in a film. The basic procedure is as follows. First the cold trap (liquid nitrogen) section is evacuated. Then this section is closed off from the pump and opened briefly (ca. 5 sec) to the degassing apparatus. The pressure is read. The process is repeated until the pressure in the trap section reaches the base pressure of the pump. This normally takes about 20 expansions, or about 30 min. This apparatus can be scaled up or down in size, and the procedure of multiple expansions can be adapted to other designs of the degassing vessel. One further advantage of this design is the anaerobic gravity feed linkage via stopcock *S2*.

Some techniques for stripping a dissolved gas from a liquid are discussed in Section 7 under gas chromatography.

7 GAS SOLUBILITY—EXPERIMENTAL METHODS

Earlier reviews [4, 79] described many different methods for determining the solubility of gases in liquids. Yet, the development of new approaches and equipment continues unabated. Some methods are simple, using commercially available glassware, and some are complex and costly, for example, the van Slyke method versus mass spectrometry. Most of the methods we describe are based on physical methods, although a few chemical procedures are also discussed. The analysis of dissolved gases is in general ignored. Also, the highly special area of gas solubilities in molten salts and alloys is not considered.

Physical methods may be divided into two broad classifications: saturation methods wherein a previously degassed solvent is saturated with a gas under conditions in which the necessary pressures, volumes, and temperatures may be determined; and extraction methods wherein the dissolved gas in a previously saturated solution is removed under conditions in which the pressure, volume, and temperature may be determined. Equilibrium saturation conditions have been attained for the gas and liquid phases by shaking a mixture of the two, by flowing a thin film of liquid through the gas, by bubbling (usually via a fritted disc) the gas through the liquid, or by flowing the gas over the liquid while it is held stationary on some supporting medium (as in gas-liquid partition chromatography). Determination of the amount of dissolved gas has been carried out by various physical and chemical methods.

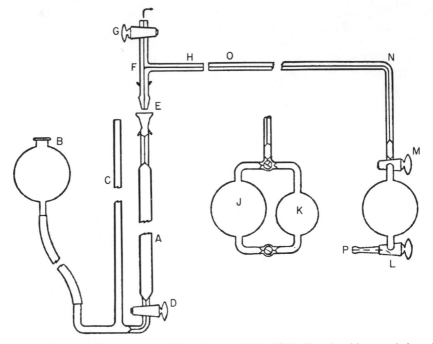

Fig. 7.2. Gas solubility apparatus of Markham and Kobe [80]. (Reprinted by permission of the copyright owner, The American Chemical Society.)

Volumetric—Atmospheric Pressure

Volumetric methods are among the most precise available—to ±0.1% in Cook and Hanson's work [33]; but they can also be rather crude. The volumetric method employs calibrated burets and bulbs in a system in which the pressure is reproducibly controlled or precisely measured. The apparatus, if small enough, is set in a water bath, or if large is placed in an air bath. Temperature control to ±0.1°C is easy for both baths, since on an air bath 10 cm of Styrofoam insulation provides thermal transfer conditions approaching that of a Dewar flask. In general, all the methods in this section are capable of precisions of ±1% or better for gas partial pressures near 1 atm.

Markham and Kobe Apparatus

Markham and Kobe [80] designed the Ostwald-type apparatus shown in Fig. 7.2. The absorption flasks consist of the two bulbs J and K. One bulb is about twice the volume of the other. Either bulb can be filled with degassed solvent by way of the three-way stopcock connecting the two flasks at the bottom. The buret system and second bulb are filled with solvent-saturated gas to atmospheric pressure, the buret levels read on a cathetometer, and the stop-

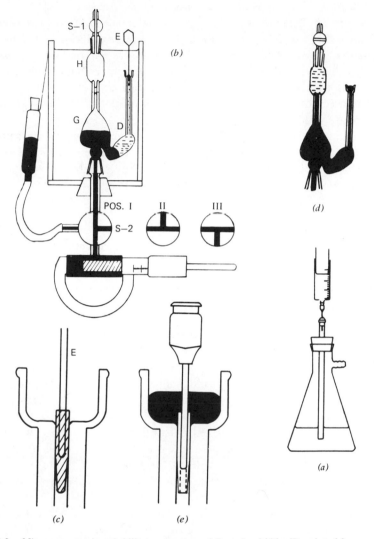

Fig. 7.3. Microgasometric solubility apparatus of Douglas [40]. (Reprinted by permission of the copyright owner, The American Chemical Society.)

cocks opened between the absorption flasks. The liquid-gas interface comes at about the point of maximum cross-sectional area of the two flasks. The flasks are oscillated through a $10°$ arc at a rate of 160 oscillations a minute. Solution equilibrium is complete in about 20 min. The leveling bulb is used to adjust continuously the gas phase to atmospheric pressure. The volume of gas absorbed is found by subtracting initial and final buret readings.

A modification of the apparatus has been used with success [28]. The burets were replaced by Truebore tubing, a closed reference pressure was used to prevent error due to changes in atmospheric pressure during the solubility determination, and a vertical arrangement of absorption flasks was used. In the vertical absorption flask arrangement, only the upper bulb could be used to hold the solvent, but the solvent dropped into a large bulb, which allowed vigorous agitation of solvent without streaming through the small-bore stopcock and prevented splashing into the upper connecting capillary tubing.

The solubilities of carbon dioxide and nitrous oxide in aqueous salt solution were reproducible to 0.2% by the apparatus [80]. However, the same absolute error in Bunsen coefficient results in a reproducibility of only 3% for a gas with a Bunsen coefficient of the magnitude of that of argon in water [28].

Microgasometric Methods

The microgasometric method originally developed by Scholander [114] uses about 4 cm^3 of sample, has an estimated precision of ±0.25 to 0.50%, and is rapid in use (equilibration times of ca. 30 min). The method has been used primarily for solubilities in pure water and aqueous solutions [40, 114, 120, 134]. Scholander's article [114] should be read for details on microgasometric analysis, although Douglas' [40] modifications are adequately described below. For very low solubilities, such as helium and neon in water and sea water, Weiss [134] increased the solvent volume to 10 cm^3. A version of this apparatus is available commercially [97].

Douglas' procedure can be followed by reference to Fig. 7.3. With H filled with water and G (5-cm^3 capacity) and D filled with mercury, degassed water is introduced into the side-arm reservoir D via a 10-cm^3 syringe and plastic-tubing-tipped cannula (Fig. 7.3e). The plug E is replaced. With G filled with mercury to the top of the capillary and H half filled with water, a conventional gas-transfer pipet filled with the pure gas is seated on the capillary and the gas meniscus pulled down to the mark on the capillary. By using the leveling bulb the micrometer is set to a zero reading, and some of the gas is pulled (using the micrometer) into the absorption chamber G. The water in H is removed, save for an indicator drop in the capillary below H. When the system has equilibrated, the gas volume V_1 is read. The degassed water is tipped into G from D, mechanical shaking begun, and the indicator drop kept at the mark by adjustments of the micrometer. When equilibrium is attained, the final gas volume V_2 is read. The liquid volume V_4 is determined by opening stopcock $S1$ and adjusting the micrometer until the liquid level reaches the mark on the capillary. The Ostwald coefficient is then simply given by $L = (V_1 - V_2)/V_4$.

If required, corrections may be made for the volume change on dissolution of the gas in the liquid. Weiss [134] estimates that for helium and neon in water this correction increases the solubility by about 0.13%.

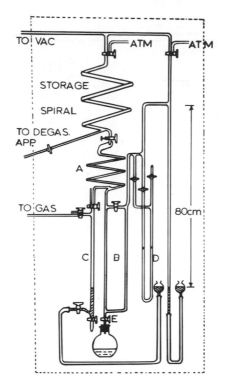

Fig. 7.4. The Morrison and Billett [89] apparatus as modified by Battino, Evans, and Danforth [6]. (Reprinted by permission of the *Journal of the American Oil Chemists Society*).

The Morrison and Billett Design

The Morrison and Billett [89] apparatus has been the basis for several modifications. One of these [6] is shown in Fig. 7.4. The entire apparatus is housed in an air thermostat as indicated by the dashed lines. Battino, Evans, and Danforth [6] replaced the storage reservoir and the absorption spiral by the more easily fabricated flat zigzag pattern of tubing. Degassed solvent [9] is introduced via gravity flow into the previously evacuated storage "spiral." Dry gas is introduced into B, C, and D. Some solvent drips through A to saturate the gas and fill the lower part of the two 10-cm³ microburets B. The stopcock connecting the two burets at the top is closed. Solvent is then dripped slowly into the absorption spiral. For low solubilities excess solvent is drawn off through E. For high solubilities extra "wet" gas is added from the 50-cm³ buret C. Ostwald coefficients from 20 to 0.01 can be measured on this apparatus to a precision ranging from ±0.25 to 1.0%. Drawbacks

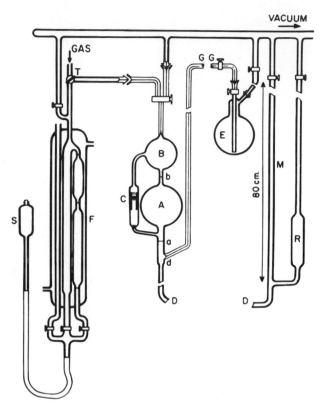

Fig. 7.5. The Dymond and Hildebrand [41] solubility apparatus. (Reprinted by permission of the copyright owner, The American Chemical Society.)

are the large volume of solvent (ca. 500 cm³) used and long times for measurements (up to 8 hr).

The apparatus described by Hayduk, Walter, and Simpson [52] is similar in design, but measurements take only 2 to 4 hr. High- and moderate-solubility versions are described, and the apparatus has a precision of about ±1%.

Apparatus of Dymond and Hildebrand

In the apparatus of Dymond and Hildebrand [41], equilibrium is attained by using an all-glass pumping system to spray slugs of solvent into the gas, causing contact as the solvent runs down the container walls in a thin film (see Fig. 7.5). The circulation of solvent through the gas speeds attainment of equilibrium. The apparatus has a precision of about ±0.5% and is readily adaptable to high and low solubilities, although the precision falls off as the solubility becomes smaller. This is due to the extreme sensitivity in these cases to the final pressure reading, since the quantity of gas absorbed is determined by gas buret and pressure readings. One special advantage of this apparatus is its ability to measure gas solubility at several temperatures on one loading. Over a moderate range Henry's law may also be checked.

Cukor and Prausnitz [36] have described a modification of the Dymond and Hildebrand apparatus [41] which has an operating temperature range of 25 to 200°C. The apparatus operates at gas pressures of 1 atm and can yield solubility data accurate to about 1%. This is an important design to consider, since most gas solubility apparatuses operate over rather restricted temperature ranges.

Cook and Hanson Apparatus

The Cook and Hanson apparatus [32, 33] is probably the most accurate and precise (reproducible to 0.1%) built to date. Unfortunately, perhaps because of its apparent complexity, Cook and Hanson have been the only workers (to our knowledge) to use this design. One difficulty is that the volume of the solvent and gas bulbs must be chosen such that 95% of the gas charged to the system is dissolved. A threefold range of the pressure (0.5-1.5 atm) partially compensates for this, but the apparatus is limited by the small range of solubilities it can accommodate for one set of gas and solvent bulbs. The apparatus has been described in great detail [32, 33], and workers in the field are urged to study Cook and Hanson's careful analysis of factors relating to gas solubility measurements. A shorter description is also available [4].

Modified Physical Adsorption Apparatus

A novel apparatus has just been described by Novak and Conway [92], which eliminates the need for complete degassing of the liquid phase, although the only gas present should be the solute gas. The method is based on the principle employed in volumetric procedures commonly used for adsorption studies on large-area porous materials in which the apparent PV relation for a given initial mass of gas is determined as a function of pressure.

If n_g is the number of moles of solute gas in the gas phase at partial pressure P_g, the following relations apply:

$$P_g V = n_g RT \tag{7.11}$$

$$n_T - n_s = n_g \tag{7.12}$$

$$n_s = K_H{}^1 P_g \tag{7.13}$$

where n_T is total moles of solute gas in the system, n_s is moles of gas in the liquid phase, and $K_H{}^1$ is the Henry's-law constant for the actual volume of solution in the system.

Appropriate substitutions in the above equations give

$$P_g V = (n_T - K_H{}^1 P_g) RT \tag{7.14}$$

The $P_g V$ product is determined as a function of P_g at about 8 or 10 pressures between 660 and 860 torr. A plot of the PV_g product against P_g is linear with slope equal to $K_H^1 RT$.

The sensitivity of the method depends on the relative magnitude of n_T and n_s. If n_T is too large, $P_g V$ will vary little with P_g. This means that the dead volume of the apparatus must be comparable to the volume of gas dissolved in the solution vessel. The apparatus should be designed with this in mind.

So far only the solubility of argon in water has been determined by this method. The argon solubility found was about 4% lower than the average value from other more conventional techniques. However, the method has several attractive features and deserves further testing.

High-Pressure Gas Solubility Apparatus

Determination of the solubility of gases in liquids at high pressure has become of increasing importance. The problems of adequate mixing of the gas and liquid phases to ensure saturation, pressure and temperature control, and sampling and measurement of the gas dissolved at high pressure present greater difficulties than in apparatuses that operate at atmospheric pressure.

The early work of Wiebe, Gaddy, and co-workers [137, 138] was successful in solving these problems. First attempts to saturate the liquid phase with gas at high pressure by stirring was slow and subject to mechanical failure. Wiebe and Gaddy solved the stirring problem by bubbling the compressed gas continuously through the liquid, allowing the excess gas to escape to the atmosphere. This is still a good stirring technique for inexpensive, nontoxic, nonflammable gases. The gas-saturated liquid was sampled and analyzed by removing a portion to an evacuated mercury-enclosed buret system. The gas was removed from the liquid at low pressure, and both liquid and gas volumes measured in the buret system at controlled temperature and pressure.

N. O. Smith and co-workers [48, 49, 96] have modernized the Wiebe and Gaddy apparatus by taking advantage of technological developments of the past 40 years. Notable improvements include use of (1) a magnetically driven, flat-bladed, turbine stirrer which creates a partial vacuum in the liquid, which draws gas into the liquid through the hollow turbine shaft and disperses the gas into the liquid by the turbine blades; (2) modern autoclave design and temperature control methods; and (3) a meniscus volume correction for the liquid in the buret measurement.

The autoclave, and the tubing and valve system connecting the compressor and compression cylinders to the autoclave, are shown in Fig. 7.6. The autoclave is a 1-gal, 316 stainless-steel container with a modified Bridgman (packless) closure. The tubing is 0.125-in.-i.d. 316 stainless steel. The autoclave can be either evacuated to about 0.03 atm or filled to a gas pressure of about 600 atm.

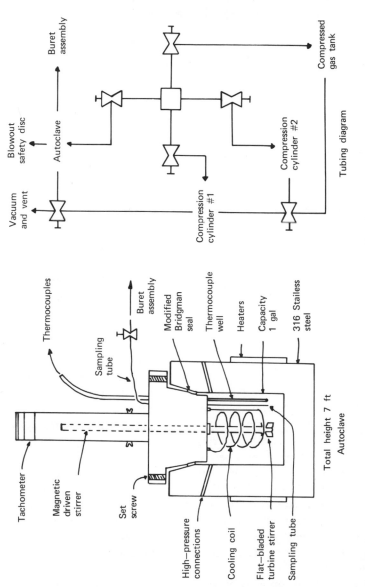

Fig. 7.6. The high pressure gas solubility apparatus of Smith and Gardiner [48, 49]. (Reprinted by permission of the authors.)

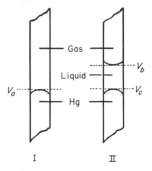

Fig. 7.7. The liquid volume meniscus correction [49]. (Reprinted by permission of the copyright owner, The American Chemical Society.)

The autoclave pressure is measured with two precision Heise Bourdon gauges with a range of 0 to 15,000 psi. The gauges are calibrated frequently against an Ashcroft deadweight tester. The maximum possible error in pressure is estimated to be 0.5 atm at 100 atm and 1 atm at 600 atm.

The autoclave temperature is controlled to $\pm 0.5°C$ between temperatures of 25 and 125°C.

To make a gas solubility measurement, the Bridgman closure is opened and the solvent is introduced directly into the autoclave. The solvent is degassed by displacement of air by the solute gas. The autoclave is sealed and charged to approximately 1000 psi with the solute gas; the gas is equilibrated with the liquid phase by stirring and then vented. The procedure is repeated, and after the second venting it is estimated that the residual atmospheric gases in the autoclave make up less than 1 ppm of the total. The autoclave is then filled with solute gas to the pressure of the solubility measurement.

The gas is allowed to equilibrate with the liquid for 8 hr or longer. Tests made approaching equilibrium from both higher and lower pressures agree within experimental error after 8 hr. Stirring with the magnetically driven, flat-bladed, turbine stirrer is stopped for a period of 15 to 30 min before sampling to allow gas bubbles to escape from the liquid phase. A 15 to 30 cm^3 sample of gas-saturated liquid is passed from the autoclave into the evacuated (0.01 torr) buret system at a rate of 0.5 to 1 cm^3 min^{-1}. The autoclave pressure is maintained at a constant value during the transfer of sample. After transfer is complete, the liquid is degassed by manipulation of the buret system. The gas and liquid volumes are measured in the buret system at atmospheric pressure.

It is important to apply a meniscus correction to the measurement of the liquid volume in the buret. There is extra liquid in both the mercury-liquid meniscus and the liquid-gas meniscus (Fig. 7.7) which should be added to the liquid volume read from the buret. In a standard, 50 cm^3, 1.18-cm-diameter buret, the correction to be added to the volume was found to be 0.36 ± 0.04 cm^3 by adding a known volume of water to a buret partially filled with mercury. Wiebe and Gaddy did not use such a correction. In the apparatus described here, neglect of the meniscus correction causes the solubility to be high by about 1.5% at 100 atm and 4% at 600 atm.

Other methods for high-pressure determination of solubility of gases in liquids involve the use of rocking autoclaves for the solubility of noble gases in salt solutions [106], the solubility of carbon monoxide and hydrogen in water up to 100 atm and 300°C [61], and the solubility of carbon dioxide in water [85, 122] and in heavy water [64] to 40 atm.

Fleck and Prausnitz [47] designed and built an apparatus for studying three-phase liquid-liquid-gas equilibria up to 1000 psia. Design criteria for the apparatus include small liquid volumes and minimal dead space, magnetic pumping of all three phases with sampling from flowing streams and analysis of all three phases by gas chromatography, good temperature control, and operating pressure up to 1000 psia.

Pressure-Drop Method

The pressure-drop method uses an apparatus in which the gas absorbed by a given volume of solvent is measured by determining the pressure change in a gas reservoir of known volume.

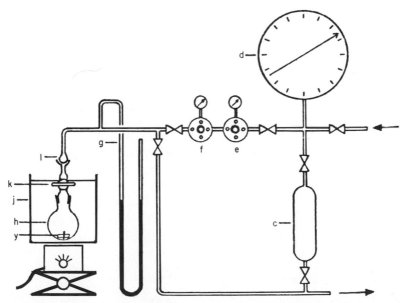

Fig. 7.8. The pressure-drop apparatus of Walters, Mortimer, and Clements [133]. (Re printed by permission of the copyright owner, The American Chemical Society.)

Apparatuses have been designed for use in all pressure ranges. This method has been used to determine the solubility of hydrocarbons in water between 0.5 and 1.5 atm [65], of hydrogen, ethylene, ethane, and propane in toluene at pressures up to 12 atm [133], and of methane in *n*-decane at pressures up to 1000 psia [63]. The error in the method has been estimated to be about ±2%, except for hydrogen solubilities which have an uncertainty of 5%. A maximum error of 5% and a probable error of 1.5% were estimated for the methane-decane system [63].

The nonideal character of the gas should be known for the pressure interval used in the apparatus. For hydrogen gas a wall adsorption correction is required to obtain satisfactory solubility data.

The apparatus of Waters, Mortimer, and Clements [133] is shown in Fig. 7.8. The apparatus components are a steel bomb c, a Heise gauge d, two regulators e and f, a manometer g, and a 500-ml flask h fitted with a stirring bar and immersed in a thermostat. A known volume of solvent is degassed at $-78°C$ and then brought to the temperature of the solubility measurement. The bomb and gauge are filled with gas to 12 atm pressure, and regulator e is adjusted to give a downstream pressure a little above the pressure desired for the solubility measurement. Regulator f is then adjusted to fill the apparatus as far as stop-cock k to the desired pressure, taking into account the solvent vapor pressure. Stopcock k is opened, and the pressure drop recorded. The solubility of the gas in the liquid is calculated in moles per liter using the equation

$$K(M) = \left[\frac{\rho_t}{V_s\rho_r} \frac{273.15V_1(P_0f_0 - P_1f_1)}{22.414(273.15 + t_g)} \right]$$
$$- \left[\left(V_2 - \frac{V_s\rho_r}{\rho_t} \right) \frac{P_2}{22.414} \frac{273.15}{273.15 + t_b} \right] \quad (7.15)$$

where ρ_t is density of solvent at thermostat temperature, ρ_r is density of solvent at room temperature, P_0 is initial gauge reading (in atmospheres), P_1 is final gauge reading (in atmospheres). P_2 is partial pressure of the gas at the thermostat temperature (in atmospheres), f_0 and f_1 are compressibilities of the gas at pressures P_0 and P_1, respectively, t_g is temperature of the gauge and band (in degrees Celsius), t_b is thermostat temperature (in degrees Celsius), V_1 is volume of gas reservoir (in liters), V_2 is volume of adsorption vessel (in liters), and V_s is volume of solvent at room temperature (in liters).

Gas Chromatography

Gas chromatography is applied to the determination of gas solubility in liquids in two ways. First, it is used to determine the amount of gas dissolved in a liquid by passing the carrier gas successively through a known volume of gas-saturated liquid to purge the liquid of the dissolved gas, through a preabsorbing column or columns to dry and sometimes concentrate the gas, and finally through the chromatograph to measure quantitatively the amount of solute gas removed from the gas-saturated solvent. Second, the theory of gas-liquid chromatography has been applied to the behavior of low concentrations of various vapors as the solute gas, in the column-supported liquid as the solvent, to obtain the vapor activity coefficient in the solvent at infinite dilution by elution chromatography and at higher concentrations by frontal analysis.

The first approach was used by McAuliffe [73b] to determine the solubility of 65 gaseous and liquid hydrocarbons in water at room temperature. Since then gas chromatography has been used in the determination of the solubility of nitrous oxide in water [16], C_1 to C_3 hydrocarbons in dodecane, octadecane, and eicosane [91], and methane-ethane-propane mixtures in water [1]. Swinnerton, Linnenbom and Cheek have developed gas chromatographic methods for the determination of dissolved oxygen, nitrogen, and argon [70, 123], hydrocarbons [124], and carbon monoxide [125] in sea water.

In addition gas chromatography has been used in the determination of liquid hydrocarbon solubilities in water and deuterium oxide [101], and of solid hydrocarbons, carbon dioxide, and sulfur dioxide in liquids at cryogenic temperatures [24, 107].

There is evidence that some [73b] gas-chromatographic solubility measurements are low [65, 101]. It is suggested that the reason for the low values may be the use of a preabsorbing column to remove water prior to the hydrocarbon entering the chromatographic column [101]. The absorption effect may be a function of sample size.

The sample size for the gas-chromatographic technique varies over a wide range. Gas-saturated liquid samples of from 1 to 50 $\mu\ell$ are often directly injected into the chromatograph or onto a suitable precolumn. The methods developed for gas analysis of sea water use samples of up to 1 ℓ. The 1-ℓ samples are purged of dissolved gas by the carrier gas, and the solute gas absorbed on a precolumn for later introduction into the chromatograph.

Flame ionization detectors are used for hydrocarbon solutes, and thermal conductivity detectors are used for oxygen, nitrogen, and argon. Carbon monoxide is first catalytically converted to methane which is measured by a flame ionization detector.

Both gas [16, 91] and liquid [101] hydrocarbon solubilities in water could not be reproduced to better than ±3% by gas-chromatographic methods. This is of marginal accuracy for studies that use the temperature derivative of solubility to obtain enthalpy and heat capacity change values for the solution process. However, a statistical study of the procedures worked out for oxygen and nitrogen analysis [70a] had a relative standard deviation of 0.48% for oxygen and a relative standard deviation of 1.23 and 1.42% for oxygen and

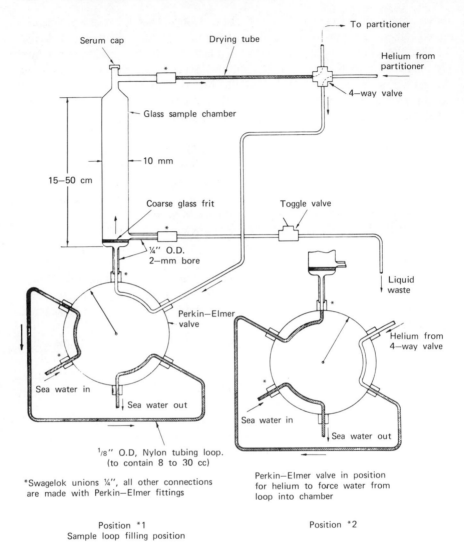

Position *1
Sample loop filling position

Position *2

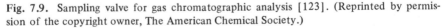

Fig. 7.9. Sampling valve for gas chromatographic analysis [123]. (Reprinted by permission of the copyright owner, The American Chemical Society.)

nitrogen, respectively, in analyzing the gas mixture concentrations in water.

Some of the features of the gas chromatographic techniques used by Swinnerton, Linnenbom, and Cheek are discussed below. Much of the success of their method for oxygen, nitrogen, and argon depends on the use of a six-way sampling valve and the gas stripping chamber shown in Fig. 7.9. The nylon tubing loop determines the sample size. In the apparatus shown various lengths

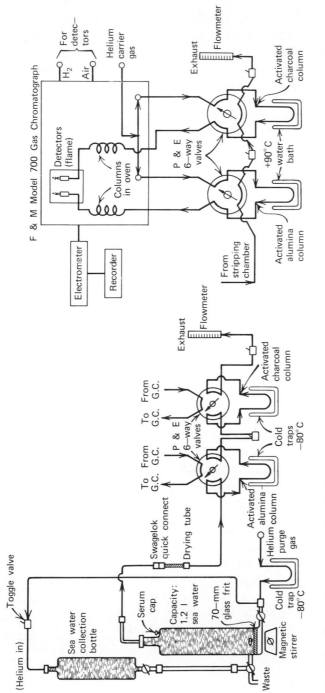

Fig. 7.10. The gas chromatographic analysis apparatus for hydrocarbon mixtures of Swinnerton and Linnenbom [124]. (Reprinted by permission of the *Journal of Gas Chromatography*, Preston Abstracting Co., Evanston, Ill.)

of tubing were used to vary the sample size from 3 to 30 ml. When the six-way valve is turned, the sample in the loop is pushed into the gas stripping chamber and bubbled with helium gas, which reduces the solute gas pressure and effectively and quickly removes the dissolved gas. A trace of wetting agent improves the efficiency of the gas stripping step, apparently because the reduced surface tension results in smaller bubbles of the stripping gas. Smaller samples of gas for calibration runs can be added directly through the rubber serum cap on the stripping chamber.

Oxygen, nitrogen, and argon [123]. Linde 13X molecular sieves are used to separate the nitrogen from the oxygen and argon. In air-saturated solutions about 5% of the oxygen peak is due to argon, and the argon can be corrected for without appreciable error. In biological fluids in which the oxygen/argon ratio may not be constant, both oxygen and argon may be determined by using a hydrogen carrier gas and passing the sample over a palladium catalyst to convert oxygen to water.

Hydrocarbons [124]. A block diagram of the apparatus used to analyze hydrocarbon mixtures is shown in Fig. 7.10. The solute gases stripped from up to a 1-ℓ sample are absorbed on two precolumns at low temperature. The first column, activated aluminum oxide, absorbs all the hydrocarbons except methane. The second column, activated charcoal, absorbs the methane. The C_2 to C_4 hydrocarbons are desorbed from the precolumn at 90°C and separated on an activated aluminum oxide plus 10% paraffin oil column. The methane is desorbed and passed through a silica gel column. By proper choice of columns, propane-propylene mixtures can be separated, and higher-molecular-weight hydrocarbons including *n*-octane and benzene can be separated. The method detects down to 1 part hydrocarbon in 10^{13} by weight.

Carbon monoxide. The carbon monoxide is absorbed on a precolumn of one-fourth charcoal and three-fourths molecular sieve. The carbon monoxide is desorbed, mixed with hydrogen, and passed over a nickel catalyst which converts the carbon monoxide to methane [102]. The methane is then detected. Ten tests on a synthetic sample gave results with a 1.3% standard deviation and 0.44% standard deviation of the mean [125].

The application of gas-chromatographic theory to obtain thermodynamic data has been under investigation for several years. The subject may appear peripheral to gas solubility, but the present theory allows the rapid determination of thermodynamic properties of dilute vapor solutes in the column-supported liquid as solvent. Several good discussions of the theory are available [29, 81]. The field is very active, and recent work is briefly summarized in the physical-analytical measurements section of the Cram and Juvet reviews [34]. A few references are given to illustrate recent trends in the field [23, 30, 37, 46, 126].

The expression

$$\gamma_2^\infty = \frac{RT}{Mf_2^0 V_g^0} = \frac{17.04 \times 10^6}{Mf_2^0 V_g^0} \qquad (7.16)$$

relates the activity coefficient at infinite dilution γ_2^∞ to the specific corrected retention volume of the solute in centimeters of carrier gas per gram of liquid V_g^0. M is the molecular weight of the liquid phase, f_2^0 is the fugacity of the pure saturated solute vapor in torr, and RT has units of cubic centimeter torr per mole.

The specific retention volume should be independent of all operational variables of the column except temperature. The equation is valid when (1) small sample sizes are used and the peak retention volumes are extrapolated to zero sample size; this ensures that the solute is in the Henry's-law region, and the extrapolation gives the solute property at infinite dilution in the liquid phase; (2) the column temperature is controlled to $\pm 0.2°C$ and the packing is uniform, which allows the retention volume to be compressibility-corrected to zero pressure drop in an isothermal column; (3) the solid support material is pretreated to remove active sites and covered with 15% by weight liquid to prevent interaction between solute and the solid support; (4) equilibrium is maintained between the vapor and the liquid at all points in the column; (5) it is assumed that there is no solute adsorption at the gas-liquid interface and that the liquid phase acts as a bulk partitioning liquid; and (6) the vapor phase is assumed ideal with no solute-solute, solute-carrier gas, or carrier gas-carrier gas interactions. Only if helium is the carrier gas, can one assume ideal behavior of the vapor phase. However, one can obtain the activity coefficient corrected for vapor-phase imperfections from

$$\ln \gamma_2^\infty = \frac{17.04 \times 10^6}{MP_2^0 V_g^0} + \frac{P}{RT}(2B_{12} - V_2^0) - \frac{P_2^0}{RT}(B_{22} - V_2^0) \quad (7.17)$$

where P is the average column pressure, P_2^0 is the vapor pressure of the pure saturated solute vapor in torr, V_2^0 is the molar volume of pure solute, B_{22} is the second virial coefficient of the solute vapor, and B_{12} is the second virial coefficient of interaction between solute and carrier gas. The value of B_{12} must often be estimated. It is of significance when the carrier gas is carbon monoxide, carbon dioxide, oxygen, or nitrogen. When the carrier gas is helium, (7.17) can be approximated as

$$\ln \gamma_2^\infty = \frac{17.04 \times 10^6}{MP_2^0 V_g^0} - \frac{P_2^0 B_{22}}{RT} \qquad (7.18)$$

The measurement of γ_2^∞ as a function of temperature allows the calculation of other thermodynamic properties. The excess partial molar free energy, enthalpy, and entropy of solution are given by the equations

$$\Delta \overline{G}^E = \Delta \overline{H}_M - T\Delta \overline{S}^E = RT \ln \gamma_2^\infty \tag{7.19}$$

$$\Delta \overline{H}_M{}^E = - \frac{\partial(\Delta \overline{G}^E/T)}{\partial(1/T)} = -R \frac{d \ln \gamma_2^\infty}{d (1/T)} \tag{7.20}$$

Recently, frontal analysis has been used to obtain activity coefficients at finite concentrations [23, 30].

Mass Spectrometry

The mass spectrometer may be used for gas solubility determinations and for solute gas ratios and isotope ratios in a solvent. The procedure is to outgas a gas-saturated solvent sample, trap the gas, and then analyze the gas by mass spectrometry.

Mass spectrometry has been used to determine the solubility of argon in liquid ammonia [35], methane, oxygen, and nitrogen in water [20], various gases in blood [103], helium in liquid fluorine [19], and solid carbon dioxide and hydrogen sulfide in hydrocarbons [24]. Mass spectrometry has also been used to determine nitrogen/argon, nitrogen/oxygen, and nitrogen isotope and oxygen isotope ratios in water and sea water [13, 14, 62].

The solubility of argon in liquid ammonia showed a mean deviation from the average of 0.88% [35]. A direct calibration of known amounts of helium gas in liquid fluorine showed an average deviation of 2.5% below 0.2% helium in fluorine, which increased to over 5% at 0.5% helium in fluorine [19]. A direct comparison of solid carbon dioxide solubility in liquid methane and ethane by mass spectrometry and gas chromatography had an average deviation of nearly 2%, with a maximum error of 9% for the lowest solubility measured [24]. For nitrogen/argon and nitrogen/oxygen ratios the precision is ±1%, and for oxygen isotope and nitrogen isotope ratios the precision is ±0.01% [14, 62].

Chemical Methods

Standard chemical analytical techniques can often be used to determine the concentration of a solute gas that has either acid, base, or redox properties. Basic gases include ammonia and methylamine, acidic gases include the hydrogen halides and carbon dioxide. Iodimetric methods have been used with sulfur dioxide, hydrogen sulfide, ozone, and chlorine. Both chlorine and hydrogen cyanide have been precipitated as silver salts. Phosgene has been determined by treatment with excess alkali and back-titration; acetylene has been determined by absorption in silver nitrate solution followed by back-titration of the acid liberated. Direct references to these chemical methods are in the Markham and Kobe review [79].

Probably the most used chemical method is the Winkler method for oxygen. The procedure requires a rigid technique to control the pH and iodide-ion concentration. In the Winkler method the dissolved oxygen oxidizes freshly precipitated manganous hydroxide to manganic hydroxide at high pH:

$$Mn^{2+} + 2OH^- = Mn(OH)_2$$

$$2Mn(OH)_2 + \tfrac{1}{2}O_2 + H_2O = 2Mn(OH)_3$$

The solution is made acidic, under which condition the manganic ion reacts with excess iodide ion. The resulting iodine (I_3^- ion) is titrated with thiosulfate:

$$2Mn(OH)_3 + 6H^+ + 3I^- = 2Mn^{2+} + I_3^- + 6H_2O$$

$$I_3^- + 2S_2O_3^{2-} = 3I^- + S_4O_6^{2-}$$

Possible errors in the Winkler method have been extensively discussed, and procedures presented which hopefully avoid these errors [22, 42, 88].

The Winkler method is often used to determine dissolved oxygen in natural waters. Most natural waters contain small amounts of nitrite ion which interfere with the determinations and result in a high value for dissolved oxygen. A sodium azide modification of the Winkler method eliminates interference from nitrite ion [119]. Standard methods [119] describe other modifications to eliminate interference when sulfite, thiosulfate, polythionate, and hypochlorite ions, chlorine, or iron(II) are present.

Commercially Available Gas Solubility Apparatuses

Biochemists, clinical chemists, and other researchers make use of commercially available versions of the Van Slyke [129], Van Slyke-Neill [130], and Warburg [132] apparatuses in gas solubility work. Many versions of these apparatuses for both macro- and microsize samples are available from laboratory equipment suppliers.

Van Slyke apparatuses extract the dissolved gas from the fluid phase at a low pressure and compress the extracted gas to a fixed pressure (Van Slyke) or to a fixed volume (Van Slyke-Neill). A Henry's-law correction for the gas unextracted at the low pressure must be applied. Extensive instructions and applications of the apparatus are available in standard clinical chemistry books [53, 98].

The Warburg apparatus can be used to measure the solubility of a gas in a liquid, but more often it is used in following the rate of either a process or a reaction that produces or consumes a gas. The Warburg method follows pressure changes for a constant volume of gas. An excellent monograph is available

[128c] which discusses in detail the use and applications of the Warburg apparatus.

A series of apparatuses to determine free, entrained, and dissolved air in hydraulic and other fluids is manufactured (Aire-Ometer, Seaton-Wilson, Inc., Burbank, Calif.). The apparatus extracts dissolved gas from a 1-ml fluid sample under vacuum, using mercury as the working medium, and compresses the extracted gas to atmospheric pressure. The principle of operation is similar to that of the Van Slyke method.

Many versions of an electrochemical device to determine dissolved oxygen are now available. The original amperometric technique [25, 26] used a gold-silver electrode system with the gold cathode separated from the sample by a Teflon membrane. The dissolved oxygen from the sample diffuses through the membrane to the cathode, where at a constant applied potential the current flowing through the cell is directly proportional to the amount of oxygen reaching and being reduced at the cathode.

Table 7.3 Characteristics of Some Commercially Available Oxygen Sensors

Cathode	Anode	Cathode Electrolyte Solution	Applied Voltage	Manufacturer
Platinum	Silver	KCl	0.8	Yellow Springs Instrument Co., Yellow Springs, Ohio
Gold	Silver	KCl	0.8	Beckman Instrument Co. Fullerton, Calif.
Platinum	Silver	KCl Phosphate buffer	0.8	Radiometer, Copenhagen
Silver	Lead	KOH	0	Precision Scientific Co., Chicago, Ill.
Platinum	Lead	KI	0	Weston and Stack, Inc. Malvern, Pa.[a]
Silver	Lead	Acetate buffer	0	New Brunswick Scientific Co. Inc., New Brunswick, N.J.
Thallium	–	–	–	Union Carbide, White Plains, N.Y.
Zirconium oxide cell	–	–	–	Westinghouse Electric Corp. Pittsburgh, Pa.

[a]This sensor is also used in the oxygen analyzer marketed by Leeds and Northrup, North Wales, Pa.

Several modifications of the cell have been proposed. When a silver-lead electrode system with potassium hydroxide as electrolyte [77] is used, the oxygen reduction is spontaneous and the current developed is proportional to the oxygen concentration.

The use of an acetate buffer as electrolyte with the silver-lead electrode system is recommended [60] when solutions with a high carbon dioxide content are involved.

The dissolved oxygen sensor response is highly temperature-dependent, but the temperature coefficient is internally compensated for in commercial apparatuses. The sensor response also depends on the nature and thickness of the membrane separating the sample from the cathode. A 10-mil Teflon membrane is commonly used. The device consumes oxygen; thus it is important to pass fresh sample over the membrane continuously to prevent oxygen depletion and to keep maximum response. Some of the sensors have small-area electrodes to minimize oxygen consumption. At least one of the oxygen sensors responds not only to dissolved oxygen but also to gaseous oxygen between 1 and 2000 torr. The oxygen sensor is not an absolute device, but requires frequent calibration against either a saturated solution of oxygen in water or against a chemically analyzed oxygen solution.

Some features and the sources of several manufactured oxygen sensors are given in Table 7.3.

At least two other electrochemical oxygen sensors have been patented and are marketed. One works on the principle that the corrosion potential generated between an exposed thallium metal electrode and a reference electrode is proportional to the dissolved oxygen concentration [21]. The electrode system is placed directly in the sample which must be free of thallium ions. No cathode membrane and no electrode electrolyte are required.

The second is essentially a high-temperature zirconium oxide-oxygen concentration cell for analyzing oxygen concentration in any noncombustible gas mixture [111]. The cell output is specific for oxygen and is not affected by carbon dioxide; however, combustible components burn at the operating temperature, resulting in an indication of less oxygen than is actually present.

The development of thin, semipermeable membranes of such materials as Teflon (Du Pont), dimethyl silicone, and silicone-polycarbonate copolymer (General Electric), and other polymers has been instrumental in the development of electrode systems for oxygen and other gases.

A combined glass-reference electrode system in a dilute bicarbonate solution responds to carbon dioxide which diffuses through a membrane causing a change in pH according to the P_{CO_2} of the sample. Combined electrode systems which respond to pH, P_{O_2} and P_{CO_2} are manufactured (Radiometer, Copenhagen; Instrumentation Laboratory, Inc., Lexington, Mass.). An oxygen-carbon monoxide analyzer is also available (Instrumentation Laboratory, Inc.).

A selective electrode for sulfur dioxide in either gas or solution, which does not respond to oxygen, carbon dioxide, carbon monoxide, nitrogen dioxide, ozone, or chlorine is available for 0 to 100 ppm concentrations (Ericsen Instruments, Ossining, N.Y.).

Miscellaneous

Some gas-liquid systems require the design of a special apparatus or allow the application of an unusual analytical technique. Some examples include the Monel metal apparatus for the solubility of fluorine, nitrogen trifluoride, and oxygen in anhydrous liquid hydrogen fluoride [109], and an apparatus for the determination of the solubility of helium in liquid fluorine [19]. There is a report of the solubility of acetylene in liquid oxygen determined by infrared spectroscopy [17].

8 TREATMENT OF DATA

In this section we discuss the effects of temperature and pressure on gas solubility, thermodynamic functions, methods of correlating data, and salt effects. The calculation of gas solubilities from experimental data was implicitly covered in Section 2.

Choice of Standard States and Calculation of the Thermodynamic Functions

The change in Gibbs energy $\Delta \overline{G}^{\circ}$, enthalpy $\Delta \overline{H}^{\circ}$, and entropy $\Delta \overline{S}^{\circ}$, are defined, respectively, for the process

$$M(\text{gas, hyp. 1 atm}) \rightarrow M(\text{solution, hyp. } x_2 = 1) \tag{7.21}$$

for 1 mol of M, where x_2 is the mole fraction of M in the solution. If we then assume ideal behavior of the gas and validity of the laws for dilute solutions, we obtain the approximate relation

$$\Delta \overline{G}^{\circ} = -RT \ln x_2(\text{sat}) \tag{7.22}$$

where $x_2(\text{sat})$ is the value of x_2 when the solution is in equilibrium with gaseous M at a partial pressure of 1 atm, estimated from the total pressure by neglecting the nonideality of the mixture of gaseous M and the solvent vapor. Subject to the same approximation holding [116], the enthalpy and entropy changes are given by

$$\Delta \overline{H}^{\circ} = RT \frac{\partial \ln x_2(\text{sat})}{\partial \ln T} \tag{7.23}$$

$$\Delta \overline{S}^{\circ} = R \left[\frac{\partial \ln x_2(\text{sat})}{\partial \ln T} + \ln x_2(\text{sat}) \right] \qquad (7.24)$$

Obviously, the entropy may also be calculated from $\Delta \overline{S}^{\circ} = (\Delta \overline{H}^{\circ} - \Delta \overline{G}^{\circ})/T$.

A convenient way to represent gas solubility data when the temperature dependence is known is to perform a least-squares fit of $-RT \ln x_2(\text{sat})$ versus T:

$$-RT \ln x_2(\text{sat}) = A_0 + A_1 T \qquad (7.25)$$

If the data are sufficiently accurate over a wide temperature range, a three-constant fit may be used:

$$-RT \ln x_2(\text{sat}) = a_0 + a_1 T + a_2 T^2 \qquad (7.26)$$

It follows that $\Delta \overline{S}^{\circ} = -A_1$ or $-a_1 - 2a_2 T$, and that $\Delta \overline{H}^{\circ} = A_0$ or $a_0 - a_2 T^2$, depending on whether (7.25) or (7.26) is used. When $\Delta \overline{H}^{\circ}$ values are available at 298.15 K, for not too large temperature intervals over which $\Delta \overline{H}^{\circ}$ may be considered constant, the mole fraction $x_2(T_2)$ at a temperature T_2 may be calculated from the mole fraction at 298.15 K with reasonable accuracy from the relation

$$\ln x_2(T_2) - \ln x_2(298.15) = \frac{\Delta \overline{H}^{\circ}}{R} \left(\frac{T_2 - 298.15}{298.15 \ T_2} \right) \qquad (7.27)$$

The Hildebrand [56] entropy of solution $\Delta \overline{S}_H$, using the approximations shown above, is given by

$$\Delta \overline{S}_H = R \left(\frac{\partial \ln x_2}{\partial \ln T} \right)_{\text{sat}, P} \qquad (7.28)$$

$\Delta \overline{S}_H$ refers to the change in entropy for the process of transferring 1 mol of gaseous M (1 atm, ideal gas) to a solution at equilibrium mole fraction x_2, above which the gas partial pressure is 1 atm. This then is a solution of *different* mole fraction for each gas and each temperature. To convert the Hildebrand entropies of solution to those for the process depicted in (7.21), it is necessary to add the entropy change for increasing the mole fraction from x_2 to $x_2 = 1$. According to Henry's law this is equal to the entropy change for compressing the gas from 1 atm to atmospheres/x_2, or just $R \ln x_2$. This gives

$$\Delta \overline{S}^{\circ} = \Delta \overline{S}_H + R \ln x_2 \qquad (7.29a)$$

Occasionally, the standard state with $x_2 = 10^{-4}$ is used. To convert from $\Delta \overline{S}^{\circ}$

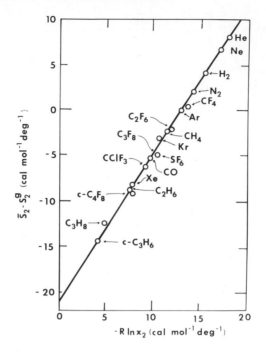

Fig. 7.11. Relation between partial molar entropy of solution and $-RT \ln x_2$ for gases in cyclohexane.

to this standard state, the following relation is used:

$$\Delta \bar{S}(x_2 = 10^{-4}) = \Delta \bar{S}^\circ - R \ln 10^{-4} = \Delta \bar{S}^\circ + 18.30 \qquad (7.29b)$$

The Temperature Dependence of Gas Solubility

The temperature dependence of the thermodynamic functions $\Delta \bar{G}^\circ$, $\Delta \bar{H}^\circ$, $\Delta \bar{S}^\circ$, and $\Delta \bar{S}_H$ are given, respectively by (7.22), (7.23), (7.24), and (7.28). Following the analysis given by Prausnitz [104], it is illustrative to break up the Hildebrand entropy ($\Delta \bar{S}_H = \bar{S}_2{}^L - S_2{}^g$) into two parts:

$$\Delta \bar{S}_H = (S_2{}^L - S_2{}^g) + (\bar{S}_2{}^L - S_2{}^L) \qquad (7.30)$$

where $S_2{}^L$ is the entropy of the hypothetical pure liquid at the temperature of the solution. The first term on the right-hand side of (7.30) is simply the entropy of condensation of the pure gas. In general, this term is negative. The second term on the right-hand side of (7.30) is the partial molar entropy of solution of the condensed solute and is given by

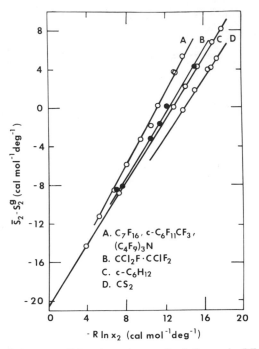

Fig. 7.12. Relation between partial molar entropy of solution and $-RT \ln x_2$ for gases in several solvents.

$$\overline{S_2}^L - S_2^L = -R \ln x_2 \qquad (7.31)$$

if we assume an ideal entropy change on mixing for the two liquids. But x_2 is always less than 1, so the second term on the right-hand side of (7.30) must always be positive. The lower the solubility, the larger the second term. We would therefore expect that, as the solubility decreases, a point is reached where the temperature coefficient of solubility becomes positive (corresponding to a positive value of $\Delta \overline{S_H}$). In general, then, sparingly soluble gases (very small x_2) show positive temperature coefficients of solubility, and readily soluble gases (relatively large x_2) show negative temperature coefficients.

In fact, this qualitative interpretation of the sign of $\Delta \overline{S_H}$ is borne out experimentally as shown in Figs. 7.11 and 7.12 [56a]. In Fig. 7.11 experimental values of $\Delta \overline{S_H}$ are plotted against the ideal partial molar entropy of the condensed solute ($-R \ln x_2$) for 18 gases in cyclohexane at 25°C and 1 atm partial pressure. Figure 7.12 contains similar information for five more solvents. The upper part of the figure (above $\Delta \overline{S_H} = 0$) corresponds to positive temperature coefficients of solubility, and the lower part to negative coefficients. The

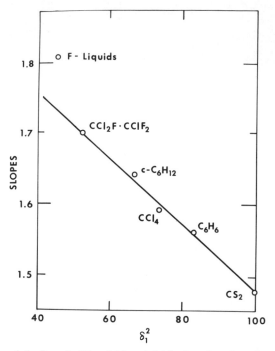

Fig. 7.13. Slopes of the lines in Figs. 7.11 and 7.12 plotted against the solvent solubility parameter squared.

range of solubilities for a zero temperature coefficient is from 3.43×10^{-3} for perfluoroheptane to 76×10^{-3} for carbon disulfide; the increased solubility (for zero temperature coefficient) corresponds to an increasing solubility parameter for the solvent. At constant x_2 the temperature coefficient of solubility for a given gas increases algebraically as the solubility parameter δ_1 of the solvent decreases. The slopes for the solvents are plotted against δ_1^2 in Fig. 7.13.

An enthalpy change corresponding to $\Delta \overline{S}_H$ can similarly be broken up into two parts:

$$\Delta \overline{H}_H = (H_2{}^L - H_2{}^g) + (\overline{H}_2{}^L - H_2{}^L) \tag{7.32}$$

where $H_2{}^L$ is the enthalpy of the hypothetical pure liquid at the temperature of the solution. The first term on the right-hand side is the enthalpy change on condensation of the pure solvent, and this quantity is negative (it may be positive for $T/T_{c_2} \gg 1$). The second term on the right-hand side is the partial enthalpy of mixing for the liquid solute and, in the absence of solvation between solute and solvent, this quantity is generally positive. Regular solution theory predicts that, the greater the difference between solubility parameters of the solute and the solvent, the greater the heat of mixing. If this difference is very large, the second term in (7.32) dominates, $\Delta \overline{H}_H$ is positive, and the temperature coefficient of solubility is positive. The smaller the difference in solubility parameters, the more the first term dominates, and then $\Delta \overline{H}_H$ is negative and the solubility decreases as the temperature increases. Specific chemical interactions between solute and solvent lead to both terms being negative, and a resulting rapid decrease in solubility as the temperature increases.

At a small partial pressure of solute, the solubility generally decreases with temperature, goes to a minimum, and then increases. The temperature corresponding to minimum solubility lies well above 25°C for most systems, although for light gases the minimum is usually observed at low temperatures. All the rare gases show a negative temperature coefficient of solubility in water at about room temperature and atmospheric pressure. The solubility goes through a minimum and then increases at higher temperatures and pressures. However, in hydrocarbon solvents at room temperature and atmospheric pressure, helium and neon solubilities increase with temperature, argon solubility shows an almost zero temperature coefficient, and krypton and xenon solubilities decrease with temperature.

For determining enthalpy changes from high-pressure phase equilibrium data, see the analysis by Sherwood and Prausnitz [116].

The Pressure Dependence of Gas Solubility

The solubility of a gas in a liquid is proportional to its partial pressure, and this empirical observation is described by Henry's law:

$$P_i = y_i P = K'_H x_i \tag{7.33}$$

where P is the total pressure, y_i is the mole fraction of i in the gas phase, x_i is the mole fraction of i in the liquid phase, and K'_H is a proportionality constant which is temperature dependent. A rigorous definition of the Henry's-law constant $K_{2,1}$ was given earlier as

$$\lim_{x_i \to 0} \left(\frac{f_i}{x_i} \right) = K_{i, \text{solvent}}$$

where f_i is the fugacity of the solute i. A more general form than (7.33) for Henry's law for solute i is

$$f_i = \phi_i y_i P = K_{i, \text{solvent}} x_i \tag{7.34}$$

where ϕ_i is the gas-phase fugacity coefficient. $K_{i, \text{solvent}}$ is not a function of composition; it is a function of temperature and to a lesser extent a function of pressure. The pressure dependence is essentially negligible at low pressures, but must be considered at high pressures.

Since

$$\left(\frac{\partial \ln f_i^L}{\partial P}\right)_{T,x} = \frac{\overline{V}_i}{RT}$$

then

$$\left(\frac{\partial \ln K_{i,\text{solvent}}}{\partial P}\right)_T = \frac{\overline{V}_i^\infty}{RT} \tag{7.35}$$

where $\overline{V}_i^\infty$ is the partial molar volume of i in the solvent at infinite dilution. We can integrate (7.35) from $x_i = 0$ to x_i with the corresponding limits of P_{sat}^L to P to obtain the more general expression

$$\ln \frac{f_2}{x_2} = \ln K_{2,1}^{(P^L\text{sat})} + \frac{\overline{V}_2^\infty(P - P^L\text{sat})}{RT} \tag{7.36}$$

which is called the Krichevsky-Kasarnovsky [66] equation. In this equation the subscript 2 refers to the solute, and 1 to the solvent. Also it is assumed that $\overline{V}_2^\infty$ is independent of the pressure. This equation is good for representing the solubility of sparingly soluble gases up to rather high pressures. Graphs of $\ln (f_2/x_2)$ versus P (P is in general much greater than P_{sat}^L) yield $K_{2,1}^{(P^L\text{sat})}$ from the intercepts and $\overline{V}_2^\infty$ from the slopes. The latter are generally in good agreement with experimental values which are quite difficult to measure. The Krichevsky-Kasarnovsky equation holds in general for small values of x_2 and temperatures far removed from the critical temperature of the solution.

The Krichevsky-Ilinskaya [67] equation adds an extra term to (7.36), which takes into account the change in activity coefficient with concentration. The equation is

$$\ln \frac{f_2}{x_2} = \ln K_{2,1}^{(P^L\text{sat})} + \frac{A}{RT}(x_1^2 - 1) + \frac{\overline{V}_2^\infty(P - P^L\text{sat})}{RT} \tag{7.37}$$

and is in general more widely applicable than (7.36). The constant A is empirically evaluated. This equation is particularly applicable to solutions of light gases such as hydrogen and helium in solutions in which the solubilities are relatively high. Orentlicher and Prausnitz [94] give values of $K_{2,1}^{(P^L\text{sat})}$, A, and $\overline{V}_2^\infty$ for hydrogen solubilities in argon, carbon monoxide, nitrogen, CH_4,

C_2H_4, C_2H_6, C_3H_8, and C_3H_6 for pressures up to 100 atm. For example, for hydrogen in CH_4 at 116 K, $K_{2,1}^{(PL\,sat)}$ = 970 atm, A = 15 ± 3 ℓ atm mol^{-1}, and $\overline{V}_2$ = 31.6 cm^3 mol^{-1}.

Estimation of the Solubility of a Gas in a Liquid

There are theoretical and semiempirical relations that are useful in correlating the solubility of a gas in a liquid with various properties of the solvent and the gas. Some of these correlations are of use in predicting the solubility of a gas in a nonpolar liquid when measurements are not available.

Already discussed and illustrated in Figs. 7.11 and 7.12 is the linear relationship between the entropy of solution and the gas solubility function $-R \ln x_2$. For solvents for which the linear relationship is known, a single measurement of solubility gives information about the sign of the temperature coefficient of solubility of the gas in that solvent.

The regular solution theory developed by Hildebrand and co-workers [55] in its simplest form allows correlation of the gas solubility with the solvent solubility parameter δ_1. The Hildebrand regular solution of two liquids can be defined as a solution that has a vanishing excess entropy of mixing at constant temperature and constant volume. In nonpolar liquids in which the molecular forces of attraction are primarily London dispersion forces, a geometric mean of the component cohesive energy densities is assumed to represent the interactions between unlike molecules. This assumption leads to a relatively simple expression for the excess thermodynamic properties of a solution in which an important term is the difference in the square roots of the cohesive energy densities for the binary solution components. The square root of the cohesive energy density is defined as the solubility parameter and is calculated from the energy change on isothermal vaporization of the liquid to the ideal gas state and from the liquid molar volume

$$\delta = \left(\frac{\Delta E_{vap}}{V} \right)^{1/2} \tag{7.38}$$

Figure 7.14 shows the logarithm of solubility at 25°C and 1 atm of several gases as a function of the solvent solubility parameter. The near-linear correlation allows estimation of solubilities of these gases in other solvents. If there is a chemical interaction between the gas and the solvent, the predicted solubility will be low. For example, the solubility of carbon dioxide in aromatic hydrocarbons is higher than predicted because of an acid-base interaction between gas and solvent. Since the extensive data needed to construct the curves of Fig. 7.14 are available only at 25°C, the correlation is rarely useful at other temperatures. The correlation is sometimes improved by plotting log x_2 against the square of the solubility parameter [56a].

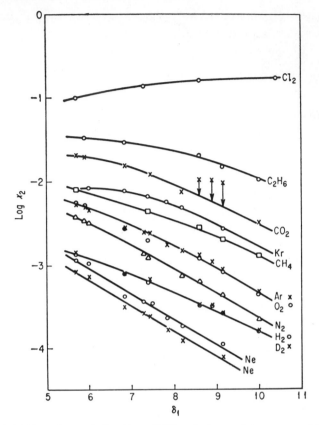

Fig. 7.14. Solubility of gases in liquids at 25°C and at a partial pressure of 1 atm against solvent solubility parameter. (J. H. Hildebrand and R. L. Scott, *Regular Solutions,* by permission of the authors.)

Prausnitz and Shair [105] and Yen and McKetta [141] have combined the theory of regular solutions and the theory of corresponding states to obtain a result that allows estimation of gas solubility in a liquid at other temperatures, provided the liquid is not near its critical temperature.

Prausnitz and Shair consider the isothermal solution of a gas in a liquid to be a two-step process. They consider the free-energy changes on (1) condensing the gas to a volume close to the partial molal volume of the gas as a solute in a liquid solvent,

$$RT \ \ln \frac{f_{\text{hyp. liq. } 2}}{f_{\text{gas } 2}} \ ,$$

and (2) dissolving of the hypothetical liquid in the solvent to saturation, RT

ln $\gamma_2 x_2$. Since the solute in the liquid is in equilibrium with the gas, the sum of the two free energies is zero.

$$\Delta G_{\text{soln}} = 0 = RT \ln \frac{f_{\text{hyp.liq.}2}}{f_{\text{gas}\,2}} + RT \ln \gamma_2 x_2 \qquad (7.39)$$

They assume the regular solution theory gives the gaseous solute activity coefficient

$$RT \ln \gamma_2 = V_2(\delta_1 - \delta_2)^2 \phi_1^2$$

where V_2 is the molar liquid volume of the solute, ϕ_1 is the volume fraction of the solvent, and δ_1 and δ_2 are solvent and solute solubility parameters, respectively. Substitution and rearrangement gives the expression

$$\frac{1}{x_2} = \frac{f_{\text{hyp.liq.}2}}{f_{\text{gas}\,2}} \exp\left(\frac{V_2(\delta_1 - \delta_2)^2 \phi_1^2}{RT}\right) \qquad (7.40)$$

for the solubility of the gas. The quantity $V_2(\delta_1 - \delta_2)^2 \phi_1^2$ may be assumed independent of temperature, since regular solution theory assumes that at constant composition ln γ_2 is proportional to $1/T$. Thus 25°C solubility data may be used to find values of V_2 and δ_2. However, the fugacity of the hypothetical liquid must be treated as a function of temperature. Values of V_2 and δ_2 for 11 gases were estimated from 25°C solubility data of Prausnitz and Shair. They are given in Table 7.4.

Table 7.4 Liquid Volumes and Solubility Parameters for Gaseous Solutes at 25°C

Gas	V_2 (cm^3 mol^{-1})	δ_2 [(cal cm^{-3})$^{\frac{1}{2}}$]
N_2	32.4	2.58
CO	32.1	3.13
O_2	33.0	4.0
Ar	57.1	5.33
CH_4	52.	5.68
CO_2	55.	6.0
Kr	65.	6.4
C_2H_4	65.	6.6
C_2H_6	70.	6.6
Rn	70.	8.83
Cl_2	74.	8.7

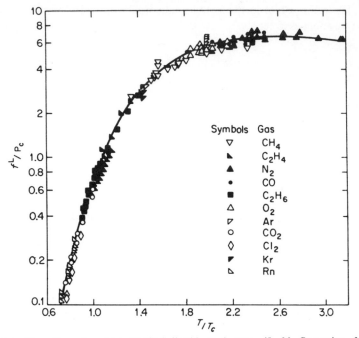

Fig. 7.15. Fugacity of a hypothetical liquid at 1 atm. (J. M. Prausnitz, *Molecular Thermodynamics of Fluid Phase Equilibria,* 1969, by permission of Prentice-Hall, Inc., Englewood Cliffs, N. J.).

They have also correlated the fugacity of the hypothetical liquid at 1 atm as a corresponding states plot (Fig. 7.15). If the gas solubility is needed at a higher pressure, the Poynting correction should be applied to the atmospheric pressure fugacity.

The equation can also be used to estimate gas solubility in mixed solvents by replacing δ_2 by an average solubility parameter for the entire solution:

$$\bar{\delta} = \sum_i \phi_i \delta_i \tag{7.41}$$

The average solubility parameter is summed for all components including the solute.

Hildebrand and Scott show that the solubility of a nonpolar gas in nonpolar solvents can be correlated with the Lennard-Jones characteristic energy ϵ/k of the solute (Fig. 7.16). Extensive data are available only for 25°C, but the relationship can be used with some confidence to estimate solubilities in solvents for which no experimental data are available at that temperature.

A useful set of reasonably consistent characteristic energy parameters for

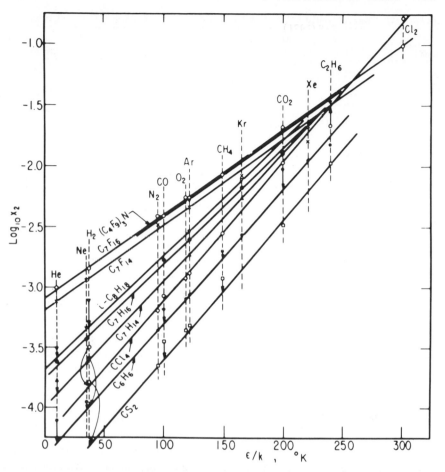

Fig. 7.16. The logarithm of the mole fraction solubility of gases in liquids at 25°C and a partial pressure of 1 atm against the solute characteristic energy ϵ/k. (J. H. Hildebrand and R. L. Scott, *Regular Solutions,* by permission of the authors.)

19 gases is compiled in Table 7.5 from several sources [45, 55, 57]. Also included is the energy of vaporization of the gas at its normal boiling point.

It has long been a useful model to consider the solubility of a gas in a liquid as a two-step process: (1) The creation of a cavity in the solvent to accommodate the solute, and (2) the introduction of a solute molecule into the cavity, followed by interaction with the solvent [43, 128b]. At present the most fruitful approach to evaluating the free-energy change associated with the model involves the use of the statistical mechanical theory of fluids of Reiss et al.

Table 7.5   Characteristic Energy and Energy of Vaporization of Some Gases

Gas	ϵ/k (K)	ΔE_{vap} (cal mol^{-1})	Gas	ϵ/k (K)	ΔE_{vap} (cal mol^{-1})
He	10	11	CO_2	200	–
Ne	35	361	SF_6	201	4080
H_2	37	176	Xe	219	2694
N_2	95	1182	$c-C_4F_8$	223	5003
CO	100	1444	C_2F_6	225	3473
O_2	118	1453	C_2H_6	236	3147
Ar	121	1380	C_3H_8	242	4029
CH_4	148	1737	C_3F_8	266	4220
CF_4	153	2720	Cl_2	300	4878
Kr	165	1920			

[108], which yields an approximate expression for the work to introduce a sphere into a fluid of spherical particles reversibly. Pierotti [99, 100] showed that the approximation for the case of hard-sphere molecules can be used to calculate the solubilities of gases in a large number of solvents from the molecular parameters of the liquids and gases. This theory can be used to determine the Lennard-Jones pair potential parameters for the solvent from the experimental gas solubilities, and the resulting parameters can be used to make good predictions of the thermodynamic functions of the solution process.

For dilute solutions it can be shown that

$$RT \ln K_H/\text{atmospheres} = \overline{G}_c + \overline{G}_i + RT \ln RT/V_1^0 \qquad (7.42)$$

where $\overline{G}_c$ and $\overline{G}_i$ are the partial molar Gibbs energy for cavity formation and interaction, respectively, K_H is the Henry's-law constant, V_1^0 is pure solvent molar volume, R is the gas constant, and T is the absolute temperature.

Reiss et al. [108] obtained the expression for $\overline{G}_c$ for a fluid of hard spheres:

$$\overline{G}_c = RT \left\{ 6 \frac{Y}{1-Y} \left[2 \left(\frac{a_{12}}{a_1} \right)^2 - \frac{a_{12}}{a_1} \right] + \right.$$

$$\left. 18 \left(\frac{Y}{1-Y} \right)^2 \left[\left(\frac{a_{12}}{a_1} \right)^2 - \frac{a_{12}}{a_1} + \frac{1}{4} \right] - \ln (1-Y) \right\} + f(P) \qquad (7.43)$$

where $Y = \pi a_1^3 \rho/6$, with ρ the number density of the fluid N/V_1^0, and a_1 is the solvent molecule hard-sphere diameter. The diameter of the cavity to be created is a_2, and a_{12} is $(a_1 + a_2)/2$, the radius of the sphere, which excludes the centers of the surrounding hard-core molecules. The term $f(P)$ is a function of pressure and the hard-sphere dimensions; it is only 0.02% of the total value of $\overline{G}_c$ [139], and so is neglected.

As a first approximation it is assumed that the free energy of interaction is equal to the molar interaction energy of the solute, or $\overline{G}_i = \overline{U}_i$. That is, the entropy and PV terms are assumed negligible compared to $\overline{U}_i$.

The interaction energy of a polar molecule dissolved in a nonpolar solvent can be described in terms of dispersion, repulsive, and inductive interactions by a Lennard-Jones [6-12] pairwise additive potential:

$$U(r) = 4\epsilon \left[\left(\frac{\sigma}{r} \right)^{12} - \left(\frac{\sigma}{r} \right)^{6} \right] \tag{7.44}$$

where σ is the value of r for which $U(r) = 0$, and $-\epsilon$ is the minimum value of $U(r)$. It can be shown that

$$\overline{U}_i = -3.555 R\pi\rho\sigma_{12}^2 \epsilon_{12}/k - 1.333 N\pi\rho\mu_2^2 \ a_1/\sigma_{12}^3 \tag{7.45}$$

For two nonpolar molecules only the first term is needed.

Terms in the equation not already defined are σ_{12} which is $(\sigma_1 + \sigma_2)/2$ and ϵ_{12} which is the geometric mean of the energy parameters $(\epsilon_1\epsilon_2)^{1/2}$; μ_2 is the dipole moment of the polar component, a is the polarizability of the solvent, and k is the Boltzmann constant. Substitution of U_i into (7.42) gives the desired equation for $\ln K_H$.

To apply the equation requires a set of consistent molecular parameters extracted from the experimental data. Extrapolation of the plot of $\ln K_H$ against rare gas solute polarizability to $a_2 = 0$ is equivalent to obtaining the solubility of a hard sphere with $a_2 = 2.58$ Å. Thus

$$\lim_{\substack{a_2 \to 0 \\ a_2 \to 2.58 \ \text{Å}}} \ln K_H = \ln K_{H,0} = \frac{\overline{G}_c}{RT} + \ln \frac{RT}{V_1^0} \tag{7.46}$$

A plot of $\ln K_H$ against a_2 is shown for the solvents n-heptane and isobutanol in Fig. 7.17.

The $\overline{G}_c$ value can be obtained with $a_{12} = (a_1 + 2.58)/2$. Then (7.46) can be solved for a_1 values.

Equation 7.42 can be rewritten as

$$RT \ln K_H - \overline{G}_c - 1.33\pi\rho N\mu_1^2 a_2/\sigma_{12}^3 - RT \ln RT/V_1^0$$

$$= -3.555\pi R\rho\sigma_{12}^3 \left(\frac{\epsilon_1}{k} \right)^{1/2} \left(\frac{\epsilon_2}{k} \right)^{1/2} \tag{7.47}$$

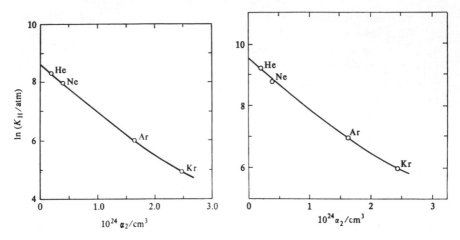

Fig. 7.17. Experimental values of $\ln K_H$ plotted against rare gas polarizabilities at 25°C in *n*-heptane (left) and isobutanol [7, 139]. (Reprinted by permission of the *Journal of Chemical Thermodynamics*.)

The term containing μ_1 disappears for a nonpolar solvent such as *n*-heptane.

Figure 7.18 shows plots of the left-hand side of (7.47) against $10^{22}(\epsilon_2/k)^{1/2}\sigma_{12}{}^3$ for the solvents *n*-heptane and isobutanol. The slope is $-3.555 R\pi\rho(\epsilon_1/k)^{1/2}$. The values of ϵ_1/k and σ_1 can then be used in the calculation of the solubility of other gases in the solvent.

The thermodynamic functions characteristic of the solution process can now be calculated. Thus

$$\Delta \overline{G}_2^\circ = RT \ln K_H \tag{7.48}$$

$$\Delta \overline{H}^\circ = -RT^2 \left(\frac{\partial \ln K_H}{\partial T} \right)_p = \overline{H}_c - \overline{H}_i - RT + a_\rho RT^2 \tag{7.49}$$

where $\overline{H}_c$ and $\overline{H}_i$ are the partial molal enthalpies associated with cavity formation and interaction, respectively, and a_ρ is the coefficient of solvent thermal expansion.

The enthalpy of interaction is assumed equal to $\overline{U}_i$. The enthalpy of cavity formation $\overline{H}_c$ is calculated from the free energy of cavity formation equation by differentiation of $\overline{G}_c/T$ with respect to T.

$$\overline{H}_c = -T^2 \left[\frac{\partial(\overline{G}_c/T)}{\partial T} \right]_p = a_\rho RT^2 \frac{Y}{1-Y} \left\{ \frac{6}{1-Y} \right.$$

$$\left. \left[2\left(\frac{a_{12}}{a_1}\right)^2 - \frac{a_{12}}{a_1} \right] + \frac{36Y}{(1-Y)^2} \left[\left(\frac{a_{12}}{a_1}\right)^2 - \frac{a_{12}}{a_1} + \frac{1}{4} \right] + 1 \right\} \tag{7.50}$$

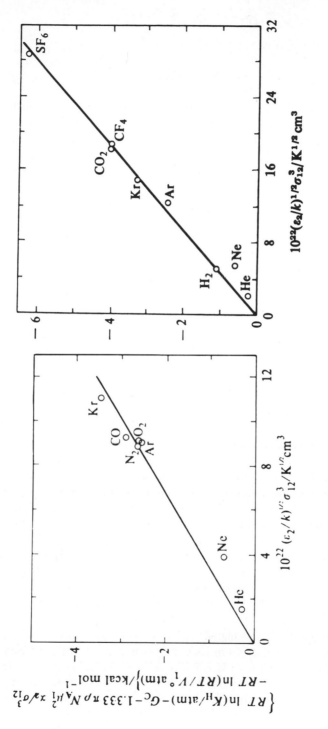

Fig. 7.18. The left-hand side of (7.47) plotted against a Lennard-Jones [6-12] potential parameter at 25°C for a number of gases in *n*-heptane (left) and isobutanol [7, 139]. (Reprinted by permission of the *Journal of Chemical Thermodynamics*.)

The partial molal entropy is obtained:

$$\Delta \overline{S}_2^\circ = (\Delta \overline{H}^\circ - \Delta \overline{G}_2^\circ)/T \tag{7.51}$$

Battino et al. [7, 139] compared calculated and experimental values of thermodynamic functions for several polar and nonpolar solvents. In general the agreement is quite good.

In 1937, Uhlig [128b] derived a linear relationship between the logarithm of the Ostwald coefficient of gas solubility and the solvent surface tension:

$$\log L = \frac{-4\pi r^2 \sigma + E}{2.3kT} \tag{7.52}$$

He assumed (1) a cavity of radius r is formed which requires energy equal to the cavity surface area times the solvent macroscopic surface tension σ, and (2) there is an energy of interaction E when the gas molecules are placed in the cavity. The basis of the relationship has been criticized, because of the use of the macroscopic surface tension to calculate the energy of formation of a microscopic cavity and because of the implication of an interaction energy that is the same for all solvents. However, the relationship has been of value in correlating the solubility of a given gas in a series of nonpolar nonassociated solvents.

Experimental evaluations of the log Ostwald coefficient-solvent surface tension relationship have been carried out for oxygen in about 20 solvents, including hydrocarbons, alcohols, ketones, esters, and halocarbons [113], nitrogen, oxygen, and air in lubricating oils and fuels [3], noble gases in hydrocarbons and substituted benzenes [27, 112], and carbon dioxide, oxygen, and nitrogen in various solvents [95]. In general the relationship is satisfactorily linear, water and polyhydric alcohols being the solvents of poor agreement. The relationship fails in the presence of surface-active agents which change the surface tension but do not change the bulk properties of the solvent [44, 50].

Burrows and Preece [18] combine the empirical relationship between surface tension and temperature, $\sigma = k/T^n$, with the ideas of Uhlig to derive an equation that predicts a linear relationship between $\log L(T/\sigma)^{1/(n+1)}$ and σ/T, where $n = $ ½ or 1. The relation holds for systems for which $\log L$ is not a linear function of $1/T$.

McGowan [75] calculated the work for the isothermal compression or expansion of a solute from its unit pressure volume to the free volume of the solvent, which he shows is equal to a function of the volume fraction of the solute at saturation. For gases his approach predicts a linear relationship between $\log L$ and $(\phi/V)^8$, where ϕ is the molecular parachor, a function of surface tension, and V is the molecular volume of the solvent. He found a good correlation for argon in 17 solvents.

Marathe and Prausnitz [78] have published an empirical equation for estimating the solubility of helium in a variety of solvents over a wide range of temperature. The equation predicts the helium Henry's-law constant with a root-mean-square error of 21% in nonhydrogen-bonded solvents and 7% in strongly hydrogen-bonded solvents (water and alcohols).

The solubility of methane in n-hexane, cyclohexane, benzene, and carbon tetrachloride has been predicted with a maximum error of 3% by Hsi and Lu [58] by applying a modified Redlich-Kwong equation of state. The method was developed for systems containing one supercritical component and includes a scheme for evaluating temperature-dependent parameters for the super-critical component.

Salt Effects

The activity coefficient of a dissolved gas is a function of the concentration of all solute species. At a given temperature the logarithm of the solute gas activity coefficient can be represented by a power series in C_s, the electrolyte concentration, and C_i, the nonelectrolyte solute gas concentration.

$$\log f_i = \sum_{mn=0} k_{mn} C_s^n C_i^m \tag{7.53}$$

It is usually assumed that, for low C_s and C_i values where there is no chemical interaction between solute species, only the linear terms are important and

$$\log f_i = k_s C_s + k_i C_i \tag{7.54}$$

Experimental measurements of the solubility of a gas in pure solvent and in a salt solution give the activity coefficient of the dissolved gas directly. The gas solute activity is the same in pure solvent and salt solution for a given partial pressure and temperature, so

$$f_i S_i = f_i^0 S_i^0 \tag{7.55}$$

and

$$f_i = f_i^0 \frac{S_i^0}{S_i} \tag{7.56}$$

where S_i and S_i^0 are gas solubilities in salt solutions and pure solvent, respectively. Combining equations gives

$$\log f_i = \log f_i^0 + \log \frac{S_i^0}{S_i} = k_s C_s + k_i S_i \tag{7.57}$$

Since $\log f_i^0 = k_i^0 S_i^0$,

$$\log \frac{f_i}{f_i^0} = \log \frac{S_i^0}{S_i} = k_s C_s + k_i(S_i - S_i^0) \tag{7.58}$$

The last term is negligible if k_i, which results from the interaction of the solute gas with itself, is small, or if the $S_i - S_i^0$ difference is small. When the nonelectrolyte solute is a gas, the last term is normally considered negligible and the equation is used in the form

$$\log \frac{f_i}{f_i^0} = \log \frac{S_i^0}{S_i} = k_s C_s \tag{7.59}$$

which is the form of the well-known empirical Setschenow equation. A salt that increases the activity coefficient of the dissolved gas salts out, and a salt that decreases the activity coefficient of the dissolved gas is said to salt in.

Debye and McAulay [38] proposed an electrostatic theory of the salt effect, which predicted nonelectrolytes that lower the dielectric constant of the solvent will be salted out, while those that increase the solvent dielectric constant will be salted in. The theory gives values of k_s which vary little with the nature of the salt and do not explain at all the salting in observed by large electrolyte ions. Conway, Desnoyers, and Smith [31, 39] have modified the continuous-distribution theory [38] and have considered a two-region model to explain salting out. They consider effects in two distinguishable regions near the ion: (1) that due to the loss of normal solvent function of the water molecule attached to the ion in the primary hydration shell, where appreciable dielectric saturation occurs, and (2) that due to the more continuous distribution of solvent and nonelectrolyte in the region of solution beyond the hydration radius arising from the different size and polarization of the nonelectrolyte and solvent molecules in the ion fields. Their results give the salting-out parameter in terms of the partial molar volumes and dipole moments of the nonelectrolytes. Ruetschi and Amlie [110] also used a two-region model which gives results in terms of dielectric decrements.

Bockris, Bowler-Reed, and Kitchener [15] modified the electrostatic expression for k_s by attempting to include dispersion forces. The dispersion forces whose signs depend on the relative polarizabilities of the nonelectrolyte solute and solvent usually lead to a salting-in contribution. The overall sign of k_s depends on the relative magnitudes of the electrostatic (salting-out) and dispersion (salting-in) terms. Trends in k_s with the sizes of molecules and ions have been explained quite well by the theory [84].

McDevit and Long [74] calculated the free energy of transfer of a nonpolar electrolyte from pure water to salt solution, assuming the neutral molecules modify the ion-water interaction in a simple manner. They obtained a limiting law expression for k_s:

$$k_s = \frac{\overline{V}_i^0(V_s - \overline{V}_s^0)}{2.3K_T RT} \qquad (7.60)$$

where $\overline{V}_i^0$ and $\overline{V}_s^0$ are partial molal volumes at infinite dilution of nonelectrolyte solute and salt, respectively, V_s is the molar volume of pure (liquid) electrolyte, and K_T is the isothermal compressibility of the solvent. The internal pressure theory of McDevit and Long relates the sign of k_s to the influence the salt has on the solvent structure. It often predicts the correct order of k_s for a given nonelectrolyte with a series of electrolytes.

If (7.60) is differentiated with respect to T and terms considered negligible omitted, one obtains

$$\frac{dk_s}{dT} \cong -\left(\frac{\overline{V}_i^0}{2.3\ RT}\right)\left(\frac{d\overline{V}_s^0}{dT}\right) \qquad (7.61)$$

which usually predicts the correct order for dk_s/dT [28].

At present the most fruitful approach in explaining salt effects appears to be the scaled-particle theory. The background of the theory was presented in the previous section. Shoor and Gubbins [117] extended Pierotti's [99, 100] scaled-particle model to obtain an equation for the solubility of a nonelectrolyte in an aqueous salt solution. Masterton and Lee [83] extended the derivation to obtain a general expression for the salt-effect parameter k_s. The scaled-particle theory gives the following expression for log S:

$$-\log S = \frac{\overline{g}_1^{\ n}}{2.3kT} + \frac{\overline{g}_1^{\ s}}{2.3kT} + \log kT \sum_{j=1}^{4} \rho_j \qquad (7.62)$$

where $\overline{g}_1^{\ n}$ is the free-energy change on cavity formation, $\overline{g}_1^{\ s}$ is the free-energy change of interaction when the solute molecule is introduced into the cavity, and ρ_j is the number density of nonelectrolyte, solvent, cation, and anion for $j = 1$, 2, 3, and 4, respectively.

The Setschenow equation is valid only at low salt concentrations. Thus an expression for k_s is obtained by differentiating the log S expression with respect to salt concentration in the limit $C \to 0$:

$$k_s = -\left(\frac{d(\log S)}{dc}\right)_{C\to0} = \left(\frac{d(\overline{g}_1^{\ n}/2.3kT)}{dc}\right)_{C\to0} + \left(\frac{d(\overline{g}_1^{\ s}/2.3kT)}{dc}\right)_{C\to0}$$

$$+ \left(\frac{d\ \log\ \Sigma\ \rho_j}{dc}\right)_{C\to0} \qquad (7.63)$$

$$= k_\alpha + k_\beta + k_\gamma \qquad (7.64)$$

where k_α, k_β, and k_γ represent the contributions to the salting coefficient from cavity formation, interaction, and number density.

Detailed expressions for k_α, k_β, and k_γ can be found in Masterton and Lee [83]. In general, the salting coefficient contributions are functions of

$$k_\alpha = f(\sigma_1, \sigma_3, \sigma_4, \phi_0)$$

$$k_\beta = f(\sigma_1, \sigma_2, \sigma_3, \sigma_4, a_1, a_3, a_4, Z_3, Z_4, \epsilon_1/k, \phi_0)$$

$$k_\gamma = f(\phi_0)$$

where ϕ_0 is salt apparent molar volume at infinite dilution, ϵ_1/k is the non-electrolyte solute energy parameter, σ_j is the diameter, a_j is the polarizability, and Z_j is the number of electrons in the ion with subscripts 1 the nonelectrolyte solute, 2 the solvent, 3 the cation, and 4 the anion.

The sign of the salting coefficient k_s depends on the relative magnitudes of k_α, k_β, and k_γ. In all cases k_α is positive. Physically, this means that the free-energy change for cavity formation becomes more positive as the concentration of the salt increases. That is, it is more difficult to form a cavity in an aqueous solution of a strong electrolyte than in water itself. The values of k_β are negative. Physically, this means that once a cavity is formed in a salt solution it is easier to introduce a nonelectrolyte molecule than it is in pure water. Thus the nonelectrolyte experiences a net attractive force when water molecules in the surrounding cavity are replaced by strong electrolyte ions. The sign of k_γ is variable, and its magnitude is negligible unless ϕ_0 is very large.

Table 7.6 gives values of k_α, k_β, and k_γ for H_2, CH_4, and SF_6 in the alkali halides sodium chloride and potassium iodide. Also given in the table are values of k_s calculated from the McDevit and Long theory and experimental values of k_s. Both theories are in remarkably good agreement, but the scaled-particle approach appears to be superior.

Table 7.6 Comparison of Observed and Calculated Salting Coefficients at $25°C$

System	Scaled Particle Theory				McDevit and Long Theory	Observed
	k_α	k_β	k_γ	k_s		
H_2-NaCl	0.126	−0.023	0.008	0.111	0.123	0.114
H_2-KI	0.115	−0.031	−0.004	0.080	0.057	0.081
CH_4-NaCl	0.203	−0.080	0.008	0.131	0.184	0.127
CH_4-KI	0.182	−0.102	−0.004	0.076	0.086	0.097
SF_6-NaCl	0.387	−0.194	0.009	0.202	0.358	0.195
SF_6-KI	0.343	−0.226	−0.004	0.113	0.166	0.145

Recently, Tiepel and Gubbins [128a] have proposed a theory based on the first-order perturbation theory of Leonard, Henderson, and Barker [68] for the thermodynamic properties of gases dissolved in electrolyte solutions. The final equations closely resemble those of the scaled-particle theory, but the hard-sphere diameters are temperature-dependent in this approach. The theory gives good agreement with experiment and correctly predicts observed trends in partial molal volumes of the solute gas in salt solutions.

Other studies of the solubility of gases in salt solutions deserve mention. The scaled-particle approach is also used by Wen and co-workers [135, 136] and by Lucas [71, 72]. The thermodynamic approach of Ben-Naim [11, 12] is of interest. Salt effect studies at high pressure have been reported [49, 96, 122], and there are new data on carbon dioxide and C_2H_4 salt effects [93].

9 PARTIAL MOLAR VOLUMES OF GASES

The partial molar volumes of gases in solution are of importance in understanding the thermodynamics of solutions. For example, for study of the high-pressure thermodynamics of mixtures, it is necessary to know the dependence of the fugacity on pressure, and this dependence is related to the partial molar volume via

$$\left(\frac{\partial \ln f_i}{\partial P}\right)_{T,x} = \frac{\bar{v}_i}{RT} \tag{7.65}$$

[See Prausnitz (Ref. 104, Section 10.5) for a discussion of this.] Hildebrand et al. [56a] have discussed the contribution of volume expansion on mixing to the entropy of solution. Although volume expansion has little effect on the Gibbs energy of mixing, it can have a significant effect on the entropy of mixing. The correction to the partial molar entropy of solution of a dilute solution requires values of solvent internal pressure $(\partial P/\partial T)_v$ and solute partial molar volume $\bar{v}_2$. Hildebrand [54, 69] and others [90, 118, 131] have used a correlation between $(\bar{v}_2 - v_{2,b})(\partial S_1/\partial V)_T$ and $R \ln x_2$, where $v_{2,b}$ is the molar volume of the gas at its boiling point, or $\bar{v}_2/v_c$ and $-(\log x_2)/v_c$, where v_c is the critical volume of the gas. These correlations are all solvent-dependent. Lyckman, Eckert, and Prausnitz [73a] use a generalized correlation between $\bar{v}_2 P_c/RT_c$ and $TP_c/c_1 T_c$, where P_c and T_c refer to the gas, and c_1 is the cohesive energy density of the solvent. Their plot shows a great deal of scatter, but the correlation is generally useful.

Experimental determinations of $\bar{v}_2$ are of two types. Most workers have preferred to use some variation of Horiuti's apparatus [59] wherein a change in solution volume is directly measured on an incremental addition of gas, that is, $\bar{v}_2 = (\Delta V/\Delta n_2)_{T,n_1}$. Masterton [82] determined the partial molar volumes of benzene, methane, ethane, and propane in water by direct deter-

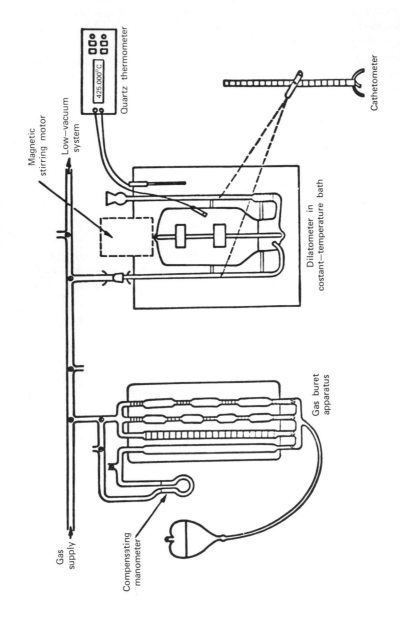

Fig. 7.19. Apparatus for the determination of partial molal volume of a gas in a liquid [127]. (Reprinted by permission of the author.)

Gas supply

Magnetic stirring motor

Low–vacuum system

425.000°C

Quartz thermometer

Cathetometer

Dilatometer in constant–temperature bath

Gas buret apparatus

Compensating manometer

mination of solution densities to 1 part in 10 million. The original article should be referred to for details. We described earlier one modification of the Horiuti apparatus in some detail. (Table VIII in Battino and Clever's review [4] catalogs references to gas-liquid systems for which partial molar volumes have been measured.)

The apparatus used by Tiepel and Gubbins [128a] for determination of the partial molar volumes of gases dissolved in electrolyte solutions is shown in Fig. 7.19. Details of the experimental method are available [127]. The dilatometer has a capacity of 500 cm^3, which is the largest that has been used by any worker. About 450 cm^3 are filled with degassed solvent, and the remainder with mercury. Two capillary side arms (i.d. ca. 0.6 mm) are affixed to the bottom. Both capillaries are calibrated with respect to uniformity and have fiducial marks for reference. Stirring is done magnetically via an internally sealed paddle. Gas is introduced from the gas buret apparatus via the left-hand capillary which has a nozzle inserted into the mercury layer. Gas volumes are determined with an accuracy of ±0.002 cm^3. Temperature control must be maintained to at least ±0.001°C for aqueous systems. For this size dilatometer, temperature control would have to be an order of magnitude better for organic solvents. Heights are read with a cathetometer to ca. ±0.01 mm. It is very important that the measurements be made at constant pressure to avoid the necessity of compressibility corrections. This is done by noting the initial pressure before the addition of gas and returning the system to this pressure by applying slight suction to one of the legs of the dilatometer. Total dissolution of the gas increment with constant stirring may require from several hours to 1 day. The overall precision with this apparatus is of the order of 1% in the partial molar volume.

10 SUMMARY

The chapter has been written as a guide to the experimental determination of the solubility of gases in liquids. A useful standard for gas solubility measurements was recommended, and some typical gas solubility values have been presented. The importance of degassing the liquid phase before a solubility measurement was emphasized, and procedures to obtain a gas-free liquid were recommended.

Gas solubility apparatuses for use over various temperature and pressure intervals were discussed. Their accuracy and precision were judged so that one may pick a technique that fits a particular need.

The effect of temperature and pressure on the solubility of a gas in a liquid were discussed, and enough theory was presented so that a worker may judge whether or not his own results are typical, or he may make reasonable estimates of the solubility of a gas in a liquid in the absence of experimental data.

Acknowledgment

One of us (R.B.) acknowledges the support in part of Public Health Service Grant No. GM 14710-05, and the other (H.L.C.) acknowledges the support in part of Professor R. S. Stein and the Polymer Research Institute while on leave of absence at the University of Massachusetts.

Note added in proof. A critical review of the solubilities of gases in water is now available [142]. An automated method using the Ben-Naim and Baer [143] apparatus has been described [144]. The Ben-Naim and Baer apparatus has been used by Wen and Hung [136] for the solubility of hydrocarbons in water, and is described in detail by Hung [145]. Murray and Riley [146 and references therein] have used it for the studies on gas solubilities in water. Many workers seem to prefer the ease of using this apparatus design.

References

1. B. Amirijafari and J. M. Campbell, *Trans. AIME,* **253**(2), 21 (1972).
2. R. R. Baldwin and S. G. Daniel, *J. Appl. Chem.,* (London), **2**, 161 (1952).
3. R. R. Baldwin and S. G. Daniel, *J. Inst. Petrol.,* **39**, 105 (1953).
4. R. Battino and H. L. Clever, *Chem. Rev.,* **66**, 395 (1966).
5. R. Battino and F. D. Evans, *Anal. Chem.,* **38**, 1627 (1966).
6. R. Battino, F. D. Evans, and W. F. Danforth, *J. Am. Oil Chem. Soc.,* **45**, 830 (1968).
7. R. Battino, F. D. Evans, W. F. Danforth, and E. Wilhelm, *J. Chem. Thermodyn.,* **3**, 743 (1971).
8. R. Battino, F. D. Evans, and M. Bogan, *Anal. Chem. Acta,* **43**, 518 (1968).
9. R. Battino, M. Banzhof, M. Bogan, and E. Wilhelm, *Anal. Chem.,* **43**, 806 (1971).
10. T. N. Bell, E. L. Cussler, K. R. Harris, L. N. Pepela, and P. J. Dunlop, *J. Phys. Chem.,* **72**, 4693 (1968).
11. A. Ben-Naim and M. Egel-Thal, *J. Phys. Chem.,* **69**, 3250 (1965).
12. A. Ben-Naim, *J. Phys. Chem.,* **71**, 1137 (1967).
13. B. B. Benson and D. M. Parker, *Deep-Sea Res.,* **7**, 237 (1961).
14. B. B. Benson and D. M. Parker, *J. Phys. Chem.,* **65**, 1489 (1961).
15. J. O'M. Bockris, J. Bowler-Reed, and J. A. Kitchener, *Trans. Faraday Soc.,* **47**, 184 (1951).
16. H. H. Borgstedt and A. J. Gillies, *Anesthesiology,* **26**, 675 (1965).
17. M. O. Bulanin, R. G. Amamchyan, V. A. Borgest, and A. I. Moroz, *Tr. Vses. Nauch.-Issled. Inst. Kislorod. Mashinostr.,* **1967**, 135; *Chem. Abstr.,* **70**, 61679v (1969).
18. G. Burrows and F. H. Preece, *J. Appl. Chem.* (London), **3**, 451 (1953).
19. W. A. Cannon and W. E. Crane, *Cryogenic Tech.,* **1968**, 178.
20. B. Cantone and S. Gurrieri, *Bull. Sedute Acad. Groenia Sci. Nat. Catania,* **72**, 681 (1960).

21. I. A. Capuano, U. S. Patent 3,218,242 (1962).

22. J. H. Carpenter, *Limnol. Oceanogr.*, **10**, 141 (1965); **11**, 264 (1966).

23. C. J. Chen and J. F. Parcher, *Anal. Chem.*, **43**, 1738 (1971).

24. H. Cheung and E. H. Zander, *Chem. Eng. Progr. Symp. Ser.*, **64**, 34 (1968).

25. L. C. Clark, Jr., *Trans. Am. Soc. Artif. Organs*, **2**, 41 (1956).

26. L. C. Clark, Jr., R. Wold, D. Granger, and F. Taylor, *J. Appl. Physiol.*, **6**, 189 (1953).

27. H. L. Clever, R. Battino, J. H. Saylor, and P. M. Gross, *J. Phys. Chem.*, **61**, 1078 (1957).

28. H. L. Clever and C. J. Holland, *J. Chem. Eng. Data*, **13**, 411 (1968).

29. J. R. Conder, in *Advances in Analytical Chemistry*, C. N. Reilley and F. W. McLafferty, Eds., Vol. 6, *Gas Chromatography*, J. H. Purnell, Ed., Interscience, New York, 1968, p. 209.

30. J. R. Conder and J. H. Purnell, *Trans. Faraday Soc.*, **65**, 824, 839 (1969).

31. B. E. Conway, J. E. Desnoyers, and A. C. Smith, *Phil. Trans. Roy. Soc. London*, **A256**, 389 (1964).

32. M. W. Cook, Ph.D. Thesis, U. S. Atomic Energy Commission, UCRL-2459, Jan. 14, 1954.

33. M. W. Cook and D. N. Hanson, *Rev. Sci. Instrum.*, **28**, 370 (1957).

34. S. P. Cram and R. S. Juvet, *Anal. Chem.*, **42**, 1R (1970); **44**, 213R (1972).

35. G. Cseko and I. Cornides, *J. Inorg. Nucl. Chem.*, **14**, 139 (1960).

36. P. M. Cukor and J. M. Prausnitz, *Ind. Chem. Eng., Fundam.*, **10**, 638 (1971).

37. R. A. Culp, C. H. Lochmuller, A. K. Moreland, R. S. Swingle, and L. B. Rogers, *J. Chromatogr. Sci.*, **9**, 6 (1971).

38. P. Debye and J. McAulay, *Phys. Z.*, **26**, 22 (1925).

39. J. E. Desnoyers and B. E. Conway, *J. Phys. Chem.*, **70**, 3017 (1966).

40. E. Douglas, *J. Phys. Chem.*, **68**, 169 (1964).

41. J. Dymond and J. H. Hildebrand, *Ind. Eng. Chem., Fundam.*, **6**, 130 (1967).

42. H. L. Elmore and T. W. Hayes, *J. Sanit. Eng. Div. Am. Soc. Civil Eng.*, **86**, (SA4), 41 (1960).

43. D. D. Eley, *Trans. Faraday Soc.*, **35**, 1281 (1938).

44. T. Enns, P. F. Scholander, and E. D. Bradstreet, *J. Phys. Chem.*, **69**, 389 (1965).

45. F. D. Evans and R. Battino, *J. Chem. Thermodyn.*, **3**, 753 (1971).

46. D. H. Everett, B. W. Gainey, and C. L. Young, *Trans. Faraday Soc.*, **64**, 2667 (1968).

47. R. N. Fleck and J. M. Prausnitz, *Ind. Chem. Eng., Fundam.*, **7**, 174 (1968).

48. G. E. Gardiner, Ph. D. Thesis, Fordham University, 1970.

49. G. E. Gardiner and N. O. Smith, *J. Phys. Chem.*, **76**, 1195 (1972).

50. J. Chr. Gjaldbaek, *Acta Chem. Scand.*, **7**, 534 (1953).

51. W. Hayduk and S. L. Cheng, *Can. J. Chem. Eng.*, **48**, 93 (1970).

52. W. Hayduk, E. B. Walter, and P. Simpson, *J. Chem. Eng. Data*, **17**, 59 (1972).

53. R. J. Henry, *Clinical Chemistry*, Harper and Row, New York, 1964, Chap. 4.

54. J. H. Hildebrand, *Proc. Nat. Acad. Sci. U.S.*, **57**, 542 (1967).

55. J. H. Hildebrand and R. L. Scott, *Regular Solutions*, Prentice-Hall, Englewood Cliffs, N.J., 1962.

56.a. J. H. Hildebrand, J. M. Prausnitz, and R. L. Scott, *Regular and Related Solutions*, Van Nostrand Reinhold, New York, 1970, Chap. 8.

56.b. D. M. Himmelblau, *Chem. Rev.*, **64**, 527 (1964).

57. J. O. Hirschfelder, C. F. Curtis, and R. R. Bird, *Molecular Theory of Gases and Liquids*, Wiley, New York, 1954.

58. C. Hsi and B. C.-Y. Lu, *Can. J. Chem. Eng.*, **49**, 134 (1971).

59. J. Horiuti, *Sci. Papers Inst. Phys. Chem. Res.* (Tokyo), **17**, 125 (1931).

60. M. J. Johnson, J. Borkowski, and C. Engblom, *Biotech. and Bioeng.*, **6**, 457 (1964); **9**, 635 (1967).

61. J. Jung, O. Knacke, and D. Neuschutz, *Chem. Ing. Tech.*, **43**, 112 (1971).

62. C. E. Klots and B. B. Benson, *J. Phys. Chem.*, **67**, 933 (1963).

63. K. T. Koonce and R. Kobayahi, *J. Chem. Eng. Data*, **9**, 490 (1964).

64. J. Kratochvil, J. Sobr, J. Matousi, and J. Pick, *Collect. Czech. Chem. Commun.*, **35**, 3761 (1970).

65. G. C. Kresheck, H. Schneider, and H. A. Sheraga, *J. Phys. Chem.*, **69**, 1316, 3132 (1965).

66. I. R. Krichevsky and J. S. Kasarnovsky, *J. Am. Chem. Soc.*, **57**, 2168 (1935).

67. I. R. Krichevsky and A. Ilinskaya, *Acta Physicochim. USSR*, **20**, 327 (1945).

68. P. J. Leonard, D. Henderson, and J. A. Barker, *Trans. Faraday Soc.*, **66**, 2439 (1970).

69. R. G. Linford and J. H. Hildebrand, *Trans. Faraday Soc.*, **65**, 1470 (1969).

70.a. V. J. Linnenbom, J. W. Swinnerton, and C. H. Cheek, U.S. Naval Research Laboratory Report 6344, 1966. Available NITS, #AD 628916.

70.b. F. A. Long and W. F. McDevit, *Chem. Rev.*, **51**, 119 (1952).

71. M. Lucas, *Bull. Soc. Chim. Fr.*, **1969**, 2994.

72. M. Lucas and A. de Trobriand, *J. Phys. Chem.*, **75**, 1803 (1971).

73.a. E. W. Lyckman, C. A. Eckert, and J. M. Prausnitz, *Chem. Eng. Sci.*, **20**, 685 (1965).

73.b. C. A. McAuliffe, *Nature*, **200**, 1092 (1963); *J. Phys. Chem.*, **70**, 1267 (1966).

74. W. F. McDevit and F. A. Long, *J. Am. Chem. Soc.*, **74**, 1773 (1952).

75. J. C. McGowan, *Chem. Commun.*, **1971**, 514.

76. W. J. Mader and L. T. Grady, in *Techniques of Chemistry*, A. Weissberger and B. W. Rossiter, Eds., Vol. I, Part V, Wiley-Interscience, New York, 1970, Chap. 5.

77. K. H. Mancy, D. A. Okun, C. N. Reilley, *J. Electroanal. Chem.*, **4**, 65 (1962).

78. V. V. Marathe and J. M. Prausnitz, *J. Appl. Chem. Biotechnol.*, **21**, 173 (1971).

79. A. E. Markham and K. A. Kobe, *Chem. Rev.*, **28**, 519 (1941).

80. A. E. Markham and K. A. Kobe, *J. Am. Chem. Soc.*, **63**, 449 (1941).

81. D. E. Martire and L. Z. Pollara, in *Advances in Chromatography*, J. C. Giddings, and R. A. Keller, Eds., Vol. 1, Dekker, New York, 1966, p. 335.

82. W. L. Masterton, *J. Chem. Phys.*, **22**, 1830 (1954).

83. W. L. Masterton and T. P. Lee, *J. Phys. Chem.*, **74**, 1776 (1970).

84. W. L. Masterton, T. P. Lee, and R. L. Boyington, *J. Phys. Chem.*, **73**, 2761 (1969).

85. J. Matous, J. Sobr, J. P. Novak, and J. Pick, *Collect. Czech. Chem. Commun.*, **34**, 3982 (1969).

86. P. Mears, *Polymer and Bulk Properties*, Van Nostrand Reinhold, New York, 1965, Chap. 12.

87. K. W. Miller and J. H. Hildebrand, *J. Am. Chem. Soc.*, **90**, 3001 (1968).

88. H. A. C. Montgomery, N. S. Thom, and A. Cockburn, *J. Appl. Chem.*, **14**, 280 (1964).

89. T. J. Morrison and F. Billett, *J. Chem. Soc.*, **1948**, 2033.

90. W. Y. Ng and J. Walkley, *J. Phys. Chem.*, **73**, 2274 (1969).

91. S. Ng, H. G. Harris, and J. M. Prausnitz, *J. Chem. Eng. Data*, **14**, 482 (1969).

92. D. M. Novak-Adamic and B. F. Conway, *Chem. Instrum.*, **5**, 79 (1973).

93. K. Onda, E. Sada, and K. Takeshi, *J. Chem. Jap.*, **3**, 18 (1970).

94. M. Orentlicher and J. M. Prausnitz, *Chem. Eng. Sci.*, **19**, 775 (1964).

95. J. O. Osburn, *Fed. Proc.*, **29**, 1704 (1970).

96. T. D. O'Sullivan and N. O. Smith, *J. Phys. Chem.*, **74**, 1460 (1970).

97. H. Otto, Kebel Scientific Instruments, 80 Swarthmore Ave., Rutledge, Pa.

98. J. P. Peters and D. D. Van Slyke, *Quantitative Clinical Chemistry*, Williams and Wilkins, Vol. II, Baltimore, 1932, Chap. VII.

99. R. A. Pierotti, *J. Phys. Chem.*, **67**, 1840 (1963).

100. R. A. Pierotti, *J. Phys. Chem.*, **69**, 281 (1965).

101. R. A. Pierotti and A. A. Liabastre, Georgia Institute of Technology Environmental Research Center, Report 0572. Available NITS., #PB21163, 1972.

102. K. Porter and D. H. Volman, *Anal. Chem.*, **34**, 748 (1962).

103. S. Possati and A. Faulconer, Jr., *Anesth. Analg. Curr. Res.*, **37**, 338 (1958).

104. J. M. Prausnitz, *Molecular Thermodynamics of Fluid-Phase Equilibria*, Prentice-Hall, Englewood Cliffs, N.J., 1969, Chap. 8.

105. J. M. Prausnitz and F. H. Shair, *Am. Inst. Chem. Eng. J.*, **7**, 682 (1961).

106. H. A. Pray, C. E. Schweikert, and B. H. Minnich, *Ind. Eng. Chem.*, **44**, 1146 (1952).

107. G. T. Preston, E. W. Funk, and J. M. Prausnitz, *J. Phys. Chem.*, **75**, 2345 (1971).

108. H. Reiss, H. L. Frisch, E. Helfand, and J. L. Lebowitz, *J. Chem. Phys.*, **32**, 119 (1960).
109. R. T. Rewick, W. E. Tolberg, and M. E. Hill, *J. Chem. Eng. Data*, **15**, 527 (1970).
110. P. Ruetschi and R. F. Amlie, *J. Phys. Chem.*, **70**, 718 (1966).
111. R. J. Ruka and J. Weissbart, U. S. Patent 3,400,054.
112. J. H. Saylor and R. Battino, *J. Phys. Chem.*, **62**, 1334 (1958).
113. P. Schlapfer, T. Audykowski, and A. Bukowiecki, *Schweiz Arch. Angew. Wiss. Tech.*, **15**, 299 (1949).
114. P. F. Scholander, *J. Biol. Chem.*, **167**, 2351 (1947).
115. A. Seidell, *Solubilities of Inorganic and Metal Organic Compounds*, 3rd ed., Van Nostrand, New York, 1940.
116. A. E. Sherwood and J. M. Prausnitz, *Am. Inst. Chem. Eng. J.*, **8**, 519 (1962); Errata, **9**, 246 (1963).
117. S. K. Shoor and K. E. Gubbins, *J. Phys. Chem.*, **73**, 498 (1969).
118. E. B. Smith and J. Walkley, *J. Phys. Chem.*, **66**, 597 (1962).
119. *Standard Methods for the Examination of Water and Waste Water*, 12th ed., American Public Health Association, New York, 1965.
120. J. B. Steen, *Limnol. Oceanogr.*, **3**, 423 (1958).
121. H. Stephens and T. Stephens, Eds., *Solubilities of Inorganic and Organic Compounds*, Vols. 1 and 2, Macmillan, New York, 1963.
122. P. B. Stewart and P. Munjal, *J. Chem. Eng. Data*, **15**, 67 (1970).
123. J. W. Swinnerton, V. J. Linnenbom, and C. H. Cheek, *Anal. Chem.*, **34**, 483 (1962); **34**, 1509 (1962); **36**, 1669 (1963).
124. J. W. Swinnerton and V. J. Linnenbom, *J. Gas Chromatogr.*, **5**, 570 (1967).
125. J. W. Swinnerton, V. J. Linnenbom, and C. H. Cheek, *Limnol. Oceanogr.*, **13**, 193 (1968).
126. Y. B. Tewari, D. E. Martire, and J. P. Sheridan, *J. Phys. Chem.*, **74**, 2345 (1970).
127. E. W. Tiepel, Ph.D. Dissertation, University of Florida, 1971, University Microfilms, Ann Arbor, Mich., no. 72-16,664.
128.a. E. W. Tiepel and K. E. Gubbins, *J. Phys. Chem.*, **76**, 3044 (1972).
128.b. H. H. Uhlig, *J. Phys. Chem.*, **41**, 1215 (1937).
128.c. W. W. Umbriet, R. H. Burris, and J. F. Stauffer, *Manometric and Biochemical Techniques*, Burgess, Minneapolis, 1972.
129. D. D. Van Slyke, *J. Biol. Chem.*, **30**, 347 (1917).
130. D. D. Van Slyke and J. M. Neill, *J. Biol. Chem.*, **61**, 523 (1924).
131. J. Walkley and W. I. Jenkins, *Trans. Faraday Soc.*, **64**, 19 (1968).
132. O. Warburg, *Biochem. Z.*, **142**, 317 (1923); **152**, 51 (1924).
133. J. A. Waters, G. A. Mortimer, and H. E. Clements, *J. Chem. Eng. Data*, **15**, 174 (1970).
134. R. F. Weiss, *J. Chem. End. Data*, **16**, 235 (1971).
135. W. Y. Wen and S. Saito, *J. Phys. Chem.*, **68**, 2639 (1964).
136. W. Y. Wen and J. Hung, *J. Phys. Chem.*, **74**, 170 (1970).

137. R. Wiebe, V. L. Gaddy, and C. Hein, Jr., with T. H. Tremearne, *Ind. Eng. Chem.*, **34**, 823 (1932).

138. R. Wiebe and V. L. Gaddy, *J. Am. Chem. Soc.*, **55**, 947 (1933); **57**, 847 (1935).

139. E. Wilhelm and R. Battino, *J. Chem. Thermodyn.*, **3**, 379 (1971).

140. E. Wilhelm and R. Battino, *Chem. Rev.*, **73**, 1 (1973).

141. L. Yen and J. McKetta, *Am. Inst. Chem. Eng. J.*, **8**, 501 (1962).

142. R. Battino, R. J. Wilcock, and E. Wilhelm, *Chem. Rev.*, (in press).

143. A. Ben-Naim and S. Baer, *Trans. Faraday Soc.*, **59**, 2735 (1963).

144. B. M. Moudgil, P. Somasundaran, and I. J. Lin, *Rev. Sci. Instrum.*, **45**, 406 (1974).

145. J. H. Hung, Ph.D. thesis, Clark University, 1968.

146. C. N. Murray and J. P. Riley, *Deep-Sea Res.*, **18**, 533 (1971).

Chapter VIII

THE SOLUBILITY OF BIOLOGICAL COMPOUNDS

T. L. McMeekin

---

1 INTRODUCTION

Precise knowledge concerning the properties of biological compounds depends on their separation, which in turn is determined by their solubility, usually in several solvents.

General relations between solubility and structure have long been recognized. The presence in a molecule of polar groups such as hydroxyl, carboxyl, and amino groups tends to make it soluble in water. In a homologous series, such as fatty acids, the lower members tend to be soluble in water, and the higher members insoluble in water and soluble in organic solvents. It has also been found that there is a relationship between the melting point and solubility. For a given solvent a substance with a higher melting point is less soluble than a substance with a lower melting point. This is particularly true if the two substances form ideal solutions in the solvent [25]. These general rules of solubility are applicable to biological compounds.

A very large number of biological compounds has been classified into a variety of classes such as proteins, carbohydrates, lipids, enzymes, hormones, vitamins, and amino acids. The solubilities and other properties of many of these substances have been compiled in a recent comprehensive handbook edited by Sober [54]. This review is limited to systematic studies on the solubilities of proteins and lipids, in which an attempt has been made to relate solubility to structure and in which generalizations concerning solubility have been developed.

2 SOLUBILITY OF AMINO ACIDS, PEPTIDES, PROTEINS, AND RELATED SUBSTANCES IN WATER AND ORGANIC SOLVENTS

Amino acids and proteins have numerous amino, carboxyl, hydrocarbon, and other groups, as well as hydrogen bonds, which are important in determining their solubility. Amino acids and proteins also contain a fundamental structural feature that sets them apart from other organic molecules. They bear an electrical charge even in neutral solutions, being dipolar ions or zwitterions. Thus the structures of amino acids are represented by the formula $^+NH_3-R-COO^-$ rather than the formula $H_2N-R-COOH$. The dipolar ionic structure is a factor of first importance in determining the solubility behavior of amino acids [16]. Amino acids have charges like salts, and side chains containing CH_2 and other residues like other organic molecules. They do not conduct an electric current at their isoelectric points, however, since they are internally ionized, and the charge of the whole molecule is zero. Aqueous solutions of amino acids, proteins, and other dipolar ions have high dielectric constants [18, 62]. The dielectric constant of a dilute solution of an amino acid or protein is a linear function of the concentration.

The importance of the dipolar structure of amino acids and proteins in determining their solubilities and interactions with other biological components was recognized by the late Edwin J. Cohn who initiated pioneer studies in this field. Solubilities of amino acids, peptides, proteins, and related compounds were determined in water, salt solutions, and mixed and pure solvents. Factors such as concentration, ionic strength, and dielectric constant were varied in determining interactions between these substances. The interpretations of these results, as well as the studies of others, have led to a more profound understanding of some of the chemical forces in physiological systems [6].

The relative solubility in two solvents is independent of the vapor pressure of the solute and of its crystal lattice energy. The solubility ratio in two solvents should then be more simply related to the chemical structure of the solute than solubility in only one solvent. Thus L-leucine is more than twice as soluble as DL-leucine in both water and 80% ethanol, but the solubility ratio of both substances in the two solvents is practically the same [10]. Solubility ratios in water and alcohol have been found to be very useful in determining the effect of structure on solubility and should be of fundamental physiological significance, since the organization of the living cell involves aqueous phases of high dielectric constant and fatty phases of low dielectric constant.

Table 8.1 Solubility of Amino Acids in Water and Ethanol at 25°C [10][a]

Amino Acid	Solubility in water N_0 (mol l[-1])	Solubility in ethanol N_A (mol l[-1])	Solubility ratio log N_A/N_0
Glycine	2.886	0.00039	−3.391
DL-a-Alanine	1.660	0.00076	−2.856
DL-a-Amino-N-butryic acid	1.800	0.0026	−2.378
D-Valine	0.706	−	−
DL-Valine	0.571	0.00128	−2.158
DL-Amino-N-caproic acid	0.0866	0.00104	−1.414
L-Leucine	0.171	0.00128	−1.622
DL-Leucine	0.0744	−	−
D-Isoleucine	0.257 [54]	−	−
L-Aspartic acid	0.0375	0.0000116	−2.999
L-Asparagine	0.186	0.000023	−3.402
D-Glutamic acid	0.0585	0.0000185	−2.992
DL-Glutamine	0.291	0.0000315	−3.466
L-Phenylalamine	0.172	0.002	−1.522
L-Tyrosine	0.0024 [41]	0.000028 [41]	−
L-Tryptophan	0.067 [41]	−	−
DL-Threonine	1.598	0.000453	−3.070
L-Cystine	0.000454	−	−

[a]Adapted from Ref. 6, p. 199.

3 EFFECT OF THE CH_2 GROUP ON SOLUBILITY

The solubility in water of monoaminomonocarboxylic acids decreases with increasing number of CH_2 groups (Table 8.1). Aminobutyric acid is an exception to this rule, being more soluble in water than alanine. If, however, solubility ratios in water and organic solvents are considered, the effect of the CH_2 group on solubility is regular and can be described by the rule that the ratio of the solubility in alcohol N_A to that in water N_0 is increased threefold for each CH_2 group for solutes with hydrocarbon chains terminating in methyl groups. This rule holds not only for amino acids, but also for amino acid derivatives such as formyl amino acids, hydantoins and hydantoic acids [34, 35]. A general statement of this rule is given by the equation log $(N_A/N_0) = K_2 + K_1 N CH_2$, where N is the number of CH_2 groups in the molecule, and K_2 is a constant depending on the nature of the polar groups. The coefficient K_1 defining the influence of CH_2 groups on the solubility varies with the organic solvent, being 0.44 for methanol, 0.49 for ethanol and acetone, and 0.53 for butanol and heptanol [35]. Values for the CH_2 group and other groups present in amino acids and proteins calculated from the ratio of the solubility in ethanol N_A and water N_0 are listed in Table 8.2.

The effect of CH_2 groups on solubility reflects the same forces that orient long-chain fatty acids in surface films [32]. A similar effect of the CH_2 group on the surface tension of amino acid solutions has been found [44].

The rule for CH_2 groups in hydrocarbon chains holds only for chains terminating in methyl groups. Thus log (N_A/N_0) is practically identical for aspartic and glutamic acids and also for asparagine and glutamine, despite the additional CH_2 group in glutamine. In other cases, such as in ϵ-aminocaproic acid, CH_2 groups between polar groups exert some effect in enhancing the solubility ratio, but not the full effect exerted by a side chain of the same length [35].

4 EFFECT OF THE CH_2CONH CONFIGURATION ON SOLUBILITY

The influence of the configuration CH_2CONH is opposite in sign to that of the CH_2 group, as is illustrated in Table 8.2 by the comparison of the solubility of glycine with glycine peptides. This configuration that constitutes the repeating pattern of the protein molecule must therefore be strongly polar. Nozaki and Tanford [41] suggest that this group has a strong preference for an aqueous environment, and that this property explains why proteins dissolve in organic solvents with helical conformations.

5 EFFECT OF THE HYDROXYL GROUP ON SOLUBILITY

The highly polar nature of the hydroxyl group is shown by comparing the

Table 8.2 Influence of Structure on the Solubility Ratio in Ethanol and Water at 25°C[a]

Group, and Substances i and K	$\left[\dfrac{Log}{(N_A/N_0)}\right]_i$	$\left[\dfrac{Log}{(N_A/N_0)}\right]_K$	$\dfrac{\Delta Log\ N_A/N_0}{N}$
CH$_2$ group			
Glycine and alanine	-3.391	-2.856	$+0.54$
Glycine and a-aminocaproic			
acid	-3.391	-1.414	$+0.49$
Glycolamide and lactamide	-0.799	-0.254	$+0.54$
CH$_2$ CONH group			
Glycine and diglycine	-3.391	-4.367	-0.98
Glycine and triglycine	-3.391	-4.965	-0.79
Hydantoic acid and diglycine			
hydantoic acid	-0.630	-1.533	-0.90
Hydroxyl group			
a-Aminobutyric acid and			
threonine	-2.375	-3.070	-0.69
Acetamide and glycolamide	-0.120	-0.799	-0.68
Alanine and serine	-2.856	-3.362	-0.51
Methionyl sulfur			
a-Aminovaleric acid and			
methionine	-1.90	-2.444	-0.54
Benzene ring			
Alanine and phenylalanine	-2.866	-1.433	$+1.40$
Dipolar ionization			
Glycine and glycolamide	-3.391	-0.799	2.59
Alanine and lactamide	-2.856	-0.254	-2.60
a-Aminocaproic acid and			
a-hydroxy caproamide	-1.414	$+1.084$	-2.50

[a] Adapted from Ref. 6, p. 206.

solubility of alanine with that of serine, and the solubility of acetamide with that of glycolamide (Table 8.2). This effect of the hydroxyl group on solubility is, however, smaller for an aromatic hydroxyl than for an aliphatic hydroxyl group.

6 EFFECT OF THE METHIONYL SULFUR GROUP ON SOLUBILITY

The sulfur of methionine lies between nonpolar groups. A comparison of the solubility of methionine with that of a-aminovaleric acid (which has the same composition except for sulfur) suggests that sulfur decreases the influence of the paraffin side chain by an amount equal to one CH_2 group, indicating that the sulfur group is polar.

7 EFFECT OF THE BENZENE GROUP ON SOLUBILITY

Since the benzene ring is composed of hydrocarbon groups, it would be expected to increase the solubility in alcohol relative to that in water. This is confirmed by comparing the solubilities of alanine and phenylalanine (Table 8.2). It appears that the benzene ring has about the same effect as three CH_2 groups on the solubility ratio log (N_A/N_0).

8 EFFECT OF DIPOLAR IONIZATION ON SOLUBILITY

Since amino acids and proteins are dipolar ions, their electrical structure is a very important factor in determining their solubility properties. The uncharged species of amino acids and related uncharged derivatives have been of particular value in determining the effect of dipolar ionization on the solubility of amino acids and peptides. Hydroxyamide isomers of amino acids

Table 8.3 Influence of Dipolar Ionization Estimated by Comparison of the Solubilities of a-Amino Acids and a-Hydroxyamides in Water (N_0) and Ethanol (N_A) [5]

a-Amino Acid	Log (N_A/N_0)	a-Hydroxyamide	Log (N_A/N_0)	Influence of Dipolar Ionization $\Delta Log(N_A/N_0)$
Glycine	−3.391	Glycolamide	−0.799	−2.592
Alanine	−2.856	Lactamide	−0.254	−2.602
Norleucine	−1.414	a-Hydroxycaproamide	+1.084	−2.498

such as glycolamide and lactamide with the same groups as the amino acids, but without dipolar ions, show solubility properties markedly different from those of the amino acids. Thus the logarithm of the solubility ratio, expressed as mole fraction in water (N_0) and in ethanol (N_A), of glycine was compared with that of glycolamide, alanine with lactamide, and norleucine with hydroxycapro-amide as shown in Table 8.3. The solubility ratio N_A/N_0 is several hundred times greater for a-hydroxyamides, illustrating the marked effect of dipolar ioniza-tion on solubility. Similar results were obtained when amino acids were compared with their uncharged hydantoic acids [34]. The effect of dipolar .onization on the solubility of amino acids was found to be essentially the same when methanol, ethanol, and butanol were used as solvents [35].

Further extended studies on the solubility of amino acids in water and in aqueous organic solvents have been reported by Nozaki and Tanford [41]. Based on the results of amino acid solubilities in water and in aqueous ethanol and dioxan solutions, the contributions of amino acid side chains and peptide unit free energies of transfer from water to organic solvent were determined. The results in these two systems were very similar particularly in higher concentrations of organic solvent. On the basis of these results and literature values for amino acid solubility, the amino acids were arranged in order of their hydrophobicity, as shown in Table 8.4.

Table 8.4 Hydrophobicity Scale Δf_t for Transfer of Amino Acid Side Chain from 100% Organic Solvent to Water at 25°C [41]

Amino Acid Side Chain	Δf_t (cal mol^{-1})
Tryptophan	3400
Norleucine	2600
Phenylalanine	2500
Tyrosine	2300
Dihydroxyphenylalanine	1800
Leucine	1800
Valine	1500
Methionine	1300
Histidine	500
Alanine	500
Threonine	400
Serine	-300

The values for hydrophobicity (Δf_t) in Table 8.4 show that the side chain of tryptophan is the most polar side chain in proteins, and that serine is the least polar. These values for the polarity of amino acid side chains are in general agreement with the previously reported values for individual amino acids. However, by evaluating the properties of the amino acid side chains, the results are of more direct value in determining the properties of proteins. It is, however, noteworthy that the values for hydrophobicity for asparagine, with one less CH_2 group than glutamine, contrary to expectations gave more negative values of Δf_t than glutamine. Somewhat similar results for the polarity of these two amino acids had been obtained previously [34]. Thus log (N_A/N_0) was found to be almost identical for asparagine and glutamine, as well as for glutamic and aspartic acids. Proline, another principal constituent of proteins, was not given a hydrophobicity value because of the relative inaccuracy of the solubility measurements.

From those results it appears that the interaction of the side chain with the solvent is approximately independent of the interaction of the backbone to which the side chain is attached. Thus the contribution of the leucyl side chain Δf_t value is essentially the same when calculated from the difference between glycine and leucine or their formyl or hydantoic acid derivatives [5]. Also, when the value for the contribution for the aspartyl side chain Δf_t was calculated by comparing the Δf_t of glycine with that of aspartic acid, the result was essentially the same as when the hydantoic derivatives were compared.

9   SOLUBILITY OF AMINO ACIDS IN PROTEIN-DENATURING SOLVENTS

In connection with the development of a theory of protein denaturation, Nozaki and Tanford determined the solubilities of amino acids and peptides in urea [38], ethylene glycol [39], and guanidine hydrochloride solutions [40].

Free-energy contributions Δf_t of the amino acid side chains and peptide units were calculated from the amino acid solubilities in these solvents. It was found that the results for amino acid side chains were roughly parallel to the effectiveness of these solvents in denaturing globular proteins. Thus effective protein denaturants, urea and guanidine hydrochloride solutions, were found to have qualitatively similar effects, though not exactly parallel, on the solubilities of the amino acids except that guanidine was about one to three times as effective as urea as a denaturant at a given concentration. Comparison of the transfer of the free-energy contributions Δf_t of the amino acid side chains and two peptide units in these two solvents is shown in Table 8.5.

The solubility of amino acids in aqueous ethylene glycol solutions was found to be essentially the same as in water, which is in agreement with the finding that aqueous ethylene glycol solutions have little effect on protein conformation.

Based on these results on amino acid solubility and other considerations, Tanford [59] has developed an important theory on the denaturation of

Table 8.5 Comparison between Urea and Guanidine Hydrochloride [40]

| | Ratio of $\Delta f_t'$ in GuHCl to Δf_t in Urea | |
For the transfer of	2 M GuHCl/ 4 M Urea	4 M GuHCl/ 8 M Urea
Two peptide units	1.91	1.52
Histidine	1.78	1.51
Glutamine	1.67	1.37
Asparagine	1.42	1.14
Leucine	1.36	1.20
Methionine	1.10	0.96
Tryptophan	1.24	1.07
Phenylalanine	1.08	0.97
Tyrosine	0.98	0.82

globular proteins. Calculations showed that the unfolding (denaturation) of a globular protein can be treated theoretically as a process resembling solution of a solid substance [58]. The major force stabilizing the globular protein structure was considered to be the hydrophobic force that arises from the unfavorable interaction between nonpolar parts of the protein molecule and water, rather than through hydrogen bonds [58]. The process of protein denaturation involves exposure of the amino acid side chains inside the native globular structure to the solvent. Denaturation is then favored when the exposed side chains interact more favorably with the solvent medium than they do with water. The difference in the action of solvents on proteins was explained in terms of free energy of transfer of the constituent parts of protein molecules as determined from the experimental solubilities of amino acids. The assumption was made that the free energy of transfer of the whole protein molecule in the unfolded state can be expressed as the sum of the free energies from the various groups in the protein molecule. Small model compounds were also utilized in determining the effect of each group in the protein molecule. The further assumption was made that the contribution of each group is independent of the rest of the molecule.

10 SOLUBILITY OF AMINO ACIDS IN SALT SOLUTIONS AND IN SOLUTIONS OF OTHER AMINO ACIDS

Amino acids are ionically charged. They therefore behave as salts and, since they have nonpolar groups, they also behave as organic compounds. Thus solutions of neutral salts tend to dissolve or precipitate amino acids. It is essential to understand the nature of these two effects in order to explain the solubility behavior of dipolar ions in salt solutions. Since the activity of

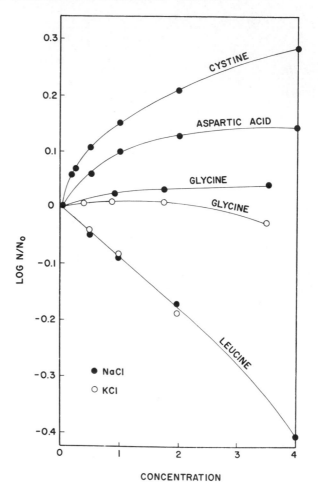

Fig. 8.1. Solubility of amino acids in NaCl and KCl. (Ref. 6, p. 238)

the solute must be the same in all saturated solutions, its relative solubility in two solvents equals the reciprocal of the activity coefficient ratio in the two solutions. If the activity coefficient γ is defined as unity for zero salt concentrations, where the solubility is S', in any other saturated solution $\log (S/S')= -\log \gamma$ [16].

Slightly soluble salts become more soluble in the presence of other salts in the solutions. The effect of each ion is proportional to the square of its valence or to the ionic strength μ.

The solubility of most uncharged molecules is decreased in salt solutions, which is generally termed the *salting-out effect*. Since amino acids are dipolar ions and have aliphatic side chains, their solubility in salts depends on the length

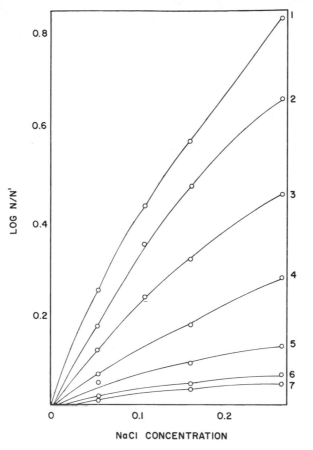

Fig. 8.2. Solubility in 80% ethanol containing NaCl of: 1, lysylglutamic acid; 2, triglycine; 3, diglycine; 4, glycine; 5, diglycyl hydantoic acid; 6, hydantoic acid; 7, *a*-aminocaproic hydantoic acid [12].

of the side chains. The results of the pioneering investigations of Pfeiffer and his associates [45, 46] on the solubilities of glycine, leucine, and aspartic acid in sodium and potassium chlorides, and those of other workers [46] on the solubility of cystine in salt solutions are shown in Fig. 8.1. Glycine, cystine, and aspartic acid which have large electrical charges and small hydrocarbon residues are dissolved by neutral salts, while the solubility of leucine with its long hydrocarbon chain is decreased in neutral salt solutions despite its electrical charge.

The effect of salts on the solubility of amino acids depends also on the dielectric constant of the solvent. Thus, in solvents of low dielectric constant, where electrostatic forces are increased, the solubilities of the amino acids are decreased. The solvent power of salt, however, becomes much greater

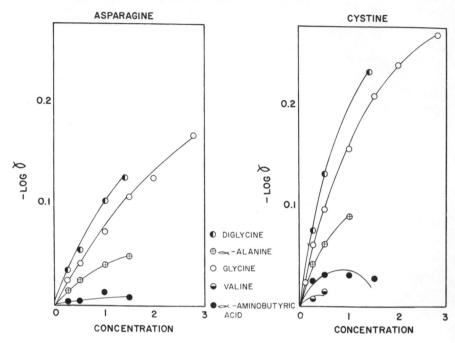

Fig. 8.3. Solubility of asparagine and of cystine in solutions containing other dipolar ions. (Ref. 6, p. 224.)

in solvents of low dielectric constant than in water [9, 12, 33]. The dipole moment of an amino acid also influences its solubility in salt solutions. The effect of salt on the solubility of glycine and a series of peptides in 80% ethanol illustrates the increased solvent effect with increasing dielectric increment of the solute (Fig. 8.2). Thus the solubility of glycine expressed as the ratio log (N_A/N_0), with a dielectric increment $\Delta = 22.6$, is 0.123, while that of triglycine, with a dielectric increment $\Delta = 113.3$, is 0.328 in 0.1 M salt, indicating that as a first approximation the solubility of the amino acids and peptides increases in the order of increasing dielectric increment.

Comparable studies in solvents of high dielectric constant may be made by determining the solubility of a relatively insoluble amino acid in a solvent of a soluble amino acid or a peptide in water. Thus the solubility of cystine was studied in solutions of glycine, a-aminobutyric acid, diglycine, and urea [12]. The results show that the solubility of cystine increased in each of these solvents (Fig. 8.3). The solvent action of glycine on cystine is essentially the same as that of sodium chloride on a molar basis. The solvent action of a-aminobutyric acid is less than that of glycine, presumably because of its longer hydrocarbon chain. Diglycine has a greater influence on the solubility of cystine than glycine, though a smaller effect than would be expected if the interaction depended only on the dielectric constant of the solution. The solvent action of urea is greater than that of the amino acids, even though it has a smaller dielectric constant, indicating that its action on cystine is different from that of amino acids.

The solubility of asparagine was also determined in amino acid and in peptide solutions [11]. The results illustrated in Fig. 8.3 show that the solubility of asparagine increased in amino acid solutions. The greatest increase is produced by diglycine, which has the highest dipole moment, and the smallest increase is in valine, which contains the largest hydrocarbon side chain. Although cystine contains two positively charged and two negatively charged groups connected by a disulfide linkage, and asparagine contains only one positively charged and one negatively charged group, the effect of amino acid and peptide solutions on the solubility of these two substances is essentially the same.

It may be concluded from these studies that the effect of the paraffin side chain of amino acids in aqueous solutions on the solubility of an amino acid is opposite in sign to that of the dipole moment.

11   SOLUBILITY OF PROTEINS

Proteins differ so profoundly in their solubility that this property has been employed as a basis for their classification. Thus globulins are insoluble in water but are soluble in salts, acids, and alkalies; albumins are soluble in water, prolamines are insoluble in water and soluble in alcohol-water mixtures, and keratins are insoluble in most solvents except strong acids and alkalies, which undoubtedly change their chemical structure.

The properties of proteins depend on the nature and number of the polar, nonpolar, ionic, and uncharged groups of which they are composed and their arrangement in space. Some proteins are similar in properties to amino acids and their derivatives. Thus silk fibroin consists of a long chain of glycyl and alanyl residues. Its insolubility and inertness resemble that of the large peptides of glycine. Also, prolamines, which have few free acid and basic groups and contain large amounts of proline, are more soluble in ethanol-water mixtures than in either water or alcohol, resembling the amino acid proline and formyl and hydantoic derivatives of amino acids [34].

Most albumins and globulins are very insoluble in organic solvents. They may be precipitated from aqueous solutions by organic solvents. Albumins and globulins may be precipitated from water without denaturation by the addition of ethanol at low temperatures, as was first shown by Hardy and Gardner [24]. Beginning in 1940, Cohn and his associates [7, 8, 14, 37, 43, 56] developed practical methods for the separation of the proteins of blood plasma by precipitation with aqueous ethanol solutions at low temperatures. Separation of the plasma proteins was accomplished by carefully controlling the temperature, protein and alcohol concentration, pH, and ionic strength.

Proteins, at their isoelectric points, were shown to be highly charged dipolar ions by Weber [60], who studied volume changes accompanying ionization. Direct proof of their dipolar ion structure was demonstrated by dielectric measurements on isoelectric protein solutions by Wyman [61] and by Oncley [42]. In view of the high dipole moments of proteins as compared to amino acids, it is not surprising that the effects of salts on the solubility of amino acids and peptides are small as compared to their effect on the solubility of proteins. The large solvent action of dilute salt solutions on globulins and the precipitation of proteins by concentrated neutral salts have been known and utilized in the separation of proteins for over 100 years [15, 31]. It was demonstrated by Mellanby [36] in 1905 that the solvent action on globulins increases with the valence of the salt ion, and that the efficiency of ions of different valences is directly proportional to the squares of their valences, which is identical to the principle of ionic strength developed by Lewis and Randall in 1921.

Numerous early studies indicated that protein solubilities were not independent of the amount of protein in the system. The careful investigations of Sörensen and Höyrup [55] on the solubility of egg albumin in ammonium sulfate demonstrated, however, that this protein has a solubility nearly independent of the total amount of protein present.

Solubility of Proteins in the Absence of Salts

Proteins differ greatly in their solubility in water. Thus egg albumin is soluble to a concentration of 40% in water, while globulins like myosin and edestin are very insoluble in water. Horse hemoglobin has a solubility of about 17 g ℓ^{-1}, while human hemoglobin has a much greater solubility. The solubility of proteins can be determined in aqueous ethanol solutions without denaturation at temperatures below the freezing point of water. In 25% ethanol at $-5°C$, the solubility of egg albumin is very low, amounting to only 0.013%, while in water it is very soluble. Horse hemoglobin has a solubility of 1.7% in water, and only 0.0036% in 25% alcohol at $-5°C$ [17]. Some of these differences in solubility may be due to a change in temperature from 125 to $-5°C$. The effect of alcohol on the solubility of proteins resembles its effect on the solubility of amino acids, but is many times greater.

Solubility of Proteins in Dilute Salt Solutions

The earlier studies on the solubility of globulins in dilute salt solutions revealed the principle of ionic strength and indicated that globulins are not chemical individuals [24, 36]. The solubility studies of Cohn and Prentiss in 1925 on oxyhemoglobin of the horse, however, indicated that this protein is a chemical individual [13]. They also showed the similarity between the action

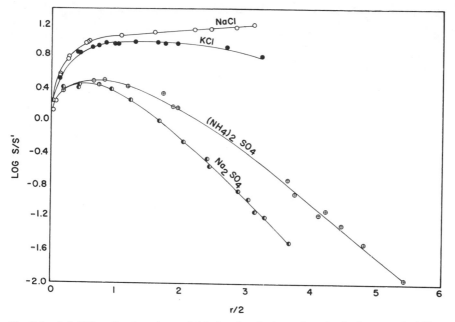

Fig. 8.4. Solubility of carboxyhemoglobin in salt solutions of varying ionic strengths [5].

of salts on hemoglobin and the solvent action of salts on relatively insoluble inorganic electrolytes of high valence.

Very extensive studies of the action of salts on oxy- and carboxyhemoglobin were made by Green [20, 21, 22, 23], who determined the solvent action of a variety of salts on this protein at low salt concentrations, and the salting-out effect at high salt concentrations. Some of her results are shown in Fig. 8.4, in which the logarithm of the solubility in salt is divided by the solubility in water and plotted against ionic strength.

The solubility of slightly soluble proteins, like that of amino acids, is minimal at their isoelectric points (zero net charge), and the addition of acid or base causes an increase in solubility. Green [23] has developed an equation for the effect of pH on the solubility of egg albumin and hemoglobin in concentrated salt solutions.

Solubility of Proteins in Concentrated Salt Solutions

In high concentrations of salt, the solubility of proteins is decreased as shown in Fig. 8.4. This effect is commonly referred to as salting out. It was first pointed out by Cohn [4] in 1925 that a linear relation is obtained when the logarithm of the solubility is plotted against the ionic strength $\Gamma/2$. The salting-out characteristics of proteins in concentrated solutions can be described by an equation of the same form as that describing the solubility of many other

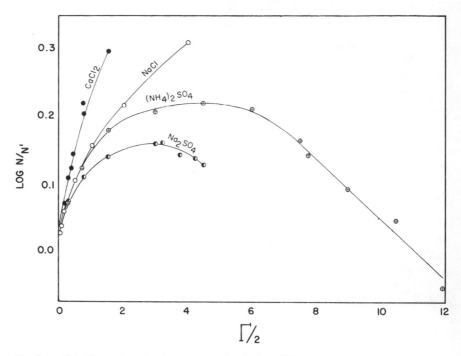

Fig. 8.5. Solubility of cystine in aqueous salt solutions [5].

molecules in salt solutions. If the solubility S of a protein is expressed in grams per liter, the equation may be written log $S = \beta^1 - K'_S \Gamma/2$. K'_S, the salting-out constant, gives the slope of the curve, while β^1 is the extrapolated intercept of the linear portion of the curve on the ordinate axis. The equation describes the data only when the solubility of the protein is small. Thus in Fig. 8.4 the solubility curves for hemoglobin in high concentrations of sodium chloride and potassium chloride do not show a significant decrease in solubility.

Values for the salting-out constant K'_S for a given protein and a given salt are essentially independent of both temperature and pH, whereas β^1 is markedly affected by pH and temperature. The value for K'_S is unique for the protein and also differs with the type of salt. The value of K'_S for horse hemoglobin in ammonium sulfate solution is 0.71, while that of fibrinogen is 1.46.

The similarity of the solubility of the amino acid cystine in ammonium sulfate with that of hemoglobin is shown by comparing Figs. 8.4 and 8.5. The solvent action of salts decreases in the order NaCl > $(NH_4)_2SO_4$ > Na_2SO_4, and the form of the curves are similar. The value of the salting-out constant for cystine in ammonium sulfate is only 0.05, as compared with the value of 0.71 for horse hemoglobin [6].

12 SOLUBILITIES IN SOLUTIONS CONTAINING MORE THAN ONE DIPOLAR ION

Dipolar ions affect the solubility of amino acids and proteins in a manner similar to other ions. Richards' [50] results on the effect of glycine on the solubility of hemoglobin illustrate this property of glycine in increasing the solubility of hemoglobin. The solubility of horse hemoglobin in water averaged 15.8 g per 1000 g of water and increased to 60 g in 1.5 m glycine. Similar results have been reported for the solubility of the amino acids cystine and asparagine in the presence of amino acid and peptide solutions by Cohn and his associates [11]. Thus the solubility of asparagine increased from 0.184 mol ℓ^{-1} in water to 0.247 mol ℓ^{-1} in 2.8 m glycine, and the solubility of cystine increased in a similar similar manner in 2.8 m glycine.

Knowledge of the combined effects of ions and dipolar ions on amino acids and proteins is essential for an understanding of physiological systems. The magnitude of these effects has been evaluated by studies on the solubility of cystine in solutions containing sodium chloride and glycine. Both glycine and sodium chloride solutions increase the solubility of cystine; however, the effect of each of these substances on the solubility of cystine is diminished when they are combined. Thus, in 2 m sodium chloride containing 2.8 mol of glycine, the solubility of cystine increased only from 0.000714 to 0.000811 M, whereas 2.8 M glycine increased the solubility of cystine from 0.000454 to 0.000779 mol ℓ^{-1} and 2.0 m sodium chloride increased the solubility of cystine from 0.000454 to 0.000714 mol ℓ^{-1} [9]. It may be concluded from these results that the interaction between dipolar ions is a function of the salt concentration, and that interaction decreases with increase in ionic strength.

13 SOLUBILITY OF LIPIDS

Lipids constitute a major and important component of living material. They comprise fatty acids and their glyceride esters, phosphatides, and related substances. Like proteins, they occur in every type of animal and plant cell. Unlike proteins, however, lipids are largely composed of nonpolar hydrocarbon groups, so that larger lipids have low solubilities in water and are soluble in a variety of organic solvents. In aqueous solutions, however, they interact with proteins to form soluble complexes, films, and emulsions which are important in physiological systems.

Solubility of Fatty Acids in Water and Organic Solvents

Since 1940 systematic studies on fatty acids and their derivatives in water and organic solvents have revealed the close correlation between their composition and solubility. The first three members of the homologous fatty acid series, acetic, propionic, and butyric acids, are soluble in water in all proportions. With an increase in the number of carbon atoms in the molecule, the solubility in water decreases, as shown by the data of Ralston and Hoerr [47] in Table 8.6.

Table 8.6 Solubilities of Fatty Acids in Water [47]

	Grams Acid per 100 g Water			
Acid	*0°C*	*20°C*	*30°C*	*45°C*
Caproic	0.864	0.968	1.019	1.095
Heptanoic	0.190	0.244	0.271	0.311
Caprylic	0.044	0.068	0.079	0.095
Nonanoic	0.014	0.026	0.032	0.041
Capric	0.0095	0.015	0.018	0.023
Undecanoic	0.0063	0.0093	0.011	0.013
Lauric	0.0037	0.0055	0.0063	0.0075
Tridecanoic	0.0021	0.0033	0.0038	0.0044
Myristic	0.0013	0.0020	0.0024	0.0029
Pentadecanoic	0.00076	0.0012	0.0014	0.0017
Palmitic	0.00046	0.00072	0.00083	0.0010
Heptadecanoic	0.00028	0.00042	0.00055	0.00069
Stearic	0.00018	0.00029	0.00034	0.00042

Extensive systematic studies have been reported by Ralston and his associates on homologous fatty acids [47, 48], primary amines [48], secondary amines [28], primary alcohols [29], symmetrical ketones [19], nitrites [27], amides [48], anilides [48], and N-n-diphenylamides [48], in a large number of organic solvents.

Skau and Boucher [53] have developed an isotherm interpolative method of calculating solubilities of missing members of homologous series. The linear equation $\log N = a + bn$ was derived from the approximate freezing-point-lowering equation and the empirical relationship between the number of carbon atoms and the heat and entropy of fusion of members of a homologous series. In the derived equation N is the solubility of the homolog, expressed in mole percent, in a given solvent at a given temperature, a and b are constants which depend on temperature and the system involved, and n is the number of carbon atoms in the dissolved compound. Log N versus n isotherms were plotted for 125 systems involving 10 homologous systems in 16 different solvents using published data for known members of homologous series. The values obtained were presented in tabular form. Only about one-third of the 584 isotherms were straight lines as required by the equation, the rest being smooth curves which are useful in predicting solubilities and in locating errors in published results.

The charts on which the tabulated solubility data of the homologous series of fatty acids were based were published later [3].

Methods for the prediction of solubility of long-chain compounds have been extended by Skau and his associates [2, 51, 53]. Their isopleth method of predicting solubilities of missing members of homologous series is based on the linear equation $1/T = a(1/n) + b$, where T is the absolute temperature at which saturated solutions of the various homologs in a given solvent contain the same mole fraction of the solute. The coefficients a and b are constants, and n is the number of carbon atoms. The validity of the isopleth method was established by the fact that $1/T$ versus $1/n$ plots based on accurate data for several systems consisted of straight lines or only slightly curved lines. The combined use of isopleth and isotherm methods of correlating solubilities makes it possible to obtain similar complete solubility data for solvents and other homologous series on the basis of a minimum of experimental measurements.

A new graphical isopleth reference method [2] of correlating solubility data was developed, which complements the isotherm and isopleth methods. This method is based on the linear equation

$$1/T = \Delta H_f'/\Delta H_f - 1/T' + C$$

derived from the approximate freezing-point depression equation in which T and T' are the primary freezing points of two analogous compounds at the same molar concentration in a given solvent. ΔH_f is the heat of fusion in calories per mole of the first compound, and $\Delta H_f'$ for the second compound, and C is a constant. The basic assumption that $\Delta H_f'/\Delta H_f$ remains essentially constant for homologous and analogous compounds in the same solvent over wide ranges of temperature was validated by constructing graphs from solubility data on long-chain fatty acids and their derivatives. Plots of $1/T_a$ versus $1/T_r$ (analogous versus reference) in most cases gave smooth, slightly curved lines and straight lines over small changes of concentrations. The previous methods for predicting solubility are applicable to homologous series only, whereas this method is also applicable to analogous compounds. By this method the complete solubility curve for a given compound can be obtained from two or three determinations if the complete solubility data are known for a single analogous or homologous compound. This method was further perfected by Skau [51, 52] for correlating and predicting solubility data for long-chain compounds. It is based on the linear relationship of $1/T_a + 1/T_r$ derived from the freezing-point-depression equation. T_a and T_r represent the temperature in kelvins at which the mole percent solubility of a compound a in a given solvent is the same as that of an analogous reference compound r in a related solvent. This method can be used for the prediction of complete solubility data from scant experimental data.

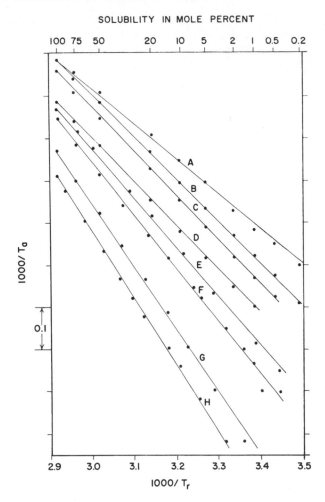

Fig. 8.6. Isopleth reference plots for: A, stearic acid in hexane; B, stearic acid in *o*-xylene; C, heptadecanoic acid in cyclohexane; D, stearic acid in chlorobenzene; E, elaidic acid in toluene; F, oleic acid in hexane; G, linoleic acid in hexane; H, oleic acid in *o*-xylene. Reference system, stearic acid in benzene [51].

A reasonably accurate isopleth reference plot can be constructed from the melting point of a compound and the temperature at which its solubility in a given solvent is about 5 mol %. Thus, with a single accurate experimental determination, an isopleth plot can be made using as a reference system the complete accurate solubility data for an analogous compound in a related solvent. This procedure should make it possible to predict the complete solubility data for C_{16} to C_{20} saturated and unsaturated fatty acids in a wide variety of solvents from one solubility determination. The results of this method of calculating solubilities are shown in Fig. 8.6 in which the solubility of stearic acid in benzene is the reference system.

Hildebrand and Scott [26] have described a general method of determining the relative accuracy of solubility data by plotting the logarithm of the mole fraction of the solute against the reciprocal of the absolute temperature. This method has been used by Bailey [1] to test the accuracy of solubility data on fatty acids and their derivatives.

References

1. A. E. Bailey, *Melting and Solidification of Fats,* Interscience, New York, 1950.
2. A. V. Bailey, J. A. Harris, and E. L. Skau, *Am. Oil Chem. Soc.,* **46**, 583 (1969).
3. R. E. Boucher and E. L. Skau, *Technical Bulletin ARS-72-1,* Southern Utilization Research Branch, New Orleans, La., 1954.
4. E. J. Cohn, *Physiol. Rev.,* **5**, 349 (1925).
5. E. J. Cohn, *Chem. Rev.,* **19**, 241 (1936).
6. E. J. Cohn and J. T. Edsall, *Proteins, Amino Acids and Peptides,* Reinhold, New York, 1943.
7. E. J. Cohn, W. L. Hughes, and J. H. Weare, *J. Am. Chem. Soc.,* **69**, 1753 (1947).
8. E. J. Cohn, J. A. Leutscher, J. L. Oncley, S. H. Armstrong, and B. D. Davis, *J. Am. Chem. Soc.,* **62**, 3396 (1940).
9. E. J. Cohn, T. L. McMeekin, and M. H. Blanchard, *J. Gen. Physiol,* **21**, 651 (1938).
10. E. J. Cohn, T. L. McMeekin, J. T. Edsall, and J. H. Weare, *J. Am. Chem. Soc.,* **56**, 2270 (1934).
11. E. J. Cohn, T. L. McMeekin, J. D. Ferry, and M. H. Blanchard, *J. Phys. Chem.,* **43**, 169 (1939).
12. E. J. Cohn, T. L. McMeekin, J. P. Greenstein, and J. H. Weare, *J. Am. Chem. Soc.,* **58**, 2365 (1936).
13. E. J. Cohn and A. M. Prentiss, *J. Gen. Physiol.,* **8**, 619 (1927).
14. E. J. Cohn, L. E. Strong, W. L. Hughes, D. J. Mulford, J. N. Ashworth, M. Melin, and H. L. Taylor, *J. Am. Chem. Soc.,* **68**, 459 (1946).
15. P. S. Denis, *Memoire sur le Sang,* Paris, 1859.
16. J. T. Edsall, in *The Chemistry of Amino Acids and Proteins,* C. L. A. Schmidt Ed., Charles C. Thomas, Springfield, Ill., 1938.
17. R. M. Ferry, E. J. Cohn, and E. S. Newman, *J. Am. Chem. Soc.,* **58**, 2370 (1936).
18. J. D. Ferry and J. L. Oncley, *J. Am. Chem. Soc.,* **60**, 1123 (1938).

19. F. M. Garland, C. W. Hoerr, W. O. Pool, and A. W. Ralston, *J. Org. Chem.*, **8**, 344 (1943).
20. A. A. Green, *J. Biol. Chem.*, **93**, 495 (1931).
21. A. A. Green, *J. Biol. Chem.*, **93**, 517 (1931).
22. A. A. Green, *J. Biol. Chem.*, **93**, 524 (1931).
23. A. A. Green, *J. Biol. Chem.*, **95**, 47 (1932).
24. W. B. Hardy and S. Gardner, *J. Physiol.*, **40**, 68 (1910).
25. J. H. Hildebrand, *Solubility*, 2nd ed., Reinhold, New York, 1936.
26. J. H. Hildebrand and R. L. Scott, *The Solubility of Non-Electrolytes*, 3rd ed., Reinhold, New York, 1950.
27. C. W. Hoerr, E. F. Binkerd, W. O. Pool, and A. W. Ralston, *J. Org. Chem.*, **9**, 68 (1944).
28. C. W. Hoerr, H. J. Harwood, and A. W. Ralston, *J. Org. Chem.*, **9**, 201 (1944).
29. C. W. Hoerr, H. J. Harwood, and A. W. Ralston, *J. Org. Chem.*, **9**, 267 (1944).
30. C. W. Hoerr and A. W. Ralston, *J. Org. Chem.*, **9**, 329 (1944).
31. F. Hofmeister, *Arch. Exp. Pharm.*, **24**, 247 (1887-1888).
32. I. Langmuir, *Chem. Rev.*, **13**, 147 (1933).
33. T. L. McMeekin, E. J. Cohn, and M. H. Blanchard, *J. Am. Chem. Soc.*, **59**, 2717 (1937).
34. T. L. McMeekin, E. J. Cohn, and J. H. Weare, *J. Am. Chem. Soc.*, **57**, 626 (1935).
35. T. L. McMeekin, E. J. Cohn, and J. H. Weare, *J. Am. Chem. Soc.*, **58**, 2173 (1936).
36. J. Mellanby, *J. Physiol.*, **33**, 338 (1905).
37. P. R. Morrison, J. T. Edsall, and S. G. Miller, *J. Am. Chem. Soc.*, **70**, 3102 (1948).
38. Y. Nozaki and C. Tanford, *J. Biol. Chem.*, **238**, 4074 (1963).
39. Y. Nozaki and C. Tanford, *J. Biol. Chem.*, **240**, 3568 (1965).
40. Y. Nozaki and C. Tanford, *J. Biol. Chem.*, **245**, 1648 (1970).
41. Y. Nozaki and C. Tanford, *J. Biol. Chem.*, **246**, 2211 (1971).
42. J. L. Oncley, *J. Am. Chem. Soc.*, **60**, 1115 (1938).
43. J. L. Oncley, M. Melin, D. A. Richert, J. W. Cameron, and P. M. Gross, *J. Am. Chem. Soc.*, **71**, 541 (1949).
44. J. R. Pappenheimer, M. P. Lepie, and J. Wyman, *J. Am. Chem. Soc.*, **58**, 1851 (1936).
45. P. Pfeiffer and O. Angern, *Z. Physiol. Chem.*, **133**, 180 (1924).
46. P. Pfeiffer and J. Wurgler, *Z. Physiol. Chem.*, **97**, 128 (1916).
47. A. W. Ralston and C. W. Hoerr, *J. Org. Chem.*, **7**, 546 (1942).
48. A. W. Ralston, C. W. Hoerr and W. O. Pool, *J. Org. Chem.*, **8**, 473 (1943).
49. A. W. Ralston, C. W. Hoerr, W. O. Pool, and H. G. Harwood, *J. Org. Chem.*, **9**, 102 (1944).
50. M. M. Richards, *J. Biol. Chem.*, **122**, 727 (1938).
51. E. L. Skau, *Am. Oil Chem. Soc.*, **47**, 233 (1970).
52. E. L. Skau and A. V. Bailey, *J. Phys. Chem.*, **63**, 2047 (1959).

53. E. L. Skau and R. E. Boucher, *J. Phys. Chem.*, **58**, 460 (1954).

54. H. A. Sober, Ed., *Handbook of Biochemistry*, Chemical Rubber, Cleveland 1968.

55. S. P. L. Sörensen and M. Höyrup, *C. R. Lab. Carlsberg*, **12**, 213 (1917).

56. D. M. Surgenor, L. E. Strong, H. L. Taylor, R. S. Gordon, and D. M. Gibson, *J. Am. Chem. Soc.*, **71**, 1223 (1949).

57. C. Tanford, *J. Am. Chem. Soc.*, **84**, 4240 (1962).

58. C. Tanford, *J. Am. Chem. Soc.*, **86**, 2050 (1964).

59. C. Tanford, *Advan. Protein Chem.*, **24**, 1 (1970).

60. H. H. Weber, *Biochem. Z.*, **218**, 1 (1930).

61. J. Wyman, *J. Am. Chem. Soc.*, **90**, 443 (1931).

62. J. Wyman and T. L. McMeekin, *J. Am. Chem. Soc.*, **55**, 908 (1933).

INDEX